T0291033

Bruno Grösel
Stage Technology

Also of Interest

Optical Imaging and Photography
Imaging Optics, Sensors and Systems
Ulrich Teubner, Hans Josef Brückner, 2023
ISBN 978-3-11-078990-4, e-ISBN 978-3-11-078996-6

Mechanical Engineering: Practical Problems
Problems and Solutions in Statics
Sayavur Bakhtiyarov, 2024
ISBN 978-3-11-132967-3, e-ISBN 978-3-11-132971-0

Metrological Infrastructure
Edited by Beat Jeckelmann, Robert Edelmaier, 2023
ISBN 978-3-11-071568-2, e-ISBN 978-3-11-071583-5
in *De Gruyter Series in Measurement Sciences*
Edited by Klaus-Dieter Sommer, Thomas Fröhlich
ISSN 2510-2974, e-ISSN 2510-2982

Technoscientific Research
Methodological and Ethical Aspects
Roman Z. Morawski, 2024
ISBN 978-3-11-117980-3, e-ISBN 978-3-11-118003-8

Page and Stage
Intersections of Text and Performance in Ancient Greek Drama
Edited by Stuart Douglas Olson, Oliver Taplin, Piero Totaro, 2023
ISBN 978-3-11-124739-7, e-ISBN 978-3-11-124802-8

Bruno Grösel

Stage Technology

—

DE GRUYTER

Author
Em.O.Univ.Prof. Dipl.-Ing. Dr.techn. Bruno Grösel
1170 Wien
Austria
bruno.groesel@tuwien.ac.at

ISBN 978-3-11-136623-4
e-ISBN (PDF) 978-3-11-136696-8
e-ISBN (EPUB) 978-3-11-136748-4

Library of Congress Control Number: 2024937061

Bibliographic information published by the Deutsche Nationalbibliothek
The Deutsche Nationalbibliothek lists this publication in the Deutsche Nationalbibliografie;
detailed bibliographic data are available on the Internet at http://dnb.dnb.de.

© 2024 Walter de Gruyter GmbH, Berlin/Boston
Cover image: Wiener Volksoper
Typesetting: VTeX UAB, Lithuania

www.degruyter.com

Preface to the 6th edition

Technical efforts do not guarantee the quality of a performance. However, a theater or opera performance without the use of stage technology equipment would be possible only to a very limited extent. Stage technology is of particular importance in theaters that work in a repertory system and possibly perform a different play every day, whereby rehearsals must still be possible between performances. Stage technology provides the necessary transport equipment for this purpose.

The technology must also create the conditions for the artist to design the stage set. Only with technical means the construction and change of stage sets can be realized in a reasonable time. Their use is usually hidden from the audience – "behind the curtain." But sometimes stage technology is also visible to the audience and enables scenic effects.

With the progress of technology, these technical stage facilities are also becoming more and more powerful, but also more complex, and require much more technical knowledge for planning and construction, but also for their operational use and maintenance.

In the narrower sense, "stage equipment" is usually understood to mean universally applicable mechanical installations in the upper and lower stages, i. e., lifting platforms, stage wagons, revolving stages, bar and point hoists, curtain devices, etc., but also mechanical devices for fire protection. In a broader sense, this term can also include lighting and sound equipment. This technical separation into stage mechanical equipment, lighting and sound technology is generally given by manufacturers of stage equipment, but also in the training and deployment of specialist personnel on stages.

This book deals with the many aspects of stage technology in the narrower sense, i. e., with mechanical equipment. Therefore, fields of expertise such as mechanical engineering, electrical and hydraulic drive technology, but also the fundamentals of mechanics must be included.

In order to provide the widest possible range of interest for persons and parties with subject-specific informations, the structure of the book has been designed to appeal to people with different levels of technical training.

In many sections, the book is certainly understandable for graduates of a compulsory school with technically oriented knowledge and practical stage experience. However, it is unavoidable to assume higher technical knowledge in some parts of the book. However, even in these parts, an attempt is made to formulate basic statements in a way that is as generally understandable as possible, admittedly here and there also by foregoing strictly scientific exactness.

The sale of the book has shown that there is obviously a need for such a compilation of stage-specific expertise. It is therefore gratifying to see that a sixth edition is now required. This one has again been revised and expanded. Firstly, new installations have been erected in the meantime that are worth mentioning, and secondly, the development of technology is so rapid that it has become essential to revise it in the light of the now existing European standard EN 17206 entitled "Entertainment Technology – Lifting

https://doi.org/10.1515/9783111366968-201

and load-bearing Equipment for Stages and other Production Areas within the Entertainment Industry – Specifications for general requirements (excluding aluminium and steel trusses and towers)." With this in mind, new text sections have been formulated and illustrations added. This is to ensure that the book corresponds to the state-of-the-art and provides up-to-date information.

I would like to express my sincere thanks to the companies that provided me with documents.

I am especially pleased that this sixth edition is now also available in English.

Vienna, May 2024 Bruno Grösel

Contents

1 Stage equipment, construction types and application criteria

In this section, the variety of stage equipment, its types of construction and criteria of use will be described in a systematic way. To begin with, the historical development of European theater should be outlined very briefly.

1.1 Historical development

The Greek amphitheater (Figure 1.1) can be regarded as the root of theater construction in the occident. The seats of the spectators were arranged in a circular ring rising towards the outside. In front of this was the playing area for the actors. It consisted of a kind of *proscenium* and a raised stage behind it, called *skene*. The proscenium and skene were already architecturally designed with columns, supports and roofing. In the center of the circle was the place for the choir, the so-called *orchestra*.

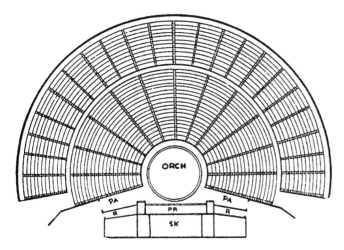

Figure 1.1: Greek amphitheater – top view. Picture reference: Leaflet 289 (see the list of supplementary literature).

The terms of the *Greek theater* are also used today, but with somewhat changed meanings. The orchestra became the area of the musicians, the *orchestra*, the architecturally designed *proscenium* became the framing of the stage opening. The component originally named skene finally became the *stage house*. *Theatron* initially referred only to the area of the audience; later the term *theater* was used for the entire building complex, i. e., the auditorium and stage.

In the *Roman theater*, on the one hand, the longitudinal orientation was retained according to the Greek model and a semicircular section was intended for the specta-

https://doi.org/10.1515/9783111366968-001

tors and a separate area for the actors. On the other hand, in the Roman *arena theater*, which in Roman times was mainly used for sporting events, animal and gladiator fights, a sandy area in the center was surrounded by an audience area. This second variant can be found in the circus and in the modern *arena theater*.

Already in the ancient theater, painted panels between the columns were used as decorations and also very simple transformations of the stage set were made with *rotating three-sided prisms*. In the *Telari stage* of the early Renaissance, this idea was taken up again. However, *mechanical stage technology* was also already used to make a god suddenly appear as *"Deus ex machina"* (Latin for "appearance of a deity using a stage machinery") with lifting platforms or fly equipment.

The theatrical activity of the Middle Ages was mainly characterized by ecclesiastical mystery and passion plays. Churches or squares served as stages. Often, the existing architecture or scaffolding created a three-part tiered arrangement for hell, earth and heaven as the stage area, or fixed stage decorations were used side by side. The play took place in several scene areas at the same time (simultaneously) or changed from one decoration section to another. This type of stage is therefore also called a *simultaneous stage*. The stage set did not have to be changed, therefore no technical equipment was required for transformations.

Secular theater events were dependent on the use of barns and simple wooden stalls or took place in halls and courtyards with the simplest stage design. It was not until the sixteenth century that so-called *playhouses* came into being.

In England, it is said that wagons with scenery built on them were pushed in front of the audience to enable the rapid change of a stage set. This could be seen as the forerunner of today's sliding stage system. In Italy, *Leonardo da Vinci* constructed the first *revolving stage* for a wedding festival performance.

The classical theater construction of the Italian Renaissance tried to transfer the forms of the ancient theater scaled down into a closed space. The *Teatro Olimpico* by the architect *Palladio* in Vicenza is an interesting example of this: by emphasizing the perspective of the stage architecture, a stage space with a special effect of depth was created.

To make better use of space, the *rank theater* (*"Rangtheater"* in German) was invented, in which up to six tiers arranged one above the other extended approximately semicircularly from one side of the proscenium wall to the other. In courtly theaters, the tiers often contained only boxes (*"Logentheater"* in German). The auditorium of this classical theater was already separated from the actual playing area, the *stage house*, by the proscenium wall and the curtain, as is common today. This design of the classical theater is still largely retained today, although the open seating arrangement is again preferred in the modern rank theaters.

While in Shakespearean theater decorative elements were used only sparingly and scene changes were often indicated only symbolically, later the general trend led to ever greater use of decorative material; this required technical means for variable stage space and scene design.

The stage system generally used in Europe until the end of the nineteenth century was the so-called *scenery stage* (*"Kulissenbühne"* in German). The stage space was delimited at the back wall by a painted *prospectus*, at the sides by painted picture surfaces, hanging or standing, arranged one behind the other, and at the top by arches or soffits (see Figure 1.2, left). Decorative elements were attached to horizontal bars hanging from the grid and could be removed from the view of the spectators by raising these bars or moved into the view of the spectators by lowering them. Standing side scenery could be moved into or out of the stage from the side on narrow movable support frames in *scenery carriages*. These scenery carriages were installed in gaps 2 to 4 cm wide across the entire width of the stage, so-called *free rides* (*"Freifahrten"* in German). Between these free rides, floor elements, some of which could be moved vertically, were installed for the sudden appearance or disappearance of parts of the picture, so-called *cassettes*. *Cassette flaps* were used to close open cassettes.

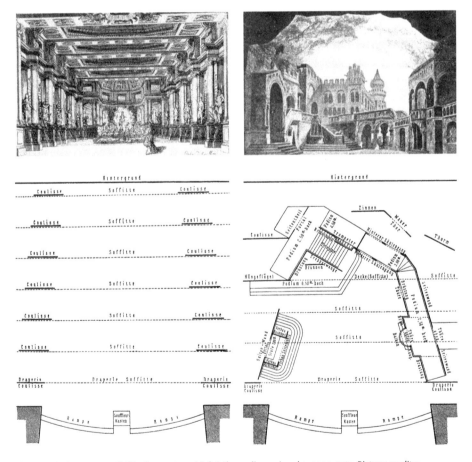

Figure 1.2: Scenery set: (left) wing system, (right) three-dimensional scenery sets. Picture credits: Leaflet 289.

A typical element of this period was also the raisable and lowerable *lattice girder* – an element of stage engineering that can sometimes also be found in modern stages, e.g., in the Staatstheater Stuttgart. This refers to a very narrow lifting platform in the width of the stage, designed as a flexurally rigid lattice girder. In order to leave a corresponding gap in the stage floor, the lifting platforms are equipped with *lattice flaps*. Narrow scenery elements can be mounted on the lattice girder and used primarily for scenic effects. Thus, a prospectus can be attached to the front or, for example, a water wave screen can be raised. This can also be used in conjunction with laterally movable screens and soffits suspended in the upper stage to vary the viewing opening of the proscenium stage (see Section 1.2). The lattice girder is moved vertically with ropes suspended from the grid.

A so-called *conversion prospectus* (*"Wandelprospekt"* in German) was sometimes used to define the back of the stage set. In this case, several stage sets were painted next to each other on a fabric web; the fabric web could then be wound up or unwound on two cylinders standing on the left- and right-hand sides of the back wall of the stage.

In the case of the *scenery stage* (*"Kulissenbühne"* in German), the stage space was thus designed using painted surface elements. The alleys between the individual sets were used by the actors for their entrance; this is why it was also called an *alley system* (*"Gassensystem"* in German). By means of perspective-painted leaflets and soffits, stage sets with great depth effect could be created. The perspective was further enhanced by a stage floor rising towards the backstage, the so-called *stage fall*.

Step by step, there was a growing need to use real stage structures and plastic decorations (see Figure 1.2, right). As a consequence, rapid transformations were no longer possible only by raising and lowering or moving light surface decoration elements as in the case of the scenery stage. More elaborate technical equipment had to be created to enable the rapid transport of heavy spatial stage structures. Thus, large *stage platforms* that could be raised and lowered, movable *stage wagons* and *revolving stages* were created. However, a sloping stage floor then proved to be impractical.

A particular innovation of development in stage technology began at the turn from the nineteenth to the twentieth century. The "Asphalia-Gesellschaft für die Herstellung zeitgemäßer Theater" ("Asphalia Society for the Production of Contemporary Theaters" and the Austrian engineer *Robert Gwinner* (Vienna, Budapest) developed the so-called *Asphalia system* towards the end of the nineteenth century. (The Greek word "Asphalia" means "safety, absence of danger.") The inventions also involved new concepts for improving acoustics and ventilation, but especially hydraulically operated stage platforms in the lower stage and lifting devices with wire ropes in the upper stage. The engineer *Fritz Brandt* (Berlin) designed a side-stage system with horizontal transformability, known as the *"Reform stage Berlin"* (*"Reformbühne Berlin"* in German). In side stages that could be separated from the main stage by sliding doors or sound-insulating curtains, stage sets could be prepared on movable stage wagons and moved to the main stage. Around the same time, *Karl Lautenschläger* (Munich) rediscovered the revolving stage, especially as a *cylinder revolving stage* with built-in lifting platforms.

1.2 Tasks of technical stage equipment

Technical equipment of stages usually includes:
- Mechanical systems for the design of the stage and auditorium in venues such as speech, music and dance theaters, opera houses, concert halls, but also in conference rooms and multipurpose halls.
- Mechanical systems for moving construction elements of the stage, for scenic design and for transformations to change the stage set.
- Mechanical systems for conveying and storage of decorative elements.
- Mechanical equipment for fire protection.
- Lighting and projection equipment.
- Acoustic systems of sound engineering.

This book deals with the variety of *mechanical equipment* mentioned in the above list, i. e., not with lighting, projection and sound equipment. These mechanical devices are used to raise, lower, tilt, move or rotate stage or auditorium elements and decorations and are driven manually, electromechanically or hydraulically. They are therefore *materials handling equipment* for the vertical or horizontal, translatory or rotatory movement of loads, and possibly also of persons. Stage technology can thus be regarded as a special discipline of general materials handling technology. This fact makes it clear that development trends in general materials handling technology are naturally reflected in stage technology, albeit in special forms. The use of these conveyors is mostly unnoticed by the audience before and after the performance or during the breaks between acts.

The way many European theaters and opera houses arrange their schedules, with different performances on consecutive days and rehearsal operations during the day, requires rapid transportation of decorative elements that often weigh several tons. Such a stage operation therefore requires universally applicable means of transport devices in the stage.

However, as already mentioned, the movability of stage elements also allows them to be used for *scenic stage design*. For example, lifting platforms can be used to design the stage floor or as supporting elements for stage structures.

For the spectator, the stage technology becomes visible when it is used during the play or during transformations on the open stage during *scenic effects*. The technical conception of a device designed for this purpose requires low sound emissions during movement and, above all, high operating speeds as well as their stepless controllability.

However, the components such as podiums, carriages, revolving platforms, etc., must also be dimensioned and designed in such a way that they not only withstand the loads in terms of strength but also exhibit very low deformation and high stiffness as a result of the loads and cannot be excited into disturbing vibrations.

Mechanical stage equipment can therefore – in brief – be used for the following purposes:

-   for *transport-related tasks*,
-   for *room design* (*interior design*) and
-   for *scenic effects*.

In general, mechanical stage equipment also includes *safety equipment* for fire protection purposes, such as fire protection curtains and smoke escape systems, since their function also requires mechanical processes.

In the enumeration of the three purposes of mechanical stage equipment, "room design" meant the design of the stage space.

In *multipurpose halls*, however, "room design" also refers to adaptation of the overall space to different uses, e. g., as a concert hall or opera house, etc. Some examples are given in Section 1.6.

1.3 Space concepts

There are different concepts for the spatial arrangement of the audience and stage area.

The *proscenium stage* as shown in Figure 1.3(a) (also called a *frame stage*) has a longitudinal orientation of the theater. The audience looks out of the auditorium onto a playing area separated by an optical constriction. This frame limiting the view is called *proscenium* or *proscenium opening*. Figure 1.4 shows the proscenium opening in the *Teatro Madrid as* seen from the stage into the auditorium.

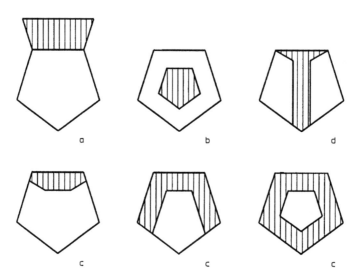

Figure 1.3: Arrangement of the stage and the audience area (according to DIN 56 920, sheet 1): (a) proscenium stage, (b) arena stage, (c) area stage (three examples) and (d) catwalk stage (dashed area – stage; white area – auditorium).

Figure 1.4: Teatro Madrid. Photo: Waagner-Biró.

In the *arena stage* shown in Figure 1.3(b), the scene area is surrounded on all sides by spectator seats; the spectators thus look from all directions onto a centrally arranged stage.

In addition to these two basic forms, stage arrangements are also possible with playing surfaces of various shapes that are not surrounded on all sides by visitor seats – referred to as *room playing areas* according to Figure 1.3(c) – as well as the *cat walk stage* typical of Japanese Kabuki theater, for example, according to Figure 1.3(d).

The so-called *thrust stage* should also be mentioned. (The English term "thrust" means "to thrust forward," "thrust into" or "to thrust out," "to stretch out.") This refers to a stage that extends into the auditorium and is surrounded by the audience on three sides.

Room areas of a proscenium stage

In the following, the stage-specific spatial arrangement possibilities in the stage house of a proscenium stage in particular will be explained in more detail:

For small stages, the stage is an area arranged on a platform construction at a slightly higher level with approximately normal room height. For larger stages, there is a large free space above the main playing area of the stage to store hanging decorations above the main stage, which cannot be seen by the audience. This gives the stage space a tower-like structure.

A significant expansion of the available space can be achieved by assigning additional rooms to the stage at the same level. In these, standing decorative elements can be provided and brought into the view of the audience when the stage set is changed.

In this sense, *side stages* can be arranged to the left and right of the *main stage*, or a *backstage* behind the main stage area, in addition to the actual playing area visible to the audience (see Figure 1.5).

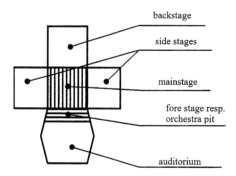

Figure 1.5: Position and identification of the stage areas in a proscenium stage. (DIN 56 920, sheet 2).

If sufficient space is available, adjacent areas at stage level can also be used as assembly areas or as rehearsal stages, as shown in Figure 1.6, for example, based on the *Copenhagen Opera House*.

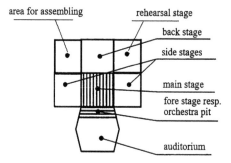

Figure 1.6: Example of further use of stage areas.

However, for appearance from below, it is also important to provide a space below the main stage. While the aforementioned area above stage level is called the *upper stage*, this area of space is called the *lower stage*, or *understage*.

Figure 1.7 shows the longitudinal section and floor plan of a theater with two side stages, a back stage, an upper stage and a lower stage. Among other things, one can see side stage wagons, a backstage wagon with turntable, double-decker platforms in the main stage, orchestra platforms in the lower stage, hoists in the upper stage, etc.

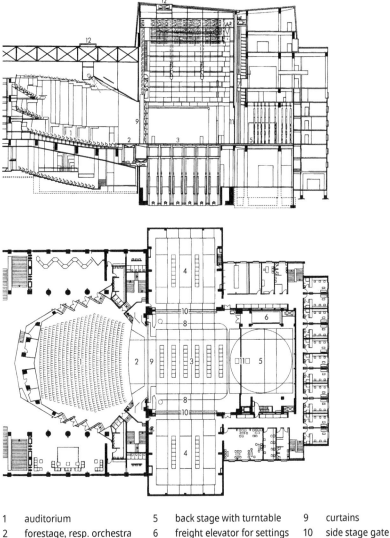

1 auditorium
2 forestage, resp. orchestra
3 main stage
4 side stages
5 back stage with turntable
6 freight elevator for settings
7 backdrop elevator and depot
8 cyclorama
9 curtains
10 side stage gate
11 back stage gate
12 flue gas outlet

Figure 1.7: Longitudinal section and top view of a theatre. Photo credit: Krupp Industrietechnik GmbH (D-Duisburg), subsequently cited as "Krupp."

The Opera de la Bastille in Paris is particularly generously equipped in terms of space. Figures 1.8 and 1.9 show that a large number of additional stage areas – including a rehearsal stage – are available both at stage level and in the basement. On both levels, *stage wagons* can be moved in ring traffic; their vertical transport is carried out with the *lifting platforms* of the main stage. A *shunting turntable* in the rear stage enables the stage wagons to be swiveled.

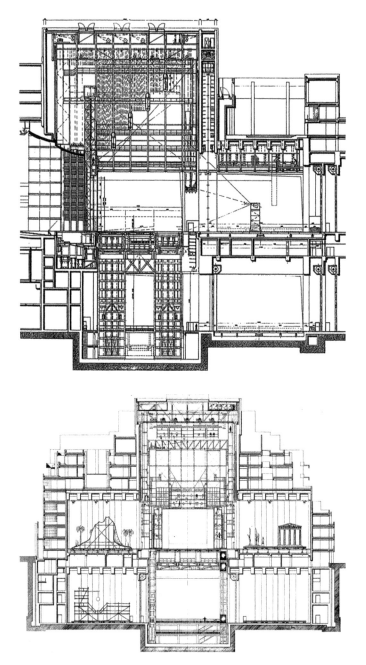

Figure 1.8: Opera de la Bastille, Paris – longitudinal and transversal section. Photo credits: BTR 5/1989, R. Biste–K. Gerling, Architektur- und Ingenieurbüro (D-Berlin).

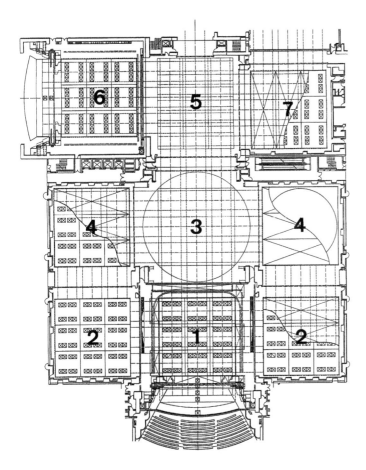

1	main stage	6	rehearsal stage
2	side stages in front	7	additional stage
3	back stage with turntable for shunting		Below these stage areas in the lower stage ad-
4	side stages behind (with a turntable stage car)		ditional stages and a turntable for shunting are
5	space for shunting and preparations		situated

Figure 1.9: Opera de la Bastille, Paris – floor plan. Photo credits: BTR 5/1989, R. Biste–K. Gerling, Architektur- und Ingenieurbüro (D-Berlin).

In opera houses and music theaters, there is an *orchestra pit* just before the proscenium opening. If several orchestra platforms are available (see Section 1.7.1), the size of the orchestra pit can be adapted to the number of musicians. In locations where performances also take place without an orchestra, this area can be used in different ways: depending on the lifting position of these podiums, this area can then serve as a playing area – a so-called *forestage* – or be assigned to the *auditorium* with additional seating, or an orchestra pit can be created (see also Section 1.7.1).

Reference should also be made to the possibility of making the floor in the auditorium in front of the orchestra podium tiltable so that it can be adapted to an inclined surface of the auditorium. This is possible, for example, in the *Kulturpalast Dresden*.

At this point, the special design of the orchestra room in the *Festspielhaus Bayreuth* (see Figure 1.10) should also be mentioned. With this arrangement, Richard Wagner wanted to ensure that the audience would not be distracted by looking at the conductor and the musicians. In addition, the sound cover in front of the stage is intended to merge the sound of the individual instruments into a muffled overall sound of the orchestra.

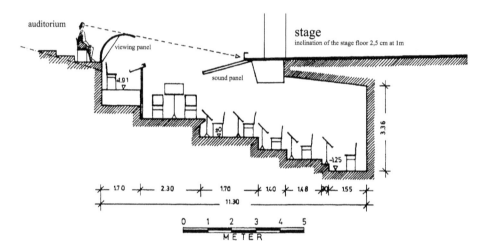

Figure 1.10: Festspielhaus Bayreuth – orchestra arrangement. Photo credit: Bosch Rexroth.

1.4 Stage systems

An essential transport-technical task is to enable a quick change of the stage set by removing decorations used on the stage from the playing area and taking already prepared complete sceneries or scenery elements from ready positions for the play action. Various stage systems serve this purpose. Various possibilities have already been mentioned in the historical review in Section 1.1; here, the most important concepts from today's point of view will be presented in summary.

Decorative elements suspended in the fly grid can be moved by lowering them into or lifting them out of the view of the audience. This allows the stage set to be changed in a simple manner.

If we disregard the scenery carriages mentioned in Section 1.1, which can be moved sideways in free runs, there are in principle the following possibilities for moving spatial decorative elements standing on the stage floor into or out of the stage area visible to the audience:

– *Vertical movement* of the entire or parts of the stage floor. This concept is named as a *lifting platform stage system* or *lowering platform stage system* ("*Versenkbühnen-system*" in German) (Figure 1.11(a)) and is actually only used in combination with other systems. As an example, we refer to the possible use of the *double-deck podiums* of the *Hamburg State Opera* (see Figure 1.13). The front podiums are double-decker podiums (see Section 1.7.1) with a distance of 10 m between the two accessible levels. Since the lower podium surface can be raised to level 0.0, two stage sets can be moved into the view of the audience as desired, depending on the podium position. However, for the stage set on the lower podium area, the upper stage equipment on the fly grid cannot be used. Therefore, simple proscenium hoists are installed below the upper podium area.

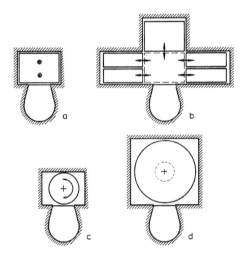

Figure 1.11: Stage systems – basic alternatives: (a) elevating stage system, (b) sliding stage system, (c) revolving stage system with a turntable and (d) revolving stage with a large transportation turntable (full or ring disc).

– *Horizontal translational movement* of the stage floor. This concept is called *sliding stage system* or *carriage stage system* (Figure 1.11(b)). It is realized in all those cases where stage wagons can be moved between main and side stage and/or main and back stage. To compensate for the level differences caused by the height of the stage wagons, compensating platforms (see Section 1.7.1) are usually installed.
– *Horizontal rotation of the stage floor* called *revolving stage system* (Figure 1.11(c)). The revolving stage can be a simple turntable (Figure 1.157(a, b)) as also used, for example, in the "*Theater in der Josefstadt*" in Vienna (Figure 1.156) or with two rotating levels in a double-decker design (Figures 1.157(c) and 1.160). However, also a foldable or dismountable turntable (Figures 1.147 and 1.148) can also be placed on the stage floor.

The stage concept should allow a turntable with a diameter larger than the width of the stage opening in the proscenium. Only then can an extensive decoration setup be accommodated on the revolving stage floor, as no excessively large side segments of fixed stage construction remain.

Figure 1.11(d) (without dashed circle) shows a giant revolving stage – also called "*transport revolving stage*". It allows several extensive decorative setups to be rotated into the audience's field of view. Figure 1.21 (*Frankfurt Opera*) and Figure 1.22 (*Linz Music Theater*) show two specific designs. They will be described in more detail later.

Figure 1.11(d) (with dashed circle) shows a large circular turntable. The circular central surface cannot be rotated and can serve as a magazine. Stage sets can be arranged side by side in a circle on the circular ring with sufficient width.

Very often, combinations of the three basic movement options can be found:
- *Platform stage with stage wagons* (*lifting-sliding stage*) (see Figure 1.12(a)). In the side stages, the back stage and, if necessary, in the lateral edge areas of the main stage, there are compensating platforms whose lift corresponds to the height of the stage wagons. Below the main stage there is a lower stage, equipped with lifting platforms of large travel range. They enable vertical transport between both levels. Thus, decorations can also be provided on stage wagons in the lower stage. Of course, these podiums can also perform the function of equalizing podiums.

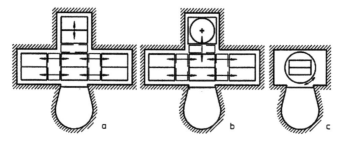

Figure 1.12: Stage systems – combined systems: (a) podium stage with stage wagons, (b) podium stage with stage wagons, in particular a stage wagon with an integrated turntable in the back stage, and (c) cylindrical revolving stage with built-in lifting platforms.

As an example, we refer to the *Vienna State Opera* (Figure 1.14). However, in this case there are no large stage wagons in a side or back stage, but only small manually movable *auxiliary stage wagons*.

The function of the lifting and sliding stage system can be explained well on the basis of this illustration: In the drawn position, the stage platforms H_1 to H_3 are raised to stage level and the platforms H_4 to H_6 are lowered to the lowest level into the lower stage. The stage wagons B_1 to B_3 are designed as *bridge wagons* (see Section 1.7.2), cover the pit above the podiums H_4 to H_6 and form a closed stage area.

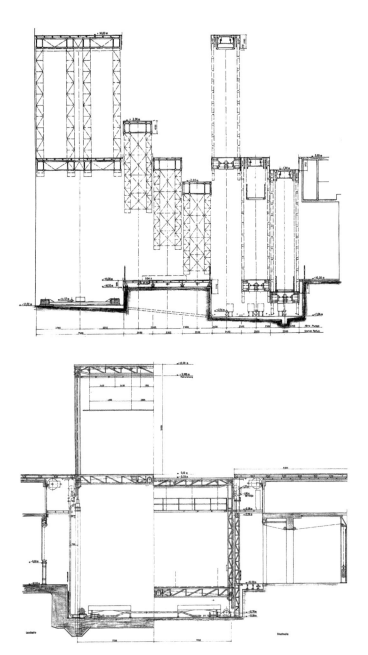

Figure 1.13: Double deck lifting platform – Hamburg State Opera: (top) lifting platforms – longitudinal section, (bottom) lifting platform with driving system – cross-section. Picture credits: SBS.

New decorations can now be brought into the main stage, for example, by lowering the podium group H_1 to H_3, moving the stage wagons B_1 to B_3 to a position above this podium group, and raising the podiums H_4 to H_6 to stage level.

If scene setups placed on podiums H_4 to H_6 are to be moved toward the prosce-nium, these scene setups can be placed on auxiliary stage carts stored in "A" and thus moved to the desired position on the stage. The equalizing wagon "E" installed in the understage can be moved to podiums H_1 to H_3 when podiums H_4 to H_6 are raised, in order to close the pit in the understage so that, for example, the above-mentioned auxiliary stage trolleys can be moved.

– *Sliding stage with turntable stage wagon*. If a turntable is installed in a stage wagon, usually the backstage wagon, as shown in Figure 1.12(b), the sliding stage and turntable systems are combined. The *Semper Opera House in Dresden* (Figure 1.73) is mentioned as an example. Furthermore, reference is made as an example to the turntable stage car in the *Opera de la Bastille in Paris*, which can be seen in the rear side stage in Figure 1.9.

– *Podium stage with a turntable to be placed on the stage floor*: In this concept, a turntable that has been stored in a disassembled or folded state is placed on the podium stage if required (see Section 1.7.3 or Figures 1.14, 1.147 and 1.148).

– *Turntable with built-in lifting platforms – cylinder revolving stage* (Figure 1.12(c)). In this system, lifting platforms are built into a rotating cylinder. By additionally using stage wagons, this solution can also combine elevating, transfer and revolving stage systems.

As an example of revolving stages with built-in lifting platforms, reference is made to the cylinder revolving stage of the *Vienna Burgtheater* shown in Figures 1.15 and 1.16. In this cylinder revolving stage, four lifting platforms in double-decker design (see Section 1.7.1) and an equalizing lifting platform "E" below each platform are accommodated. Furthermore, two bridge trolleys (see Section 1.7.2 and Figure 1.133) are installed in the revolving cylinder at stage level. This allows a closed level to be formed at each position of the lifting platforms, at stage level (0.0 m) and in the lower stage at level – 8.82 m. The lifting platforms can then be positioned at these levels in the lower stage. Magazined auxiliary stage carts can then be moved on these levels in the lower platform.

If we compare this cylinder revolving stage system with the podium stage shown in Figure 1.14, we can see that there are also lifting podiums – in this case 4 instead of 6 – and bridge carriages – in this case 2 instead of 3. In the lower platform, instead of the horizontally movable compensating carriage, there are four compensating platforms, which, depending on which platforms are raised, form a closed level at level – 8.82 m.

If decorations are made according to the cycle method described for Figure 1.14, the installation of this podium system in a cylinder revolving stage still offers the possibility of turning the revolving stage by 180° to turn the raised podiums back to the proscenium, for example.

However, four *special solutions of revolving stage systems* should also be mentioned as examples:

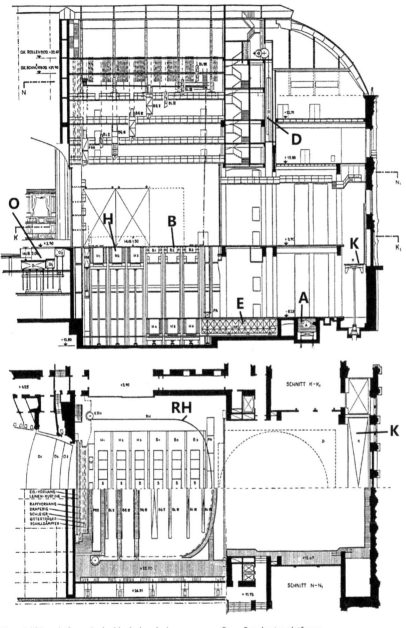

H	6 lifting platforms in double decker design	O	3 orchestra platforms
	H_1–H_3 in highest position,	PA	backdrop elevator
	H_4–H_6 in lowest position	K	elevator for road trucks to position them at stage
B	3 stage wagons (designed as bridge wagons)		or understage level
E	equalizing wagon (can be positioned below a lifting	A	storage for stage auxiliary trolleys
	platform being raised)	RH	cyclorama
D	rotary disc deposited on stage (rotary disc foldable)		

Figure 1.14: Vienna State Opera – longitudinal section and floor plan. Picture reference: Merkblätter für sachgerechte Stahlverwendung – Issue 289.

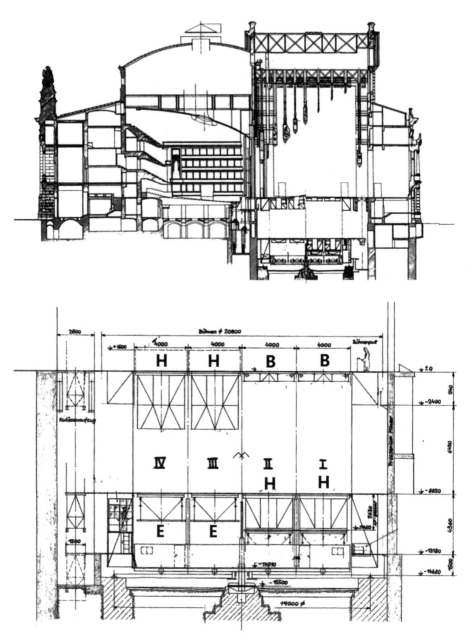

H   lifting platform
E   equalizing podium
B   stage wagon as bridge carriage

Figure 1.15: Vienna Burgtheater: (top) longitudinal section, (bottom) cylinder revolving stage. Photo credit: Waagner-Biró.

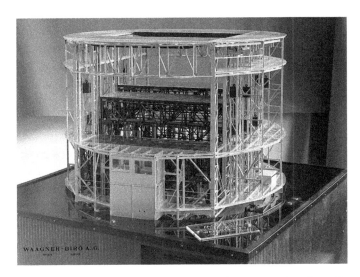

Figure 1.16: Vienna Burgtheater – cylindrical revolving stage with podiums. Photo: Waagner-Biró.

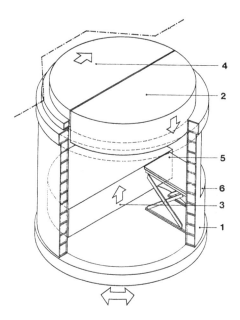

1	bearing structure	4	rotating segment for equalizing
2	lifting platform A	5	lowerable loading platform
3	lifting platform B	6	opening for transportations

Figure 1.17: National Theatre London – schematic sketch of the cylindrical revolving stage. Picture credits: BTR 3 1977.

- Figure 1.17 shows the principle of the revolving cylinder stage of the *National Theatre London*. Two lifting podiums with a semicircular shape are installed in the revolving cylinder. The opening at stage level created by lowering a podium can be closed by a disc rotatably mounted in the cylinder. This semicircular disc thus assumes the function of the movable bridge carriage just described as a rotatable element.
- Figures 1.18 and 1.19 show another special solution: In the *Vienna Volksoper*, a rotatable *core disk* is mounted on a cylindrical lifting podium, and this unit is surrounded by a *ring disk*. Thus, independently of each other, rotational movements of ring and core disk and a lifting movement of the core cylinder can be superimposed. For scenic effects, for example, the core disk can move upward by superimposing a rotation of the disk with a lifting motion of the core cylinder, while the ring disk rotates in the opposite direction.

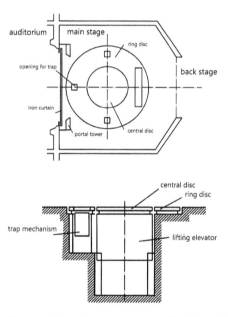

Figure 1.18: Revolving stage system of the Vienna Volksoper: (a) floor plan, (b) central disc and ring disc. Picture credit: Waagner-Biró.

- *The Theater am Goetheplatz in Baden-Baden* (Figure 1.20) is also equipped with a core and ring disc. The core disc is designed as a centrally mounted cylinder revolving stage analogous to Figure 1.157(c) and contains two traps, the ring disc runs with circular rails on a large number of support rollers (fixed rollers) with Vulkolan lining.
- Particularly noteworthy is the concept of the *large revolving stage* in the *Frankfurt am Main Opera House* shown in Figure 1.21. The revolving stage includes not only

Figure 1.19: Revolving stage system of the Vienna Volksoper – liftable cylinder with central disc. Photo credits: Waagner-Biró.

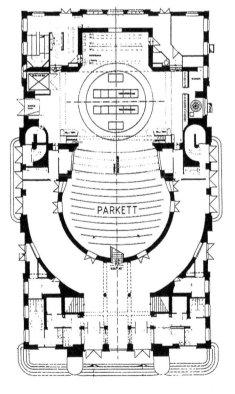

Figure 1.20: Revolving stage of the "Theater am Goetheplatz in Baden-Baden" with central disc and ring disc. Picture credits: BTR special issue 1995.

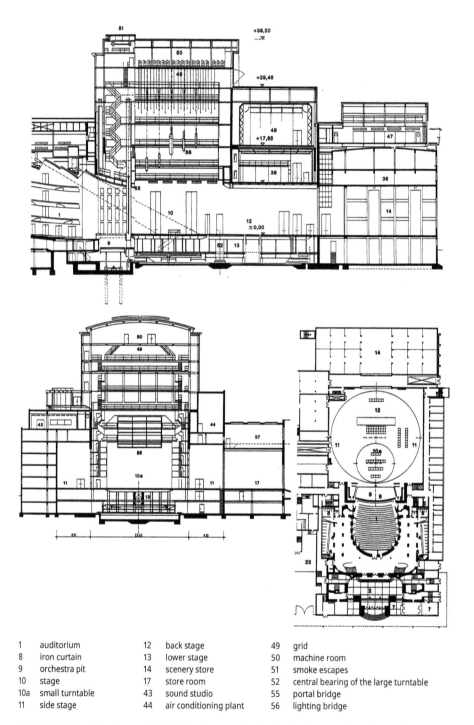

1	auditorium	12	back stage	49	grid
8	iron curtain	13	lower stage	50	machine room
9	orchestra pit	14	scenery store	51	smoke escapes
10	stage	17	store room	52	central bearing of the large turntable
10a	small turntable	43	sound studio	55	portal bridge
11	side stage	44	air conditioning plant	56	lighting bridge

Figure 1.21: Revolving stage of the Frankfurt Opera – longitudinal an transversal section and top view. Picture credits: BTR 2/1992.

the main stage but also the backstage (and small side-stage areas). A very large turntable with a diameter of 37.4 m has a small turntable with a diameter of 16 m built in eccentrically. The large turntable is used primarily for preparation and staging for repertory and rehearsal operations, while the small turntable can be used in the function of a normal revolving stage. The small turntable roughly corresponds to the size of the stage and can be partitioned off from the rest of the large turntable by means of protective gates.

– With the opening of the *Musiktheater Linz* in spring 2013, there is now another modern example of a large revolving stage in Austria (see Figure 1.22): A double-deck cylinder revolving stage with a diameter of 32 m (upper deck 32.0 m at level 0.00, lower deck 31.3 m at level –3.95 m) is installed as *transport revolving stage*. Depending on the rotational position of this transport revolving stage, there is then either a podium stage with three lifting podiums of 15 m by 4 m each or a double-deck cylinder revolving stage as a *performance revolving stage* with a diameter of 15 m in the main playing area.

As a result of the very low construction height of 6 m (there still had to be room for a garage below the stage and the groundwater level was not to be undercut), the transport disc must be supported, on the one hand, by a circular ring rail and, on the other hand, by an additional bearing in the form of a ball bearing slewing ring arranged in the center.

The performance turntable (play turntable) is also mounted on a circular rail and a ball bearing slewing ring.

Due to the low overall height, the podiums had to be inserted in lateral guide elements that can also be moved so that a telescopic highest position of +4.15 m and a lowest position of –3.95 m can be achieved. Compensating platforms underneath these podiums create level surfaces at the bottom when the play podiums are raised. More details about the podiums can be found in Section 1.7.1, revolving stages are described in more detail in Section 1.7.3.

There are also two side stages and a backstage area, with one side stage acoustically separable so that it can be used independently as a rehearsal stage.

Furthermore, a total of six *stage wagons* are available – three in the side stage and three in the back stage. A maximum of three carriages can also be moved to the transport turntable and lowered to level 0 with the podiums, but they can also be moved to the rehearsal stage so that rehearsals can also be held in the rehearsal stage with original decorations.

Figure 1.22 also shows that a *scenery store* and a *drop store* are located near the stage; both are described in Section 1.5.

Rehearsal stage

If the spatial concept allows and the operational management requires it, a *rehearsal stage* is located near the stage, and it should be possible to transport stage wagons to it.

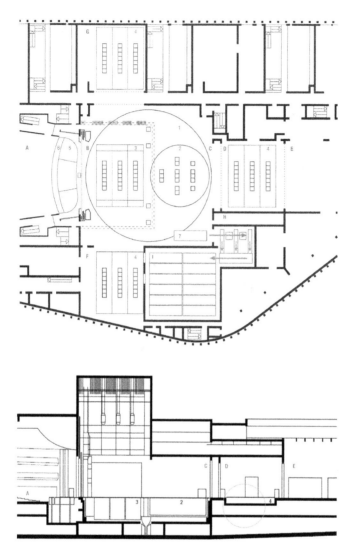

A	auditorium	1	turntable for transportation (transport revolving stage)
B	stage area for performance		
C	stage	2	turntable as acting area (performance revolving stage)
D	back stage		
E	area for preassembling	3	lifting platforms
F	side stage	4	stage wagons
G	rehearsal stage for opera	5	orchestra platforms
H	drop store	6	pasarelle (movable crossover)
I	scenery store with elevator for transport pallet	7	transport pallets for sets and sceneries

Figure 1.22: Musiktheater Linz – floor plan and longitudinal section. Picture credits: Landestheater Linz.

The *Copenhagen Opera* (Figure 1.6), the *Paris Opera* (Figures 1.8 and 1.9) and the *Musiktheater Linz* (Figure 1.22) can also serve as examples for this.

Television studios

The requirements profile for stages in TV studios is fundamentally different. There is no need for understage equipment at all. Instead, a particularly flat and smooth studio floor with a very high load-bearing capacity is required, as mobile technical equipment must be able to move freely. For this reason, wooden floors such as those used in theaters are not possible. A lighting grid of special design is common as an upper-stage facility in the ceiling area. In addition to catwalks for operation and maintenance, a light grid ceiling is formed from ceiling panels with longitudinally and transversely oriented slots, which offers the possibility of moving telescopic suspension tripods for spotlights, possibly also via switch systems, and of using point hoist winches as hoists (Sections 1.8.3 to 1.8.5). In addition, a cyclorama system is usually available for limitation of the playing area (Section 1.8.6).

Stages in ships

Special requirements are necessary for stage equipment installed in ships, as a ship is subject to lurching movements due to wave action.

In rough seas, this can result in additional accelerations of more than $1g$ (g = $9.81\,\mathrm{m/s^2}$) in all directions. This means that objects lying on the ground can lift off. Of course, theater can no longer be performed in such situations, but stage elements must be appropriately secured. For example, stage wagons must be clamped so that the wheels cannot come off the rails. If "spiral lift" elements are installed for lifting platforms, these must also be locked so that the folding spindle elements do not lose their cohesion. The new I-Lock series could provide the necessary positive locking. However, special chains from Serapid are also possible (see Section 1.7.1).

However, since additional forces due to wave action can also occur during play operation as a result of ship movements, the bars of bar hoists, for example, must be guided to prevent oscillation, and adhesion drives of stage wagons that rely on their own weight must be avoided.

Figure 1.23 shows the stage of a cruise ship. It has been equipped with stage wagons, turntables, balancing platforms, orchestra podiums with spiral lifts, etc. Lighting and sound technology are integrated in the computer control system so that stage technology sequences can be called up completely automatically.

1.5 Transport and storage systems

The previous section dealt with transport systems for changing stage sets. Decorative elements or entire stage sets that are not currently in use are stored in off-stage areas if

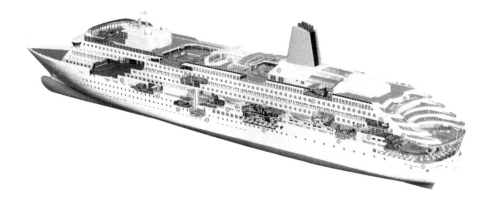

Figure 1.23: Cruise liner "Aurora", Meyer Werft – Papenburg/Germany. Model of the ship and view into the theater hall. Photo credit: Waagner-Biró.

possible. However, decorative elements must also be stored in or outside the venue for longer periods of time.

Drop store magazines in the stage area
It is particularly important to create storage facilities for the long and therefore difficult-to-transport drop rolls in the immediate vicinity of the stage. *Drop storage facilities* of various types are used for this purpose:

There is often a storage room near the stage with cantilever racking that can be operated from an elevator. The stage personnel can use the elevator to drive up to the desired rack height and store or retrieve the drop rolls. A storage room located to the side of the stage should be arranged so that the drop rolls can be brought in or out parallel to the proscenium without swiveling.

As can be seen in Figure 1.24, a *drop rack storage system* with a transport elevator has been installed in the backstage area of the *Musiktheater Linz*. For this purpose, the drops are placed in trays with internal dimensions of 17.16 m × 0.72 m × 0.40 m (L×W×H).

A similar solution is the drops store of the *National Theater in Munich* (Figure 1.26). It is a rack storage system located next to the stage with a storage and retrieval machine.

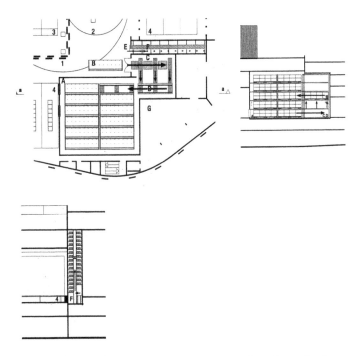

A	store for sceneries	F	elevator for drops
B	transport pallet for sceneries	G	workshops
C	access to the store for sceneries/roller conveyor	1	turntable for transportation
		2	turntable for performances
D	elevator platform	3	lifting platforms
E	access to the drop store	4	stage wagon/side stage right/back stage

Figure 1.24: Store for sceneries and drops at the "Musiktheater Linz". Photo credit: Landestheater Linz.

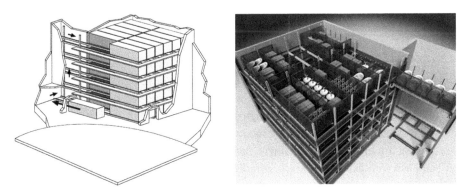

Figure 1.25: Scenery storage at the Musiktheater Linz. Picture credits (left): Landestheater Linz, Picture credits (right): Unitechnik Systems GmbH.

On the left and right, 14 cantilever racking levels are available on top of each other. The special feature of this store is that each of the *tub-like container units* consists of five individual tubs, one behind the other. In the rack, they must therefore be connected with top frames to form a load-bearing overall unit. During retrieval, they are placed on a transport carriage consisting of five articulated carriages and will be separated from the top frame, which will remain in the drop store. The carriages run on swivel rollers and can thus be steered around curves at narrow space conditions on the stage.

Figure 1.26: Store for drops at the National Theater in Munich: (left) rack with stored container tubs and trolleys, (center) spreader with top frame, (right) spreader with top frame locked with the containers. Picture credits: BTR 4/1989.

An automated store system for drops was installed at the *Hanover Opera House*. There are 37 containers measuring 18.90 m × 0.60 m × 0.56 m available for storing the drops for about 30 performances. Each container weighs about 800 kg and can be loaded with 1300 kg. The containers are manipulated in the drop store with a stacker crane, they can be moved on the stage with swivel castors.

Another interesting solution is the paternoster system installed by Schenck Handling in the *Berliner Ensemble* for storing drops and other long items (see BTR 2/2001).

An operationally interesting solution was realized in the *Paris Opera de la Bastille*. In Figure 1.8 (longitudinal section), a high two-row drop storage area with 2 × 20 = 40 storage areas can be seen in the front section of the backstage, where the drop-rolls can be stored in tub-like storage containers. Between the compartments, a container elevator equipped with passenger cabs is installed, which can be lowered to stage level. However, during normal drop changing operations, the containers are placed on a wagon bridging the stage width that can accommodate two containers and is installed at the level of the first work gallery. Figure 1.27 shows the drop storage area with these handling devices and, in particular, this bridging wagon like a crane bridge, which can be moved from the proscenium position to the wall of the back stage. This was intended to make it possible to carry out these operations independently of assembly work taking place at stage level. However, the bridge wagon is often hindered in its mobility during operation because of hanging decorations.

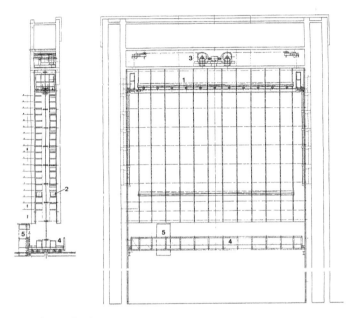

1   elevator for drops
2   container for drops
3   machine room
4   trolley for drops
5   mobile elevating working platform

Figure 1.27: Store for drops at the Paris Opera. Photo credit: BTR 5/1989.

The storage room can also be located below the main stage and served by an elevator designed as a *dropstore elevator* as in the *Semperoper Dresden* according to Figures 1.73 and 1.74 (position 29).

Another possibility is to place the prospect rolls also in the lower stage, but in a lifting podium of the main stage. In this case, not the handling elevator, but the rack is moved vertically and the manipulation of unloading and loading can be done from the stage level. However, this requires moving relatively heavy masses. In most cases, the last stage platform – viewed from the auditorium – is designed as a drop store elevator (see also Section 1.7.1, Figure 1.93). Such a drop store elevator is installed, for example, in the Zurich Opera House and can be seen in Figure 1.28 (position 7).

An extension of the storage capacity of a drop lifting podium without additional podium drives is possible by choosing a system according to Figure 1.29, as it was im-

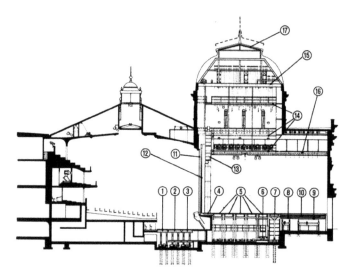

1	auditorium podium with chair wagon	9	back stage podium II
2	orchestra podium I with drag floor	10	stage wagon with turntable
3	orchestra podium II with drag floor	11	iron curtain
4	equalizing stage wagon with tiltable stage floor	12	main curtain
5	5 lifting platforms with tiltable stage floor	13	portal bridge
6	table size trap mechanism, movable underneath the	14	3 fly galleries
	stage floor	15	grid – main stage
7	podium as drop store elevator with tiltable stage floor	16	grid – back stage
8	back stage podium I	17	dome as smoke escape

Figure 1.28: Zurich Opera House – longitudinal section through stage and auditorium. Photo credit: MAN Gutehoffnungshütte AG (D-Oberhausen), hereafter referred to as "MAN" for short.

plemented in the *National Theater in Warsaw*. The drop storage system is designed as a three-part lifting shelf, shelf 3 is equipped with a lifting drive, and shelf 1 or 2 can be optionally coupled to shelf 3 and moved.

Drop store elevators are also installed in the *branch theater of the Bolshoi Theater in Moscow* (see Figure 1.255) and *Minsk National Academic Grand Opera and Ballet Theaters* (see Figure 1.94). With reference to Section 4.3 (Figure 4.20), it should also be noted that the proscenium lifting platforms in the Bolshoi branch theater are driven by roller screw spindles (planetary spindles).

Scenery store in stage area

In the completely newly built Musiktheater Linz, a large decoration magazine was integrated into the building (see Figure 1.25).

The scenery store offers a total of 55 storing positions for containers measuring 8.2 m × 2.39 m × 2.16 m (L × W × H) and with a payload of 2.5 t. The storing positions are arranged in 5 levels, each with 11 storage bays. The containers can be moved in longitudinal and transverse direction by means of roller and chain conveyors in such a

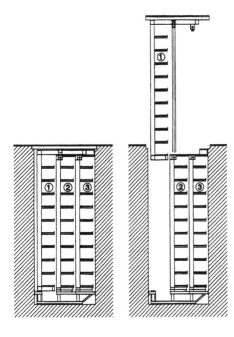

Figure 1.29: Warsaw National Theater. Three-part drop store elevator. Picture credits: Bosch Rexroth.

way that clockwise rotation is possible. The individual containers are then transferred within the floor until the desired container reaches the transfer area to the elevator.

During storage, the containers are placed on a pallet truck and inserted into the elevator via a roller and chain conveyor system. The removal of a container from storage takes up to 50 minutes, depending on its location in the store. In daily operation, the storage system is only used outside performance and rehearsal times and after the end of a performance. Preferably, the retrieval orders are entered in the evening and processed during the night.

Thus, a logistics concept applied in airport technology was realized in the area of stage technology. The containers are stored chaotically, i. e., a specific storage location is not assigned during storage, but a free storage location is simply selected.

Scenery depots spatially separated from the venue
However, because of the usually far too little space available in theaters and opera houses, decorations also have to be stored outside, often in scenery depots several kilometers away. This means that the operation of a large stage often requires not only transport solutions in the sense of internal conveying and storage technology, but also transport technology through the use of special transport vehicles approved for public transport for the transport of decorative elements between the theater and the scenery depot.

For this purpose, solutions common as general transport and storage technology, such as those used in production and distribution. Decorations have to be transported or manipulated with special vehicles and stored in racks. Favorable transport conditions should be provided for loading and unloading the road transporters and for bringing the decorations into and out of the venue. Often, the stage and road are at different levels, so that suitable means of transport are required to overcome this difference in height. In Figure 1.14, a large *scenery trolley elevator* – marked "K" – can be seen on the back wall of the backstage of the *Vienna State Opera*, on which special trailers for transporting decorations can be placed without a towing vehicle and moved vertically.

If the lifting device for the transport vehicle cannot be accommodated within the building, an *external truck-mounted lifting platform* can also be installed if space per-

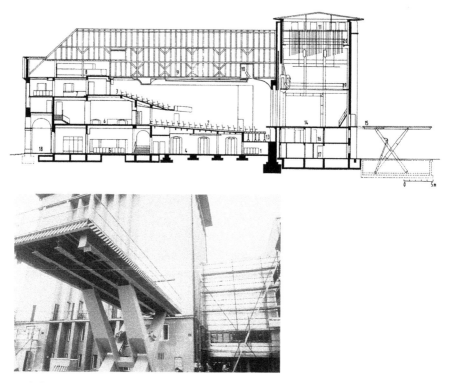

1	buffet	11	grid
2	auditorium parquet	12	portal bridge
3	auditorium balcony	13	orchestra pit
4	foyer / ground floor	14	stage
5	cloak room	15	truck elevator
6	foyer / parquet	16	cloakroom area
7	foyer / balcony	17	area for technicians
8	air conditioning plant	18	arcade hall
9	attic floor	19	fly gallery
10	lighting bridge	20	working galleries

Figure 1.30: Theater am Kornmarkt, Bregenz – Docking ramp at the rear of the stage building: (top) transversal section, (bottom) photograph of the docking ramp. Picture credits: Brochure 1995/issue 1, Meyer Stahl- und Anlagenbau - Nütziders.

Figure 1.31: High-rack store with 4-way forklift in the logistics center of the Württemberg State Theater in Stuttgart. Picture credits: Grösel.

mits, as was done, for example, at the *Theater am Kornmarkt in Bregenz* (see Figure 1.30): A platform of size 12 m × 3.8 m with a lifting capacity of 16 t (statically 32 t) can be raised 5.6 m above street level as a scissor platform. In addition, the lifting platform can be moved approx. 0.8 m horizontally to the building in order to bridge a structurally conditioned gap to the building and dock with the building.

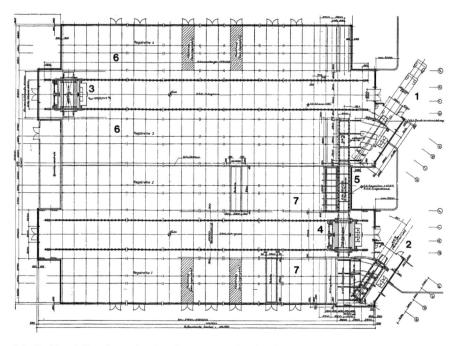

1, 2   Docking stations for truck trailers for storage and retrieval
3, 4   Rail-bound storage and retrieval unit (stacker-crane)
5   Roller conveyors for cross transport
6   Racks served by stacker crane 3
7   Racks served by stacker crane 4

Figure 1.32: Decoration warehouse in Munich/Poing – floor plan. Picture credits: BTR 4/1989.

The logistics center of the *Württemberg State Theater in Stuttgart* should be mentioned as an example of a central store: it was put into operation in June 2006 and is responsible for the entire logistics services of the three divisions of opera, drama and ballet. The store extends over three floors. In the assembly area, an in-house locksmithery, carpenter's workshop, turner's workshop, etc., are used to manufacture, assemble, fit and install scenery. Scenery is manufactured, assembled, adjusted and tested for load-bearing capacity. In a high rack store (see Figure 1.31), decorations can be stored on four levels on a total of 560 scenery carts. The scenery carts offer a surface area of 8.0 m×2.3 m and can accommodate a weight of 4 t. The scenery carts are manipulated with a 4-way forklift (total weight of the special forklift is 18 t).

A modern solution in terms of transport and storage technology has been realized in *Munich*. The decorations are transported and stored in special *decoration containers* that can be loaded from the long or short side. *Transport trolleys* with roller conveyors on the loading area are used to move the containers inside the venue. Likewise, the *truck trailers* are equipped with roller conveyors on their loading area. In Poing near Munich a three-story rack storage facility with four storage rows and two storage aisles is available as a scenery depot to accommodate a total of 457 containers. Storage and retrieval are carried out automatically by two rail-mounted storage and retrieval machines. Figure 1.32 shows a floor plan of the store. Figure 1.33 explains the container manipulation at the venue and Figure 1.34 that in the depot. This container concept implemented in Munich eliminates reloading and sorting operations; just as the containers are loaded on the stage, they are also unloaded when they are used again for the play. This naturally results in major advantages in terms of operational organization.

Figure 1.33: Decoration magazine in Munich/Poing – pallet manipulation at the venue: (top) road truck at the docking station and off-loading of the container to the slewable roller conveyor, (bottom) loading of a container on the truck. Photos: SHS Fördertechnik (D-Darmstadt).

If containers are to be used for transporting decorations in international guest performance operations, standardized ISO containers should generally be used. However, there are operational disadvantages due to the smaller dimensions and more difficult manipulation (restricted access).

Figure 1.34: Decoration magazine in Munich/Poing – pallet manipulation in the decoration magazine: (top, left) loading of a container for sceneries; (top, right) transport trolleys with roller conveyors; (bottom, left) loading of the storage and retrieval machine (stacker crane); (bottom, right) the three-story rack storage facility. Photos: SHS Fördertechnik.

1.6 Design of multipurpose rooms and halls

In a theater designed exclusively for performance purposes with a proscenium stage, the auditorium cannot be changed besides minor variations in the orchestra area. Multipurpose halls and modern experimental and arena stages usually offer more extensive design possibilities; in particular, the seating arrangement can be varied by means of podiums and grandstands.

Flexible design of the auditorium
An interesting example of great flexibility is the *Schaubühne Berlin* (see Figures 1.35 and 1.36). Its purpose was to create a theater space whose use is not predetermined by stage

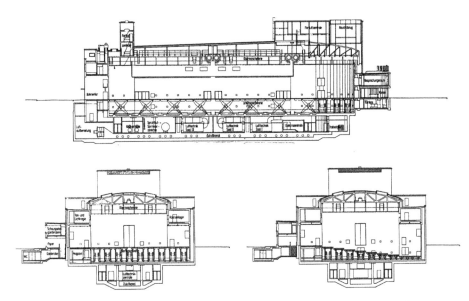

Figure 1.35: Schaubühne Berlin: (top) longitudinal section; (bottom, left) transversal-section; (bottom, right) transversal-section with lifting platforms in different positions. Picture credits: BTR 2/1982.

equipment and grandstands in a fixed spatial arrangement. For this reason, the *entire auditorium floor* is equipped with 76 hydraulically driven *scissor-type lifting platforms* (see Section 1.7.1). In the roof area, a light grid ceiling is available as a working floor, primarily for the installation of point hoists and light fittings.

Figure 1.37 shows the *Kuwait Conference Center* – a facility meanwhile destroyed in the acts of war. It is a multipurpose space that could be used both as a theater and as a conference center. To *transform the auditorium*, it too is equipped with lifting podiums so that a level or stepped floor can be created.

Figure 1.38 shows the hall of the *Kassel conference center*. Regarding the folding spindles for the podium drives, please refer to Section 1.7.1.

The *Berlin Congress Center* is equipped with a *grandstand system* that can be lowered from the hall ceiling. Figure 1.39 shows the grandstands both in the lowered position for theater, concerts, lectures and congresses and while being raised for rearrangement for banquets, balls and similar events. (The lifting chains are pulled into the hall ceiling after the lowering process.)

In the *Stadthalle Reutlingen*, the hall can be used in 2 variants: as a slope with a stepped floor or as a horizontal surface. The drive of the platform and the drive of the steps serve for this purpose. The area of the hall slope is 19.40 m × 22.50 m, that of the steps approx. 1 m × 19.40 m. Figure 1.40 shows the hall with seating in the "hall slope" variant.

The mechanism for forming the hall incline can be seen in Figure 1.41: The lifting bars are lifted 4.45 m at the end by a *push chain* and are supported in the final position

Figure 1.36: "Schaubühne" Berlin: (top) modification of the floor with scissor lifts in different stroke positions, grid over the entire space for flexible use of the upper machinery; (bottom) scissor lifts in the lower stage. Photos: Otto Vogel KG, Theaterbühnenbau; Stahl-, Metallbau GmbH & Co. (D-Berlin), hereinafter referred to as "Vogel" for short.

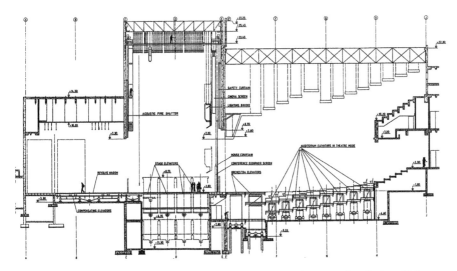

Figure 1.37: Kuwait Conference Center – longitudinal section through stage and auditorium. Main stage with double-decker podiums, backstage with equalizing platforms and turntable stage wagon, auditorium with orchestra podiums and hall podiums. Drawing: Waagner-Biró. BTR 2/1989.

Figure 1.38: Kassel conference center: (left) festival hall with lifting platforms, (right) platform drive with folding spindles ("Spiralift"). Picture credits: Sächsische Bühnen-, Förderanlagen- und Stahlbau-GmbH (D-Dresden), hereafter referred to as "SBS".

by the pendulum supports. The pendulum supports are passive, i. e., they are not driven separately but unroll only. The plate, built by the lifting bars, is lifted slightly, then bolts move into the track of the pendulum supports and the slab is lowered until the bolts stop the pendulum supports in a defined end position. The time required for lifting the floor is approx. 4 min.

The formation of the stages is carried out by electric driven spindle drives.

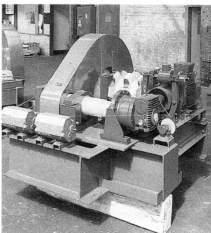

Figure 1.39: Congress Center Berlin – grandstand system that can be lowered from the hall ceiling: (top) lowered position, lifting chains not yet unhooked and raised; (bottom, left) when raising the tribune; (bottom, right) chain hoist. Photos: Krupp.

A similar transformation of the hall is also taking place in the *"Giant Amber Concert Hall" in Liepaja in Latvia*. Waagner-Biró has developed a special articulated kinematic system that uses a drive to manage both the inclination of the support system and the formation of steps. This compensates for the displacement of the podiums caused by pivoting so that they are lifted vertically. The drive is provided by worm gear screw jacks.

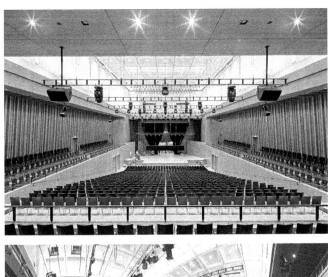

Figure 1.40: Stadthalle Reutlingen – different inclined positions of the floor. Picture credits: SBS.

The situation can be seen in Figure 1.42. The podium in front is an orchestra podium with spiral lift drive.

Different use of stage and audience area

Variable design options are also available at the *New Concert Hall in Athens*, allowing the building to be used for concerts, theater performances and congresses (see Figure 1.43).

A particularly elaborate hall transformation technique was realized at the *Royal Opera House Muscat* in *Oman*. The house can be used as an opera or concert hall. Both variants are shown in Figures 1.44 and 1.45. The core element for the transformation is a self-propelled concert shell with a total deadweight of 525 t, which is moved to the main stage in concert mode and to the backstage in opera mode. Twenty eight Demag wheel blocks (eight of them are powered) serve as the travel gear. The concert shell also houses an organ (approx. 30 t), two acoustic ceilings and 22 acoustic flaps.

The two proscenium boxes with three floors are moved to different positions depending on their use. For concert mode, they are rotated by 15° and retracted by 800 mm.

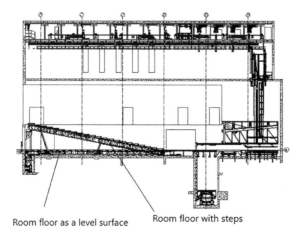

Room floor as a level surface   Room floor with steps

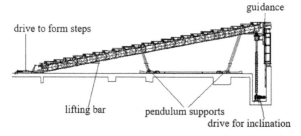

drive to form steps

guidance

lifting bar   pendulum supports

drive for inclination

Figure 1.41: Reutlingen City Hall – mechanism to realize different inclinations of the hall floor for forming the hall slope. Picture credits: SBS.

Columns 14.8 m high are also moved outward by the same amount using linear drives. The size of the portal opening can be changed in width by sliding walls (rack and pinion drive) and in height by a panel hoist (built as tubular shaft hoist – see Figures 1.219 and 1.220).

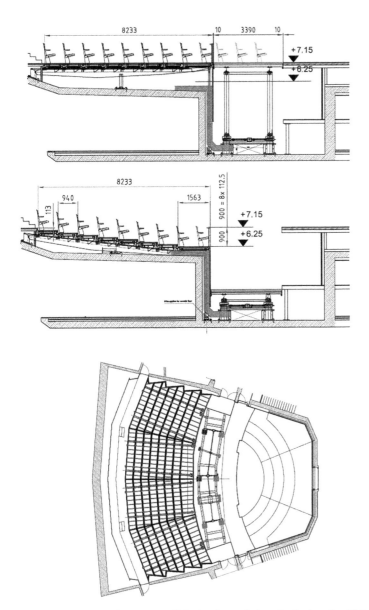

Figure 1.42: *"Giant Amber Concert Hall"* in Liepaja, Latvia. Longitudinal section: Plain and stepped position and floor plan. Photo credit: Waagner-Biró.

In the forestage area, there is a double-decker podium and two orchestra podiums, the main stage is equipped with upper stage machinery consisting of bar hoists, point hoists and panorama hoists; chain hoists are available in the side stages. Appearances from below from the understage are possible.

The auditorium enables 1100 seats in the opera version and 850 as a concert hall.

Figure 1.43: Megaro Musikis Athinon, Athens, concert hall in some room modifications. Photos: Bayerische BühnenBau GmbH, hereafter referred to as "Bayerischer BühnenBau" for short (see also BTR 4/1993).

Figure 1.44: Royal Opera House Muscat – usage for opera. Photo: Courtesy of Royal Opera House Muscat.

Special uses in the stage area

Now examples are mentioned which provide a design of the stage area for various special performances.

Worth mentioning is the technically elaborate special equipment of the *Friedrichstadtpalast in Berlin* (Figures 1.46 and 1.47), which was designed according to the model of the large show theaters in Las Vegas. In the proscenium area, a circular podium can be used to raise a *water basin*, an *ice arena* or a *circus ring*.

Similar to the Friedrichstadtpalast in Berlin, the *Shanghai Culture Square* was created as an event center that is a tailor-made solution for the needs of large-scale musicals and variety events (see Figures 1.48, 1.49 and 1.50).

In addition to the main stage, there are two side stages and a backstage at level 0.00 but also rooms below it in the understage.

Below the side stages and backstage, 3 round functional wagons with a diameter of 18 m are parked:

– A *turntable* with one core and two ring discs, the inner disc can be raised by 0.6 m, the middle ring by 0.3 m.
– A *water truck* with a water basin of 14 m diameter and 0.4 m basin depth.
– And an *ice truck*, with an ice surface 17.5 m in diameter.

These functional carts can be moved to a central lifting platform by a friction wheel drive and lifted 5 m to stage level at a maximum speed of 0.25 m/s by means of 30 push chains (see Section 4.2) and scissor guides.

The opening in the main stage can be covered by bridge carriages (one is adapted to the curve of the functional carriages). Furthermore, side stage carriages (19 m × 4.5 m) are available. All stage wagons are driven by horizontal pull/push chains (see Section 1.7.2). In addition, there are 13 equalizing platforms to play on the stage in var-

Figure 1.45: Royal Opera House Muscat: (top) usage as theater, (bottom) usage for concert. Photos: Courtesy of Royal Opera House Muscat.

ious constellations. The largest equalizing platform is located in the backstage and has the dimensions of 19 m × 19 m.

Variable room division

There is also often a desire to divide a large multipurpose hall into several smaller rooms with partition walls. The *"Congress Center" in Frankfurt am Main* offers an interesting example of this.

The "Great Hall" can be divided into 15 smaller rooms by means of 22 soundproof mobile partition walls (see Figure 1.51, left). These partitions are designed as steel tele-

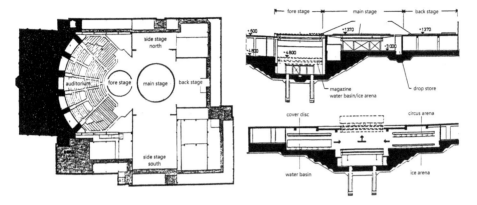

Figure 1.46: Friedrichstadtpalast Berlin: (a) floor plan, (b) longitudinal section through proscenium, main stage and backstage, (c) cross-section through proscenium. Picture credits: BTR 6/1985.

Figure 1.47: Friedrichstadtpalast Berlin. Circular lifting platform of the forestage with ice arena. Photo: SBS.

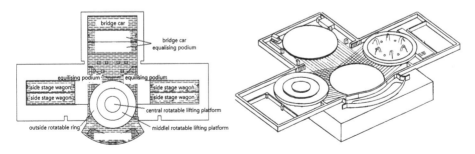

Figure 1.48: Shanghai Cultural Square. Stage areas at stage level and in the understage. Image credit: SBS.

Figure 1.49: See notes to Figure 1.50.

Figure 1.50: Central lifting platform with push-chain drive and scissor guide. Picture credits: SBS.

scopic lifting walls that can be lowered from the hall roof if required and are supported laterally in eight guide columns. They provide a sound insulation value of 45 dB.

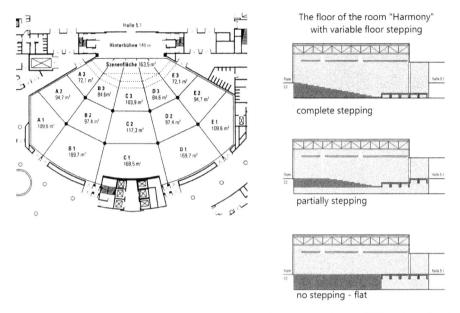

Figure 1.51: Congress Center Frankfurt: (left) possible hall subdivisions, (right) possibilities of stepping the hall floor. Picture credits: Messe Frankfurt, BTR 5/1997.

The telescopic lifting walls are driven by hydraulic cylinders and block and tackle hoists, and the guide columns protrude above the roof when raised – architecturally integrated into the roofscape. They are raised or lowered by pole-changing three-phase motors (see Section 2.2).

The floor of the hall can be stair-stepped in the spectator area with podiums (Figure 1.51, right). As the floor of a multipurpose hall, it must also be able to be travelled by 1.5 t forklifts, so a 40-mm thick beech plywood floor with a corresponding substructure was to be provided to support a 22 kN wheel load.

Figure 1.52 shows three examples of room design.

Universally usable event halls

Large multipurpose halls are shown in Figure 1.53 (*Madrid Sports Palace*) and Figure 1.54 (*Arena Rockódromo Madrid*). They can be used for sporting events, but also for musical events and theater performances. The hall in Figure 1.53 has a capacity of 15 000 visitors.

The hall in Figure 1.54 can accommodate up to 10 000 spectators. The grandstand system is arranged on three levels, has a modular design, and the individual stands can be extended or retracted telescopically.

Regarding the grandstand facilities of both multipurpose halls, please refer to Section 1.7.4.

Another interesting example is the *Koninklijk Theater Carré Amsterdam*, which is designed to allow use as both a theater and a circus.

Figure 1.52: Congress Center Frankfurt – three examples of room design. Photo credit: Messe Frankfurt.

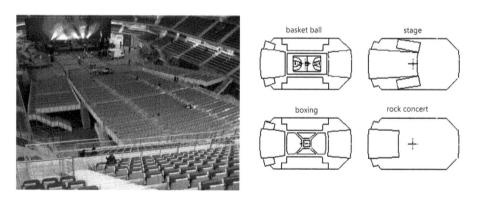

Figure 1.53: Sports Palace Madrid: (left) use as a concert hall, (right) other uses. Photo: Waagner-Biró.

The floor was equipped with six lifting podiums that allow for three hall configurations, circus with a low-set ring, theater with a raised stage area, and a flat hall area for general use.

Figure 1.54: Arena Rockódromo Madrid – tennis use. Photo: Waagner-Biró.

A tiltable platform of the same size is mounted on the hall podium. When lowered to its lowest position, the stage podium can be loaded with 2000 kg/m^2 for circus operations, and with the usual 500 kg/m^2 in other lifting positions. In theater operation, the hall podium is raised by 60 cm and a stage fall is formed with the tilting platform. The hall podium is surrounded by a horseshoe-shaped strip podium 90 cm wide, which is also equipped with an attached tilting platform and with which an aisle of variable height can be formed. On the other side, there are one large and three small orchestra podiums. All podiums are equipped with "Spiralift" drives (see Section 1.7.1), the tilting platforms with spindle drives.

Figure 1.55 shows a model of the podium system and a photo of the factory assembly of the liftable core platform.

Figure 1.55: Koninklijk Theater Carrè Amsterdam: (left) model, (right) factory assembly. Picture credits: SBS.

In the following, a multifunctional hall *"Reyno de Navarra" in Pamplona in Spain*, completed in 2011, is described. It is intended to be used as a circus, a concert hall, a theater or as a playing field for sports events (see Figures 1.56–1.59).

The hall can accommodate up to 12 000 spectators. Spectator stands can be extended from three side wall areas. As a special feature, a podium construction can also be used

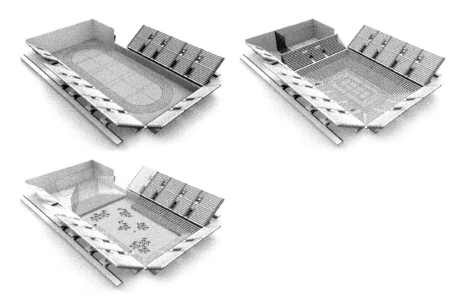

Figure 1.56: Reyno de Navarra-Pamplona, Spain. Examples for alternative usage: (top left) sports hall of largest extension, (top right) main hall with all rostrums, (bottom) main hall with stage, rostrums and hall seating. Picture credits: Waagner-Biró.

Figure 1.57: Main hall as sports field; refer to Figure 1.58.

to separate the hall into a large and small event space. This podium construction consists of four primary podiums of 11 m × 14 m each (44 m × 14 m in total) designed as double-decker podiums, which are installed within the extendable grandstands and can be extended towards the main hall or towards the secondary hall. However, they also house four secondary podiums that can be raised, then swiveled around a vertical axis and from which rostrums can also be extended, which are then available for the main hall or the secondary hall, depending on the swivel position.

The primary podiums can be loaded with 1000 kg/m^2 at the upper deck and can be raised to a height of 5.3 m above hall level. The secondary podiums can be raised about

Figure 1.58: Variants of use of the Reyno de Navarra shown as an example. Photo: Waagner-Biró.

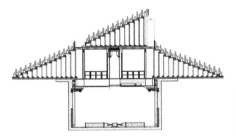

Figure 1.59: Schematic representation of the primary and secondary podiums and their built-in rostrums. Photo credit: Waagner-Biró.

the same height above the primary podiums. If primary podiums are raised to only 1 or 2 m, they can serve as a stage area for theatrical performances.

The seats in the stands can be folded up or down by motor drives.

Figures 1.56–1.58 show examples of different usage variants.

Figure 1.59 schematically shows the primary podiums with the rostrums extended on both sides, and above them the pivoting secondary podium unit with the bleachers tilted to the left.

With regard to retractable stands and auditorium seating, please refer to Section 1.7.4.

1.7 Technical equipment of the lower stage – understage equipment

The term *understage* has already been explained in the section on *"Spatial concepts"*. From a technical point of view, the presence of an understage is extremely important, even if it does not contain any special technical equipment. This is because a level below stage level enables appearances from below or exits downward through openings in the stage floor. This section deals with technical equipment whose installation position is essentially at or below stage level, i. e., lifting platforms, stage wagons and revolving stages.

1.7.1 Lifting platforms

Classification of lifting platforms
Classification according to the purpose of use:
– Podium on the main stage (acting area).
– Equalizing podium.
– Orchestra podium.
– Trap elevator.
– Drop store elevator.

Structure according to the number of levels and arrangement of podiums:
– Simple podium.
– Double-decker podium with fixed spacing of two floors.
– Double-decker podium with variable spacing of two floors (special design: drag floor).
– Primary podium with attached secondary podiums.
– Checkerboard stage (chessboard stage).

Structure according to the location of the drive:
– Drive is stationary.
– Drive travels with the podium (climbing podium).

Variants in terms of drive-design:
– Rope winch.
– Chain drive (with tension or push chain).
– Rack and pinion drive.
– Spindle drive.
– Hydro cylinder.
– Wedge drive.

Drive variants with regard to energy supply:
– Electric.
– Hydraulic.

Guidance of the podium:
– In sliding guides or roller guides.
– With scissors.

More detailed description of podiums for various uses
Stage platforms in general

Stage elevator platforms are podiums that are installed in the area of the main stage and are usually designed with regard to the technology used so that they can also be used scenically (see Sections 1.3 and 1.4). They usually extend over the entire width of the stage with a rectangular ground plan and are arranged one behind the other along the longitudinal axis of the stage (see, e. g., Figures 1.7 and 1.14).

The podium surface can be built nonadjustable in horizontal position or with inclined stage surface (stage fall) (see Figures 1.79 and 1.255). In this case, the upper playing surface of stage podiums can be designed to tilt by tilting the podium floor about a horizontal axis oriented parallel to the proscenium. By tilting all podiums by the same angle and moving the lifting podiums to corresponding height positions, a *stage fall* as shown in Figure 1.60 can be produced instead of a horizontal stage plane. In the past, such an inclined stage floor was often used with a rise of approx. 1.5 to 5 cm per meter of stage depth, i. e., with an inclination of the stage level of 1.5 % to 5 % or with approx. 1° to 3°. This results in better viewing conditions for the audience in the stalls, but it can also enhance the perspective effect in the stage set (see Section 1.1). A podium with a *tilting cover* is also shown in Figure 1.61 (*Vienna Raimundtheater*).

Figure 1.60: Inclinable floor of stage platforms. Stage fall formed by the stage platforms at the Zurich Opera. Photo: MAN.

Figure 1.61: Tiltable podium. Single podium at the Vienna's Raimundtheater. Photo: Waagner-Biró.

Figure 1.62 shows a special design for *a tiltable floor* of a *lifting platform*. The special articulated kinematics ensures that the edges of the podium remain within the dimensions of the floor plan (in horizontal position) during the tilting movement.

Tiltable podiums with a rectangular surface can usually be tilted by axial positions in the longitudinal direction of the podium. In the case of podiums of approximately square shape, especially checkerboard stages (see later), it is often also possible to tilt the playing surface in any direction. Then the stage surface formed by several podiums can be tilted in any spatial position.

Stage podiums – double-decker podium with fixed distance between upper and lower level

Lifting platforms in the main stage area are often designed as the so-called *double-decker* platforms, i. e., they offer two accessible levels.

In case a deep understage pit is present, usually at large podium stages, a single podium raised to stage level would have an empty space of great height extension below the podium surface, so that appearances from below through openings on the podium playing surface would not be possible. A double-decker podium with a second level located below the normal playing surface ideally offers the possibility of appearances from below. Usually, the two accessible levels have a distance of about 2.5–3.5 m. On this lower podium level, for example, trap elevators can then also be used flexibly. Figure 1.63 shows the double-decker podium in the revolving stage of the *Graz Opera* in the height position where the lower podium level coincides with the stage level.

If double-decker podiums are to be used in the sense of a lifting platform system as shown in Figure 1.11(a), the distance between the two podium surfaces must of course be very large. Figure 1.13 shows lifting podiums of the *Hamburg State Opera* with a dif-

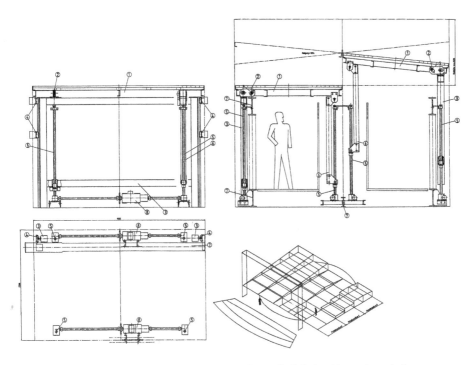

1	platform of the secondary podium	5	spindle hoist with rotating spindle
2	mechanism for tilting	6	supporting pillar
3	guided hoisting frame	7	steel structure of the primary podium
4	guides of the hoisting frame	8	driving motor

Figure 1.62: Tiltable floor of a stage platform – Patent Waagner-Biró. Photo credit: Waagner-Biró.

Figure 1.63: Double-decker elevating stage of the Graz Opera. Photo: Waagner-Biró.

ference in level between the two floors of 10 m. Since hoists of the upper stage cannot be used for a stage set built on the lower playing surface, bar hoists are mounted below the upper podium playing surface.

Stage podiums – double-decker podiums with lower level as a drag floor

If the pit is not deep enough so that a double-decker podium could not be lowered deep enough, the podium can possibly be built with a *drag floor*. A drag floor also provides a second accessible surface when the podium is raised. However, if the podium is lowered completely, the drag floor will be deposited on the ground floor of the pit. As a result, the upper playing surface can be lowered further; however, the distance between the two podium surfaces is reduced. In this case, special safety precautions must be taken to ensure that no one is on the drag floor in this operating case, being endangered by the reduction in free height.

Figure 1.64 shows the drag-floor podium in the *Valencia Opera House* (Spain). In the upper position, the drag floor is extended and the secondary platforms are raised; in the lower position, the drag floor is set down in the understage and the secondary platforms are retracted down.

The orchestra platforms I and II in Figure 1.28 in the *Zurich Opera House* are also designed with drag floors.

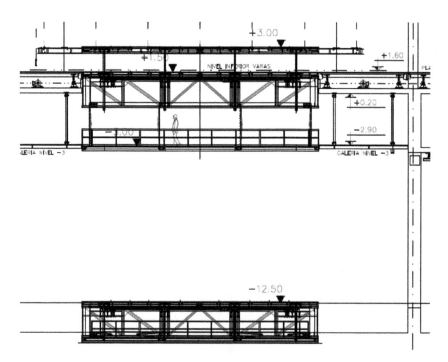

Figure 1.64: Double-decker lifting platform with drag floor in the Valencia Opera House (Spain). Picture credit: Waagner-Biró.

Stage podiums – double-decker podium as primary podium with relatively to it movable secondary podiums

More possibilities to design the stage scenery by using the podiums can be achieved by double-decker podiums so that the lower level – the *primary podium* – occupies the width of the stage, but the upper level is divided into roughly square *secondary podiums* which can be independently adjusted in height relative to the primary podium. Examples include the *Genoa Opera* (Figure 1.65), the *Oslo Opera* (Figure 1.66) and the *Vilnius Opera and Ballet Theater* (Figure 1.68).

Figure 1.65: Primary and secondary podiums – Genoa Opera (bottom pulley blocks of the movable point hoists, framework of a lighting hoist). Photo: Waagner-Biró.

The primary and secondary podiums in the *Oslo Opera House* are shown as a sectional view in Figure 1.66.

The four main podiums have a size of 16 m × 4 m and are moved by 2 hydraulic winches suspended on 8 ropes (see Figure 1.67). The use of a hydraulic drive was chosen to allow the podiums to be lifted via the pressurized hydraulic storage system even in the event of a power failure. The maximum lift is 12.5 m, and the maximum speed is 0.7 m/s. Although the podiums are suspended with ropes and this normally requires bolt locks due to the elasticity of the ropes, a different solution was chosen in this case: During the podium travel, any height differences due to different rope elongations are compensated by a fine control system by moving pulleys on each podium side via a hydraulic cylinder until the podium is back in balance. As soon as the podium travel is interrupted, friction

Figure 1.66: Primary and secondary platforms at Oslo Opera. Photo credit: Bosch Rexroth.

brakes on the deflection pulleys engage and block the ropes. This decisively shortens their effective length and sufficiently increases the system stiffness.

The secondary podiums of size 4 m × 4 m are guided in four roller-guided high-precision lifting supports each. Inside the supports are hydraulic cylinders that can be moved in such a way that inclined positions are also possible. The maximum stroke is 2.5 m, the maximum lifting speed 0.3 m/s.

A podium stage with 5 primary podiums (17.4 m × 3.0 m) and 3 secondary podiums (5.75 m × 2.98 m) per primary podium was installed in the *Lithuanian National Opera* and *Ballet Theater in Vilnius*. The arrangement can be seen in Figure 1.68. The primary platforms can be lowered from a height of 0.0 to −2.9 m, the secondary platforms are 0.665 m above the primary platforms in their lowest position and 3.215 m above them in

Figure 1.67: Oslo Opera – drive unit of the primary platforms. Photo credit: Bosch Rexroth.

their highest position. The hoist is driven by rope drum winches. The winches for the primary platforms are installed at the bottom of the lower platform at a depth of 8.0 m, while the winches for the secondary platforms are located on the primary platforms; they serve also as a tilting drive with the possibility of forming a slope to either side. The primary podiums are moved at a maximum speed of 0.1 m/s, the secondary podiums at 0.2 m/s.

Stage platforms – double-decker podium with movable lower level relative to the upper level

In the *Copenhagen Opera House*, 4 double-decker podiums with dimensions of 16 m × 4 m are installed. The basic structure of the podium with the upper level can be moved from +5 m to –5 m (stroke 10 m), the lower level relative to the upper level from –5 m to –2.9 m (stroke 2.1 m). The podium drive consists of two winches per podium narrow side, which are electronically synchronized, the lower level is moved with suspended spindles. The podiums can be seen in Figure 1.69. In addition, the illustration on the right shows a drop store elevator with dimensions of 22 m × 2 m that can be raised by 7.5 m by means of a spiral lift.

Stage podiums – double-decker podiums with movable upper and lower level

Double-decker platforms of special construction are installed in the new *Musiktheater Linz*. The three stage podiums are located in the large transport revolving stage opposite the performance revolving stage and are lowered to stage level in the basic position (see Section 1.4 and Figure 1.22). The double-decker podiums can be raised to a maximum of 3.95 m so that the lower level is part of the stage floor. When a stage wagon is positioned exactly above a stage podium, the primary podium can be lowered by the stage wagon

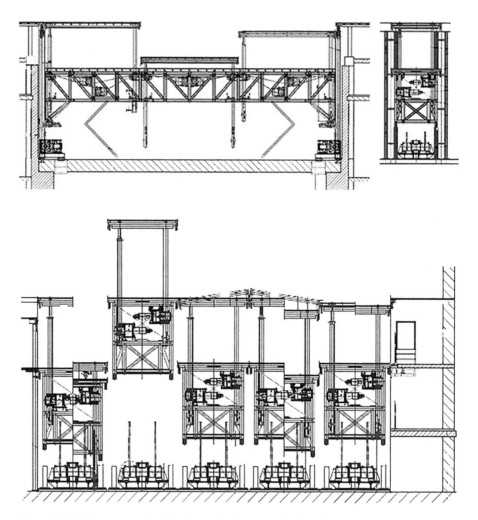

Figure 1.68: Podium stage at the Lithuanian National Opera and Ballet Theater Vilnius. Picture credit: SBS.

height of 0.2 m so that the stage wagon surface is at the same height ±0.00 as the adjacent stage floor. Each stage podium consists of a primary and secondary podium as a double-decker podium. The primary podium also has a equalizing podium built into it. Possible podium positions are shown in Figures 1.70 and 1.71.

The *primary podium* consists of a U-shaped steel structure with welded main beams at the base and so-called towers to which the guide rails for lifting frames are attached. It has dimensions of 15 m × 4 m and an overall height of about 5.6 m, the total lift being 4.15 m. Each primary podium hangs in a four 6-strand rope pulley block and is moved by two winches with two ropes each located in the transport turntable. The maximum lifting speed is 0.3 m/s. The two drive winches are electronically synchronized. The podium can be locked in 7 lifting positions.

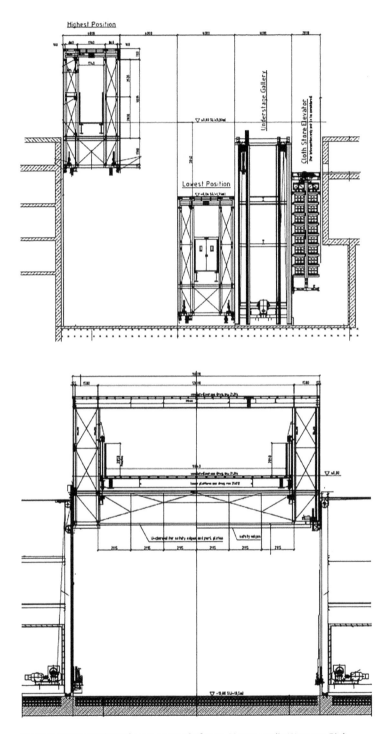

Figure 1.69: Opera Copenhagen stage platforms. Picture credit: Waagner-Biró.

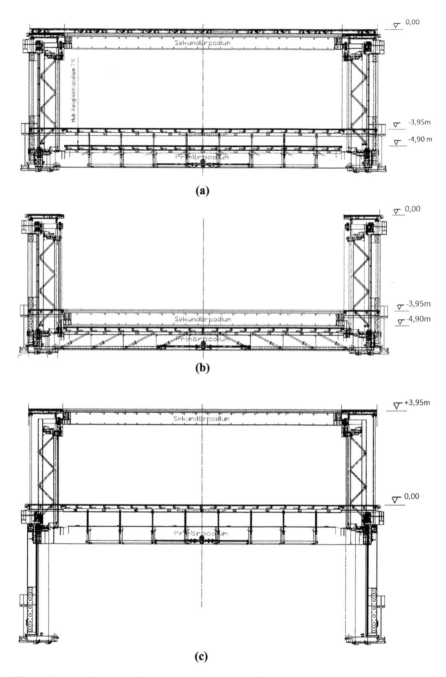

Figure 1.70: Double-decker podiums in the Musiktheater Linz – section across the stage: (a) primary podium in lowest position, secondary podium with attached stage wagon in uppermost position, balancing podium raised, (b) primary podium and secondary podium in lowest position, equalizing podium lowered, (c) primary podium and secondary podium in uppermost position, equalizing podium lifted. Image credit: Waagner-Biró.

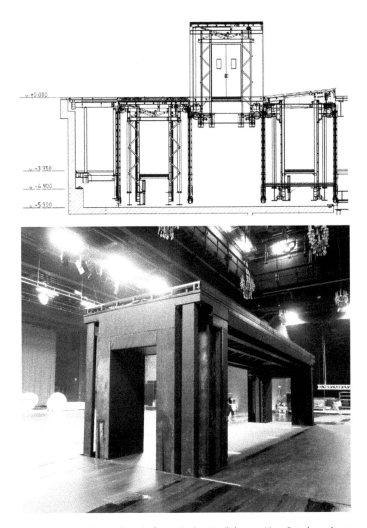

Figure 1.71: Double-decker platforms in the Musiktheater Linz. Cut along the stage and photo. Photo credit: Waagner-Biró.

For space reasons, an unusual solution had to be found for the *secondary podium*, which can be moved inside of the primary podium. The secondary platform is a deck that can be tilted via spindles (max. 10 % incline) and is attached on so-called lifting frames. The tilting mechanism in the lifting frame is designed in such a way that the pivot point is at the front edge of the wooden deck and the rear end is lifted by the screw jacks. The two lifting frames are moved vertically with 4 pieces of pulley blocks, each with 4 falls (electronically synchronized). Since these winches are integrated in the transport turntable, the associated winch drives must be released when the primary podium is moved and the secondary podium is locked; they are put under tension with a low torque so that the drive ropes of the secondary podium remain slightly tensioned.

The secondary podium can thus be moved within the primary podium 3.95 m relative to the primary podium with a maximum lifting speed of 0.6 m/s. For the secondary podium, there is only one lockable position in the upper end position and a fixed stop at the bottom when the lifting frame rests on the primary podium.

When the primary podium is lifted, the secondary podium is locked during this time and when the secondary podium is moved, the primary podium is locked. Simultaneous movement of the primary and secondary podiums is not provided.

The lower level of the primary podium (12 m × 4 m) is accessible and adjustable in height and is designed as a compensating podium to allow level surfaces at different positions of the primary and secondary podiums despite the little depth of the lower stages. The lifting of this podium for level compensation takes place without payload and amounts to 0.71 m and is accomplished with horizontally lying spindles with push rods attached to the primary podium. In the uppermost position, the push rods are vertical (see Figure 1.70(a, c)).

In order to be able to tilt the entire stage area generated by the double-decker podiums, tiltable surfaces are also arranged to the side of the tiltable secondary podium above the towers of the primary podium. In the central area of the secondary podium, seven *traps* are situated to allow appearances from below. Generally, at stage level and therefore also in the secondary podiums, guide rails are integrated in the wooden stage floor so that stage trolleys can be guided in longitudinal and transverse directions.

This podium concept was further developed for the *Berlin opera "Unter den Linden"*. It is a double-decker podium in which the two accessible levels can be spaced differently. This is achieved by means of three movable construction elements that can be moved by two rope winch groups located in the understage (each group to the left and right of the narrow side of the podium) (see Figure 1.72).

The podium consists of the upper and lower platforms and lateral lifting supports in which the two platforms can be moved in a guided manner, as well as two stationary lateral locking supports integrated into the lower platform structure to guide the lifting supports, which can also be moved.

One winch group – the main drive (platform drive) – moves either the lifting supports with the upper platform locked in them or, in the case of lifting supports locked stationarily in the latch supports, only the upper platform. For this purpose, the lifting ropes are guided from the rope drum to the upper end of the latch support, then via deflection pulleys to the lower end and then to the upper end of the lifting supports and at the end attached to the upper platform.

The second winch group – the auxiliary drive (drag floor drive) – is used to move the lower platform. The hoisting ropes are also guided through the latch and lifting supports and are attached to the lower platform.

If the lifting supports and the upper platform are coupled together and the lifting supports can be moved freely in the locking supports, winding up the ropes on the drum causes the platforms with the lifting supports to be raised to the uppermost "positive position" 8.67 m. The lower platform is moved along by simultaneous winding of the ropes

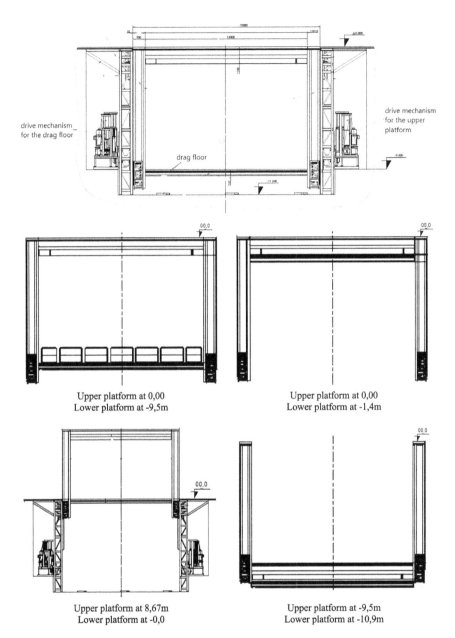

Upper platform at 0,00
Lower platform at -9,5m

Upper platform at 0,00
Lower platform at -1,4m

Upper platform at 8,67m
Lower platform at -0,0

Upper platform at -9,5m
Lower platform at -10,9m

Figure 1.72: Opera "Unter den Linden" – Berlin. Double-decker lifting platforms consisting of lifting supports, upper platform and lower platform (drag floor) guided in latch supports, below listed figures show extreme positions. Photo credit: Waagner-Biró.

on the auxiliary drive, and is also raised relative to the upper platform if the winding is faster. Of course, the lower platform can also be moved within the lifting supports when the upper platform is stationary.

If the lifting supports are coupled to the latch supports – that means they are stationary – unwinding the ropes on the drums of the main drive allows the upper platform to be lowered within the lifting supports as a "negative lift" to a maximum depth of the upper platform at level –9.5 m. The lower platform can also be lowered a little further to allow a level in the lower platform when the upper platform is completely lowered.

Figure 1.72 shows extreme positions of the platforms. Of course, many intermediate positions are possible, including a slightly lowered position of the upper platform to level –0.33 m in order to lower the turntable carriage to level 0.00.

These special platforms are installed in the main stage as podium 1, 2 and 3 and have a size of 15 m × 3 m each. The upper platform alone can be moved with 0.5 m/s, together with the lifting supports with 0.3 m/s, the lower platform also with 0.3 m/s.

Stage elevator platforms – podiums in a checkerboard arrangement with one deck or in a double deck arrangement

Even greater flexibility is offered by podiums which are not designed as rectangles in stage width but squares or rectangles arranged in a chessboard pattern. Examples of *chessboard* stages (checkerboard stages) include the *Semper Opera in Dresden* (Figures 1.73, 1.74 and 1.75) and the small *Akzent Theater in Vienna* (Figure 1.76). In the latter, the checkerboard platforms cover not only the stage area but also the front part of the auditorium. In terms of fire protection, this resulted in complex measures in the area of the iron curtain.

Another example of a checkerboard stage was also realized in the *National Theater Budapest* (see Figure 1.77). A centrally arranged tubular telescope serves as a guide, allowing a total lift of approx. 5 m. The inner tube is moved by means of two Biflex toothed chains as a space-saving "Omega drive." The drive is attached to the outer tube (see Figure 1.78).

"Biflex toothed chains" are flexible chains toothed on both sides, which enable the drive of any number of shafts with an s-shaped wrapping. The name Omega drive was chosen by the manufacturer because the shape of the chain guide corresponds to the Greek letter omega.

Another example of a checkerboard stage is the *branch theater of the Bolshoi Theater in Moscow*. The stage consists of 20 podiums of size 3 m × 3 m with a stage slope of 4 %. Each podium is driven by 4 spindles and a central motor. A longitudinal section of the entire stage is shown in Figure 1.255. A small lift is only used to lower the ballet trolley to stage level.

Stage platforms built into a cylinder revolving stage

However, stage platforms can also be installed in large revolving stages. An example is the *cylinder revolving stage* of the *Vienna Burgtheater* (Figures 1.15, 1.16), which is equipped with four lifting platforms.

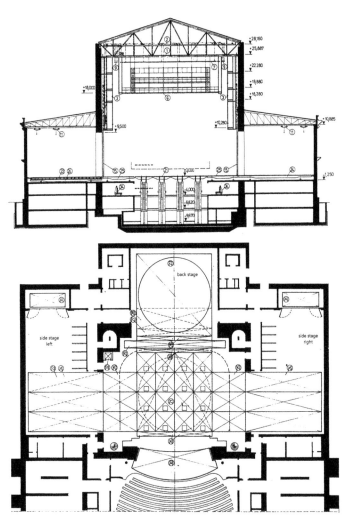

1	grid, working floor	17	divisible lifting curtain
2	grid for pulleys	18	sound curtain
3	working galleries and connecting links	19	smoke escape of the mainstage
4	stage lighting bridge (see Figure 1.74)	20	smoke escape of the auditorium
5	portal towers	21	lifting platforms at the main stage
6	stage lighting bridge	22	side stage wagons
7	drop hoists	23	back stage wagon with turntable
8	lighting bar hoists	24	equalising platforms at the side stage
9	cyclorama bar hoists	25, 26	equalising platforms at the main stage
10	hoists at back stage	27, 28	equalising platforms at the back stage
11	hoists at side stage	29	scenery storage elevator
12	point hoists at the fore stage	30	orchestra podium
13	safety curtain	31	elevators for sceneries
14	back stage gate	32	elevator
15	side stage gate	33	transport and storage racks for sceneries
16	curtain	34	trap mechanism for persons

Figure 1.73: Semperoper Dresden – cross-section and floor plan. Picture credits: BTR 4/1979.

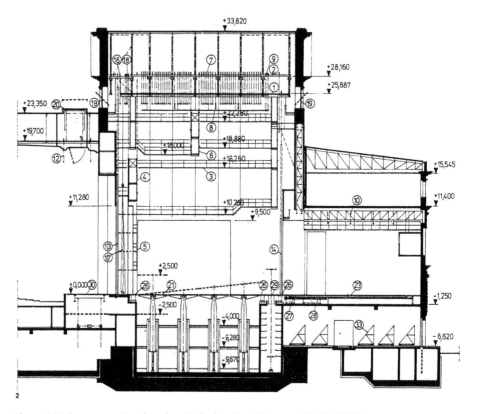

Figure 1.74: Semperoper Dresden – longitudinal section. Picture credits: BTR 4/1979.

Figure 1.75: Chessboard podiums – Semperoper Dresden. Photo: SBS.

Figure 1.76: Chessboard podiums at Theater Akzent in Vienna – podiums in different lifting positions on the stage and in the auditorium. Photos: Waagner-Biró.

Figure 1.77: Chessboard platforms at the National Theater in Budapest. Photo credit: Bosch Rexroth.

Figure 1.80 shows a historical photograph of a large cylinder revolving stage with podiums in checkerboard arrangement in factory assembly for the *National Theater Belgrade*.

In the "Großes Haus" of the *Düsseldorf Schauspielhaus*, as can be seen in Figure 1.81, a cylinder revolving stage has been installed in which there are four rectangular double-decker podiums (podiums 5–8) and a circular segment podium (podium 4). Correspondingly, there is a suitably shaped podium (podium 3) on the spectator side in front of it. The four podiums 5–8 are driven by winches and can be tilted in two directions, the circular segment podium and the two orchestra podiums (podium 1 and 2) are moved by rack and pinion drives.

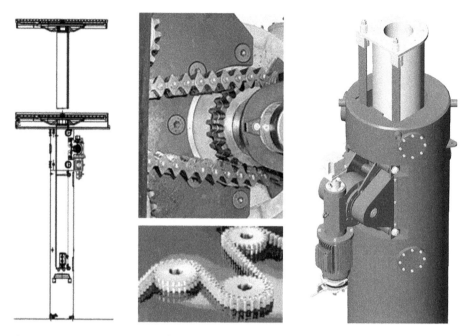

Figure 1.78: Guidance and drive unit of the chessboard podiums in the National Theater Budapest. Picture credits: Bosch Rexroth.

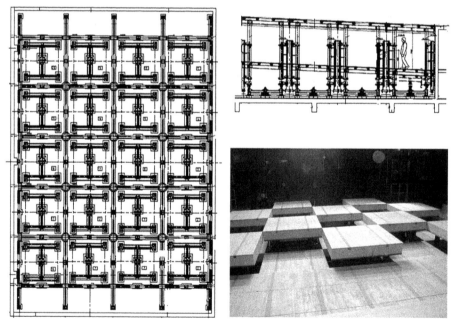

Figure 1.79: Bolshoi Branch Theater, Moscow. Checkerboard stage platforms. Floor plan and longitudinal section of stage. Photo credit: Waagner-Biró.

Figure 1.80: National Theater Belgrade. Revolving stage with checkerboard platforms. Photo credit: Waagner-Biró (Historical photograph).

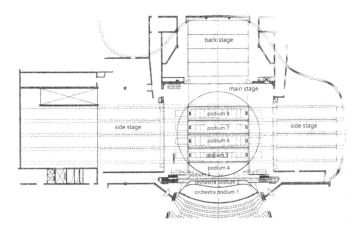

Figure 1.81: Schauspielhaus Düsseldorf. Picture credit: SBS.

Special constructions, such as in the *Smetana Theater in Prague* (Figure 1.82), make it possible to cover the entire circular area of the stage floor with the podiums.

In the *National Grand Theatre Beijing – Drama Theatre* (Figure 1.83), a large cylinder revolving stage with a diameter of 16 m and a height of 18.4 m is installed. It contains 13 square checkerboard podiums (2.5 m × 2.5 m) and 2 rectangular podiums (12.5 m × 2.5 m and 7.5 m × 2.5 m). The podiums are moved by hydraulic cylinders. In addition to the external *hydraulic accumulator pressure station* with a rotary hydraulic transmission also pressure accumulators for a maximum of two complete strokes are situated in the rotating platform (see Figure 1.151).

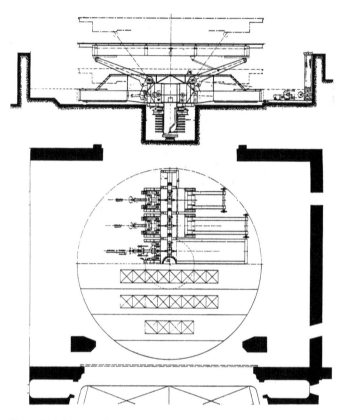

Figure 1.82: Smetana Theater in Prague. Cylinder revolving stage (the platforms cover the entire circular area). Picture credit: Waagner-Biró.

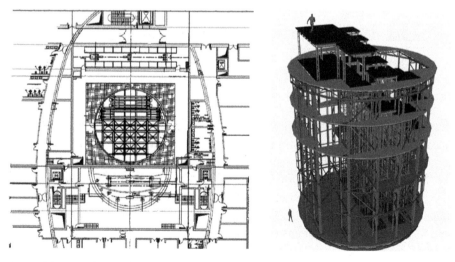

Figure 1.83: National Grand Theatre Beijing – Drama Theatre. Picture credits: Consortium Bosch Rexroth - SBS Bühnentechnik GmbH.

Equalizing elevators

Equalizing elevators are platforms with only a very short lifting distance that serve to compensate for differences in level, as they are required when moving stage wagons due to their overall height, because the stage wagon leaves behind an area of lower level at its original location. In Figure 1.73 (*Semperoper Dresden*), compensation platforms can be seen in the side stages and in the backstage, but also on the sides of the main stage.

Orchestra elevators

Orchestra elevators are podiums in the area of the orchestra and adapt their ground plan shape in one or more parts to the ground plan area of the orchestra pit. Figure 1.84 clearly shows the contours of the orchestra podiums, which are raised to form a large proscenium area, at the *Aalto Theater in Essen*. Figure 1.260 shows the orchestra platforms of the *Großes Festspielhaus in Salzburg*.

Figure 1.84: Stage of the Aalto Theater in Essen. Podium surfaces clearly visible on the stage floor. Photo: Krupp.

In opera houses there are orchestra pits for large orchestras measuring 150 to 160 m^2, even up to 180 m^2 (e. g., *Muziektheater Amsterdam*).

With the help of these orchestra podiums, the forestage area can be flexibly designed. Depending on the number of lowered podiums, the size of the orchestra area can be varied. If orchestra podiums are moved to the level of the auditorium, this area can be provided with rows of seats. If orchestra podiums are raised to stage level, the playing area of the main stage can be enlarged into the auditorium. In this way, the performance can also be moved in front of the proscenium into the auditorium. In the *Zurich Opera House* (Figure 1.28), in addition to two orchestra podiums, there is also a so-called *stalls podium* with chairs installed.

Table and people traps

The group of lifting podiums also includes table elements in the stage floor of the main stage, which are mainly used to make people or objects disappear or appear and are called table traps or person traps depending on their size.

Figure 1.85 shows such movable *table traps* as installed in the *Hanover Opera House*, as scissor lift elevators with travel shaft cladding and electrical and hydraulic supply via energy chains. Another table trap is shown in Figure 1.86.

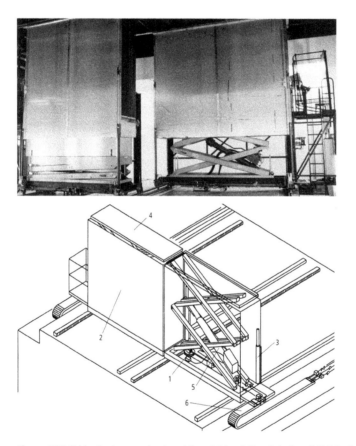

Figure 1.85: Table size trap mechanism: 1 Travel drive, 2 Travel shaft wall, 3 Cylinder actuation of travel shaft wall, 4 Lifting platform, 5 Cylinder drive of scissor lift table, 6 Energy chain. Picture credits: Bosch Rexroth. Photo: R. Budde.

Small traps, which are mainly used to transport performers, are referred to as person traps. Figures 1.87–1.89 show a trap system for persons (lowering device for persons), which can be used at variable positions, as installed in the *Musiktheater Linz*. They can be moved to various deployment positions of the stage below the stage surface. Their mode of operation is shown in Figure 1.87. The basis is a steel structure standing on

Figure 1.86: Movable table size trap mechanism. Picture credits: Bosch Rexroth.

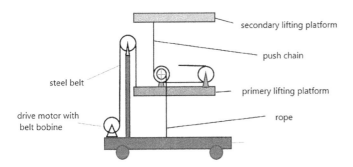

Figure 1.87: Functional scheme of a trap elevator for persons.

swiveling wheels, on which a winch for steel belts is located. The primary platform can be raised or lowered in sliding guides by winding or unwinding steel belts in a bobbin-like manner with this winch.

On the primary platform a rope drum sits on a shaft, which is set in rotary motion during this lifting process, as the rope end is attached to the base frame. On the same shaft the drive sprocket for a "Serapid" push chain is situated, which moves the secondary platform. The secondary platform is guided by double scissors. This combination of the two lifting movements makes it possible for the person to be lifted through the platform construction (through the floor) to stage level.

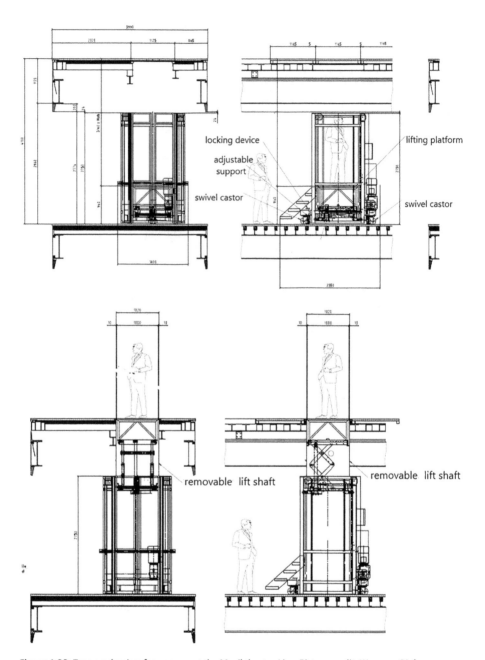

locking device

adjustable support

swivel castor

lifting platform

swivel castor

removable lift shaft

removable lift shaft

Figure 1.88: Trap mechanism for persons at the Musiktheater Linz. Picture credit: Waagner-Biró.

The base unit has a height of approx. 2.7 m and offers protective walls made of Plexiglas around the person as travel shaft cladding; a pluggable lift shaft can be placed above.

Figure 1.89: Trap mechanism for persons in the Musiktheater Linz. Photo credit: Waagner-Biró.

Figure 1.90: Movable trap mechanism for persons. Picture credits: Bosch Rexroth.

There exist also movable person trap devices suspended in the podium structure, as shown in Figure 1.90.

Figure 1.91 shows a trap-system developed by Teco-Bühnentechnik Wiesbaden (Bosch Rexroth) installed in the *Grand Theatre de Luxembourg*. It can be used multifunctionally, as it is also possible to combine several systems for persons with and without intermediate bridges to form larger podium areas, enabling so-called "tandem" or "triplet" operation. With a single lowering, an area element of 1 m × 1 m can be moved. In triplet operation, a total area of 1 m × 9 m can be produced with three traps and the additional installation of 2 × 3 intermediate aluminum bridges between the two stage elevators. In principle, each person trap can be controlled individually; in the combined operation described, they are of course moved in synchronous operation and form a table trap.

Figure 1.91: Two trap mechanism for persons linked by an intermediate bridge. Grand Theatre de la Ville de Luxembourg. Photo credit: Bosch Rexroth.

In order to be able to raise a position-variable table or person trap, an element of the stage floor with the same surface area must be removed. This can be done with removable cassette elements or with so-called *"retracting slides."* This refers to a stage floor element that can be lowered by means of a lever system and moved laterally in horizontal guide rails fixed below the stage floor. If the opening is so long that the retractable slide cannot find sufficient space in the longitudinal direction, it may be necessary to arrange slide elements one above the other. Figure 1.92 shows retractable slides installed in the podium, where the floor element to be opened is magazined below the stage floor.

Drop store elevators
Specially designed lifting platforms on the stage can also serve as store for prospect rolls, as already explained in Section 1.5 with reference to Figures 1.24 and 1.25. Figure 1.93 shows such a *drop store elevator* with hydraulic cylinder drive with shaft synchronization. (The term "synchronization shaft" is explained in more detail with Figure 1.118.)

Backdrop storage elevators can also be seen in Figure 1.255, which shows a longitudinal section of the *branch theater of the Bolshoi Theater in Moscow.*

A particularly complex understage system is installed in the *National Academic Grand Opera and Ballet Theater of the Republic of Belarus in Minsk* (Figure 1.94), which should be mentioned here because of the large number of storage lifting platforms.

Behind the proscenium, in 4 rows, there are 20 double-decker podiums in a checkerboard arrangement, each with an area of 3.2 m × 3.0 m. Behind this, in the 5th row, there is a 16 m × 3 m lifting podium, and in the 6th to 8th rows, as a backstage podium, there are three prospect lifting podiums 25 m × 0.9 m, each with 8 shelves.

Figure 1.92: Trap with retracting slide. Picture credits: SBS.

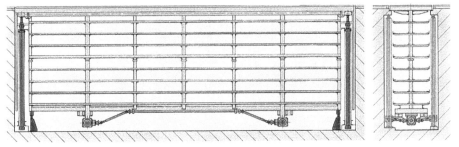

Figure 1.93: Backdrop storage elevator. Picture credits: Bosch Rexroth.

In the backstage, a double-decker podium measuring 17.2 m × 15.1 m can be used to store a turntable with core and ring disc (diameter 12.0 and 14.5 m, respectively) and a *ballet trolley*. All podiums are hydraulically operated.

As shown in Figure 1.94, a fixed platform drop of 4 % slope is provided, the covers of the checkerboard platforms can be inclined ±6.5°.

Types of podium drives

There are numerous possibilities for the technical realization to design the drive of lifting platforms. The selection of the drive variants must be based on the structural circumstances, the intended use of the podiums and, of course, the financial conditions.

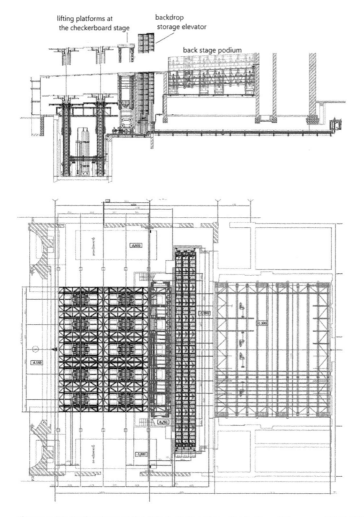

Figure 1.94: Understage at the Minsk National Academic Grand Opera and Ballet Theater. Photo credit: SBS.

General

The description of individual drive variants should be preceded by the following basic considerations:

– Most often, in the case of rotating drives, electric motors are used. The use of hydraulic motors is also possible. It should be noted that for safety reasons, two independently acting brakes must also be used for lifting platforms or an element with self-locking from movement (see Section 5.1).

– Apart from small lifting platforms, very large dead weights and payloads have to be moved. Therefore, in many cases it is favorable to partially compensate the lifting loads with counterweights in order to reduce the drive power. This makes sense especially for drives with electric or hydraulic motors. If the podium has a dead weight of E [kN] and the maximum payload is Q [kN], the lowest required lifting power results if the counterweight $G = E + Q/2$ is selected, as it is also usual for common house lifts. In this case, when the podium is empty and when it is loaded with the maximum payload, only a weight of $Q/2$ needs to be lifted in each case, since when the podium is empty the counterweight is heavier by $Q/2$ and when it is fully loaded the podium is heavier by $Q/2$.

For drives with single-acting hydraulic cylinders (plunger cylinders), the dead weight of the podium must allow lowering and counterweights are therefore generally not used. (Please, also refer to the text for Figure 1.95.) It is, of course, also possible to replace the force effect of a counterweight by that of a hydraulic cylinder.
– There are solutions in which the drive is installed in a fixed position and solutions in which the drive travels with the podium (*climbing drive*). If stage platforms are to be used for scenic purposes, the drive unit must not emit any disturbing noise. In the case of such drives, it is therefore generally better to locate them in the basement of the understage, since a drive unit that travels with the stage comes very close to the playing surface when the podium is raised. Expensive sound insulation measures are then often unavoidable, whereby airborne sound insulation is possible, but disturbing structure-borne sound transmission can usually only be prevented with great difficulty.

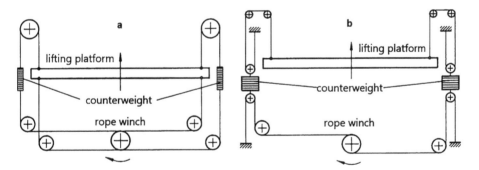

Figure 1.95: Platform drive with a rope winch: (a) with hoist rope and counter rope (counterweight heavier than dead weight of the podium), (b) counterweight only partially compensates the deadweight of the podium (or without a counterweight).

Podium dimensioning

Since podiums are part of the stage surface, they must be able to withstand the same load as the fixed stage floor when not in motion – if necessary in a locked state – which is usually 500 kg/m². Since this results in very high total loads for large platform areas, it is common practice to use smaller total loads for dimensioning the hoist drive when moving the platforms. Minimum values can be found in relevant standards.

In this sense, a distinction is made in the industry between "static" and "dynamic loads" for podiums – the designation "stationary" and "moving load" would be more accurate, since in mechanics "dynamic" loads refer to mass forces acting from acceleration processes.

Podium interlocks
As mentioned above, the lifting drive can also be designed for lower loads than it must be taken into account for the rest position. Thus, podiums must be locked in several height positions if the load bearing elements are only calculated for the lifting movement with the required safety, but larger loads must be considered in the rest position.

However, if podiums hang in elastic suspension means, such as ropes or chains, interlocks are also required, since their elasticity would also cause the height position to change when the load conditions change and the podium could also be very easily excited to vibrate under rapidly changing loads (ballet, etc.). Interlocks must also be provided for drives with hydraulic cylinders, since hydraulic fluid is also elastic due to its compressibility (see Section 3.8).

As an alternative to positive-locking latches, where only certain operating positions can be approached, clamping elements acting via frictional locking (hydraulic cylinder with clamping head, Figure 2.14(d)) may also be used.

If podiums are supported by steel components whose elasticity is negligible (threaded spindles, racks, special pressure chains, etc.) locking devices can be omitted.

Drive with ropes (see also Section 4.1)
In the case of a podium with *rope drive*, the podium hangs on ropes, possibly also in block and tackle systems depending on the size of the podiums or the deadweight and payload mass. In the basement of the lower platform there is a winch for lifting or lowering the podium. By mechanically coupling the drives of several podiums or using modern control technology, their synchronous operation can be achieved.

With regard to the type of rope guide, a variety of options is possible. Only two variants are described here.

One design is shown in Figure 1.95(a). In order to be able to move the heavy masses at a sufficiently high speed, but still with low drive power, the lifting load is partly compensated by counterweights, as already explained and often used in stage technology. In Figure 1.95(a), the counterweights are tied into the hoisting ropes, but they can also hang on extra ropes. If the counterweight masses are chosen to be equal to the deadweight plus half the payload mass, the podium or counterweight load will predominate, depending on the size of the payload, and the resulting hoisting load will be at most equal to half the payload, as already described (see also Section 3.6.3). The rope drums must accommodate upward-pulling hoisting ropes and downward-pulling counter ropes, with the corresponding counter rope being unwound when one rope is wound up.

If the counterweight is selected so that the podium mass always predominates, the counterweight rope can be omitted, as shown in Figure 1.95(b). In addition, in this example, the travel of the counterweight is halved by the pulley gear ratio, which results in counterweights with double the weight. For small podiums, of course, counterweights can be omitted.

Since rope platforms are generally mechanically locked in their position of use, the hoist control system of the winches is designed in such a way that these locking bolts can each be inserted with sufficient clearance into a perforated bar and afterwards the platform is lowered onto the locking bolts. Likewise, the podium must be lifted slightly for unlocking.

The rope drums for winding and unwinding the ropes are usually driven by electric motors. Of course, the use of hydraulic motors is also possible, as is the case, for example, with the podiums installed in the cylinder revolving stage of the *Residenztheater in Munich*. Figure 1.96 shows the arrangement in a drawing, Figure 1.97 shows a photograph.

However, this system has another special feature. Since the central hydraulic pressure station with the pumps is located outside the stage area due to the excessive noise level and for space reasons, the supply of pressurized fluid to the podiums on the revolving stage has to be carried out partly via a hydraulic rotary feedthrough. However, a large part of the instantaneous fluid demand during working movements is taken from piston accumulators located together with the nitrogen cylinders on the rotating platform, so that a *hydraulic rotary transmission* with a small nominal diameter is sufficient.

An analogous solution was chosen for the cylinder revolving stage in the *National Grand Theater Beijing* (Figures 1.83 and 1.151).

To synchronize several podiums, their rope drums were mechanically coupled in earlier times; today, electronic synchronization control is preferred.

A podium suspended from ropes can also be moved by pulling the ropes with the help of a hydraulic cylinder instead of winches, as shown in Figure 1.98 and built, for example, for stage podiums in the *Festspielhaus Bayreuth* and for table-size traps in the *Staatstheater Kassel*.

Drive with chains (see also Section 4.2)
In this case, the podiums do not hang on ropes but on chains, *single* or *multiple chains* depending on the load size as a *chain drive*. These chains are driven by sprockets.

Also with this solution, efforts will be made to balance the deadweight and payload masses as far as possible by means of counterweight masses. Figure 1.99 shows the electric chain drive and the counterweights hanging from one end of the chain in the *Tyrolean State Theater*.

Figure 1.100 shows the drive concept for the lifting platforms installed in the revolving stage of the *Vienna Raimundtheater*. The drive sprockets are arranged just below the stage level at the side of the podiums. The podium hangs on one end of the chain, the counterweight on the other end. The four drive sprockets of a podium are driven in a mechanically synchronized manner by a unit mounted in the basement of the lower stage via cardan shafts with transfer gearboxes. As this example shows, such podiums can

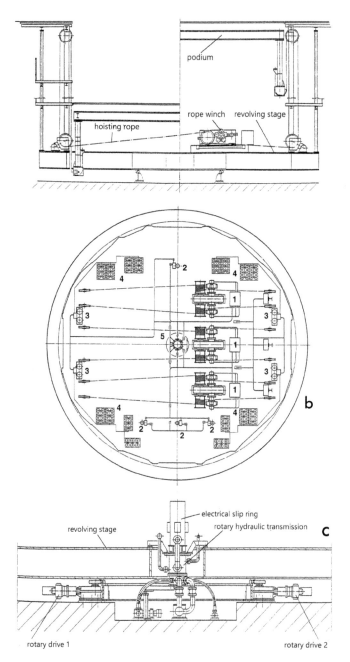

Figure 1.96: Podium drive with rope winches in the revolving stage of the Munich Residenztheater: (a) schematic of the revolving stage, rope drive of the podiums, hydrostatic winch drive, (b) top view of the revolving stage: 1. Drive mechanism of the platforms, 2. Drive of the table size traps, 3. Piston accumulators, 4. Nitrogen accumulators, 5. Hydraulic rotary feedthrough, (c) electrical and hydraulic feed to the revolving stage with electrical slip ring body and hydraulic rotary feedthrough – drive of the revolving stage, central bearing with a slewing ball bearing. Picture credits: Bosch Rexroth.

Figure 1.97: Podium drives with rope winches in the revolving stage of the Residenztheater in Munich. Photo: Bayerischer BühnenBau, Bosch Rexroth.

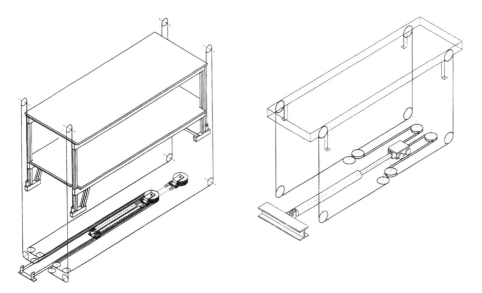

Figure 1.98: Podiums suspended from ropes with cylinder drive: (left) stage platforms at the Bayreuth Festival Theater, photo credit: Bosch Rexroth, Teco; (right) table-size trap elevator at the Kassel State Theater. Photo credit: Bosch Rexroth, MAN-GHH.

also be advantageously installed in revolving stages. On the one hand, the cardan shafts should not run with too high speed in order to exclude noise and disturbing vibration excitations (see Sections 3.10 and 4.5) and, on the other hand, the torques to be transmitted should not be too high in order to avoid expensive and heavy cardan shafts. It is there-

Figure 1.99: Podium drive with chains in the Tiroler Landestheater, Innsbruck. Photo: Waagner-Biró.

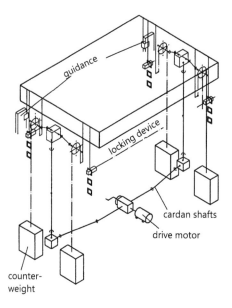

Figure 1.100: Schematic drawing of the podium drive with chains in Vienna's Raimundtheater. Photo credit: Waagner-Biró.

fore favorable to accommodate part of the transmission between motor and sprocket in the gearbox at the drive motor, and the rest at the transfer gearboxes to the sprockets.

In Figure 1.101, the lifting platforms of the *Genoa Opera* can be seen and the heavy roller chains are clearly visible.

When using a chain drive, it should also be noted that as a result of the so-called *polygon effect* (see Section 4.2.2), when the sprockets are driven at a constant angular velocity, fluctuations occur in the translational motion of the podiums, i. e., oscillations in the longitudinal direction of the chain, but the chain is also excited to transverse oscillations. This nonuniformity has a greater effect the lower the number of teeth of the sprockets. However, for large lifting loads, the use of heavy chains with relatively

Figure 1.101: Podium drive with roller chains in the Genoa Opera – primary podiums in various height positions, secondary podiums all in lowest position. Photo: Waagner-Biró.

large pitches is necessary, so that space considerations may make it necessary to select relatively small numbers of teeth. However, this polygon effect hardly has a disturbing effect at the usual working speeds in stage operation. In special cases, e. g., when, for geometric reasons, deflection sprockets have to be provided in addition to the drive sprockets, their polygon effects can be intensified by overlapping due to unfavorable positional assignment of the two sprockets. In this case, it can be advantageous to use a stationary circular segment on which the chain rolls instead of a rotating sprocket wheel as a deflection sprocket wheel, as was done in Genoa.

In the solutions shown so far, the chains are loaded as tensile elements. However, so-called *push chains* – as they have been available since some years – can also absorb compressive forces and therefore offer further design solutions. (Such push chains are also suitable for driving stage wagons; see Section 1.7.2.) The Serapid push chain is described in more detail in Section 4.2. Figure 1.102 shows chain drives in the *BMW World in Munich* as an example of the use of Serapid push chains for lifting platforms.

Push chains can also be used for scissor lift platforms as an alternative to hydraulic cylinders, spindle drives or spiralifts (see "scissor-type lift platforms" in this chapter).

Rack and pinion drive – Gear drive (see also Section 4.4)
When using racks, there are basically two possibilities:
-   Drive with *lifting/lowering rack.* A drive unit is installed on the floor, below the podium travel path. Stationary gears are driven via transfer gearboxes and cardan shafts, which engage in vertically moving racks attached to the podium. Thus, there is a requirement to allow the racks to travel the length of the total stroke into wells below basement level. Figure 1.103 shows the podium drives at the *Zurich Opera*

Figure 1.102: BMW World Munich. Photo credit: Waagner-Biró.

House and at the *Graf Zeppelinhaus in Friedrichshafen*. As an alternative to the rack with *involute gearing* (*involute toothing*), it is possible to use *pin gears* (see Sections 4.4.1) as, for example, in the *Theater der Stadt Essen* (Figure 1.104).

– At this point, it is also worth mentioning a special solution of a drive with a stroke-moved pin rack implemented in the *Ludwigsburg Theater* (Figure 1.105). To reduce the drive power, the podiums are supported by pneumatic springs. In the lowest position, the air is compressed to approx. 10 bar.

– *Climbing drive.* In the second variant (Figure 1.106 – *Mozarteum Salzburg*), the racks are installed in a fixed position in the stage and driven climbing pinions together with the drive unit travel with the podium. It has already been pointed out that this type of construction is very unfavorable in terms of sound. However, if this is not an issue in terms of the operating conditions, this can be a very economical solution.

Figure 1.103: Podium drive with lifting racks: (left) Zurich Opera House podium system, (right) detail of involute toothing, Graf Zeppelinhaus Friedrichshafen. Photos: MAN.

Figure 1.104: Podium drive with stroke-moved racks in the Theater of the German town Essen. Photo: Krupp.

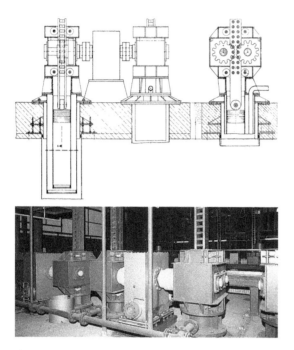

Figure 1.105: Podium drive with lifting racks. Special solution with pneumatic stroke load reduction. Picture credits: Bayerischer BühnenBau.

Figure 1.106: Climbing drive via racks – Mozarteum Salzburg. Photo: Waagner-Biró.

Spindle drive (see also Section 4.3)
Just like the rack and pinion drive for podiums, there are also design variants for the spindle drive with a stroke-moved or stationary spindle. In addition, however, there is the alternative of driving the spindle or the nut. For a podium drive, this results in the usual possibilities shown schematically in Figure 1.107:

– In the variant shown in Figure 1.107(b), the drive system is installed stationary at the bottom of the room; the *nuts* of the spindles are *driven* via transfer gears and cardan shafts. In this case, there must be *wells* into which the spindles attached to the podiums can enter when the podium is lowered. Thus, the nuts rotate, the spindles perform only a translational movement.
– Alternatively, the spindles can also be mounted in a stationary and nonrotatable position similar to the racks. The entire drive unit for the rotary movement of the nuts must then be located on the lifting platform. In this case it is a matter of a *climbing drive* (Figure 1.107(c)).
– A third option for spindle drives is to install a stationary drive system on the floor as in the first option, but to *rotate the spindles* instead of the spindle nuts. The nuts are then integrated into the podium construction to prevent them from rotating and are moved up or down with the podium by the spindle rotation (see Figure 1.107(a)). This design can also be seen in Figure 1.108.

In Figure 1.107, the spindles are supported on the foundation and take over the podium loads as compressive force. In principle, the spindles can also be suspended above and are then loaded in tension. However, since in this case the spindles are not anchored in the lower foundation of the lower platform but in the area of the platform floor, this arrangement is less favorable in terms of structure-borne sound transmission. Suspended spindles are used, for example, to move the lower deck on the double-decker platforms at the Copenhagen Opera House (see Figure 1.69).

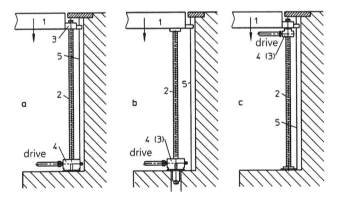

Figure 1.107: Drive system of the platforms with spindles: (a) stationary drive, rotating spindle, lifting spindle nut, (b) stationary drive, rotating spindle nut, lifting spindle, (c) drive travelling with the podium, rotating spindle nut, stationary spindle; 1 podium, 2 spindle, 3 nut, 4 spindle gear, 5 guide rail.

Figure 1.108: Podium drive with lifting spindles – Hamburg Academy of Music. Photo: MAN.

Drives with *sliding spindles* represent very simple economic solutions. Sliding spindle drives can be designed to be self-locking if a correspondingly low helix angle of the thread is applied. This means that no matter how large the loads on the platform surface, they cannot cause the platform to lower, even if no brake is applied to the drive (see Section 4.3). However, self-locking also means accepting high power losses, and the greater the losses, the more energy is converted into heat. High travel speeds are therefore not possible for thermal reasons, and the duty cycle may only be short. In addition, spindle drives can become relatively noisy at high speeds during the lowering movement. For this reason, their use is usually restricted to podiums not intended for scenic use, i. e., mainly orchestra or equalizing podiums. High power losses can be avoided by using multistart spindles with a large pitch (large helix angle of the thread), but the self-locking effect just described is then lost.

As an alternative to the sliding screw drives just described, in which the axial forces between the screw and nut are transmitted via sliding surfaces, *ball screw drives* or *planetary screw drives* can also be used, such as in the *Tampere Theater in Finland* (Figure 1.109). In ball screws, balls – similar to the balls in a ball bearing – serve as transmission elements between the screw and the nut; in planetary screws, the central screw is supported by planetary rotating spindles built into the nut (see Figure 4.20). With these components, high stroke speeds can be realized with low thermal stress, but the effect of self-locking can be ruled out because of the high efficiency. The lifting platforms for back drops shown in Figure 1.255, for example, are also actuated by roller screw drives.

Interlocks are not required on screw drives due to deformation and vibration considerations. For screw drives that are not self-locking in motion, the drive must be equipped with two independently acting brakes.

Of course, it can also be useful in a podium drive with planetary spindles to provide counterweights to reduce the load on the planetary spindle and the drive power. If this is

Figure 1.109: Podium drive with planetary spindles – vertical spindles for secondary podiums, Tampere Theater, Finland. Photos: SKF Multitec GmbH (D-Dreieichen-Sprendlingen).

not possible for reasons of space, a hydraulic counterweight mechanism can be used, as has been done, for example, in the Royal Opera House in Stockholm. Figure 1.110 shows schematically the principle of the system. In this case, 90 % of the podium's dead weight is counterbalanced.

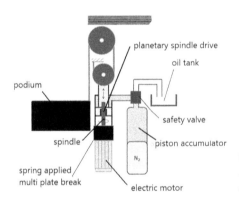

Figure 1.110: Scheme of the hydraulic counterbalance system in the Stockholm Opera House. Picture credits: Novoscen, Rexroth.

Folding spindle drive

A machine element that has already been widely used is the so-called *spiralift*. This is a *folding spindle drive* in which the spindle is formed as a tube from a horizontal and vertical steel strip. When pushed together, the two strips are stored in packages and, during the lifting process, are positively connected to each other and helically assembled. Due to its relatively large diameter, the spindle formed in this way can absorb compressive forces. However, it must be ensured that sufficient compressive force is maintained in every operating situation to ensure the cohesion of the spindle formed from the metal strips; otherwise, the form fit of the vertical strips engaging in grooves in the horizontal (gently inclined) strips would be lost. The principle of a folding spindle is shown in Figures 1.111 and 1.112. Figure 1.113 shows the installation situation in the *Festspielhaus Freiburg*, Figure 1.114 in the multifunctional hall of the *Stavanger Konserthus*, where

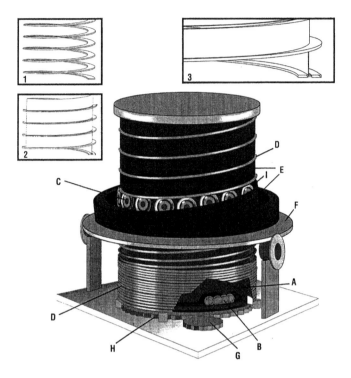

A	rotating storage cylinder for the steel belt D driven via G–H	E	vertical steel belt for spindle formation
		F	Rotatable support ring for storing the steel belt E
B	roller bearing		
C	guiding support rollers	G	drive pinion
D	horizontal guide steel belt for building up of the folding spindle	H	drive wheel
		I	joining area of belt D and E

Figure 1.111: Mode of operation of a folding spindle: 1. steel belt D; 2. steel belt D combined with steel belt E; 3. detail of the joining groove to fix the position of E and D. Photo credits: Waagner-Biró, Paco Corporation.

there are three lifting platforms with spiral lift drive in the orchestra pit to form an orchestra pit or extend the audience stalls.

Recently, Gala Systems Inc. (producer of "Spiralift") also offered the I-Lock system as a new series. In this system, the two steel belts are mechanically interlocked with each other by providing the vertical belt with a row of holes and the horizontal belt with a toothing that fits into the row of holes. This forms a compact tube.

Podiums driven with wedges

Podiums with a very small stroke, as may be the case with *compensating podiums*, can also be moved with simple wedge gears as *wedge drives* by shifting mechanically coupled wedges.

6″ Spiralift cross sectional view

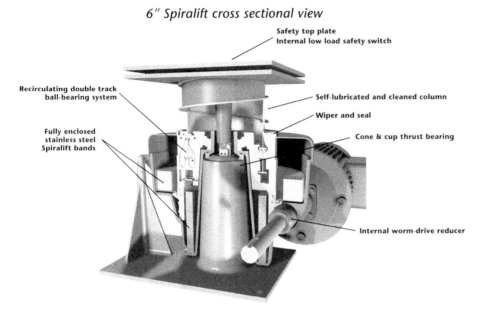

Figure 1.112: Folding spindle of the "new generation". Photo credit: Gala Theatrical Equipment.

Figure 1.113: Drive system for platforms with "folding spindles" – Spiralift. Festival hall Freiburg. Photo credit: Waagner-Biró.

Figure 1.114: Stavanger Konserthus – orchestra platforms with Spiralift. Photo credit: Waagner-Biró.

In the *Kammerspiele Paderborn*, a equalizing podium with an area of 9 m × 7.5 m and a stroke of 334 mm is designed as a *wedge-traction podium*. As can be seen in Figure 1.115, wedge carriages with rollers are pushed in and out of inclined planes at the podium and ground floor with four drive motors via spindle drives.

Guided podiums driven by hydraulic cylinders (see Section 2.3)
Hydraulic cylinders drives can exert large force effects and the hydraulic fluid is taken from pressurized hydraulic accumulators. The installation of counterweights can therefore generally be omitted. In addition, double-acting cylinders can be avoided and simple plunger cylinders can be used, since the empty podium can also be lowered under the effect of gravity. Hydraulic cylinders for operating podiums have been used for a very long time. Originally, water or a water–oil emulsion was used as the hydraulic medium; in modern plants, mineral oil-based or flame-retardant hydraulic fluids are normally used.

If a podium is driven by two or more hydraulic cylinders, it is generally not possible to achieve sufficiently accurate synchronous operation of the cylinders purely hydraulically (i. e., without electronic control), unless so-called cylinder flow dividers are used, such as those used for the orchestra podiums in the *Gothenburg Opera House*.

In this case, the lifting cylinders of a podium are not directly supplied with hydraulic fluid from the pump or accumulator station, but each lifting cylinder is supplied with the fluid quantity by a metering cylinder. By mechanically coupling the metering cylinders, their synchronism is forced and therefore the same volume flow is allocated to each lifting cylinder (see Figure 1.116).

Electronic control systems with precise path detection and combined use of hydraulics and electronics naturally also enable synchronous movements of several hy-

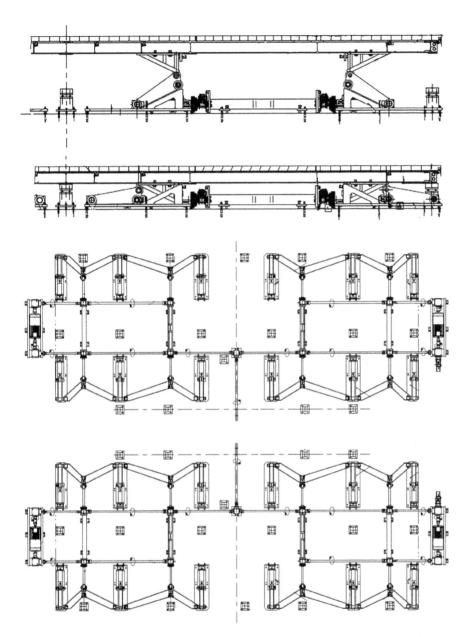

Figure 1.115: Compensating podium in the backstage in the Kammerspiele Paderborn designed as a wedge-driven podium: podium in highest and lowest position. Picture credits: SBS.

draulic cylinders without additional mechanical systems. Nowadays, synchronization of several podiums is also achieved in this way.

Within a podium, however, synchronization is forced in many cases with a *mechanical synchronization system.*

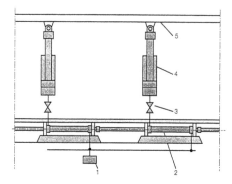

Figure 1.116: Cylinder flow divider for orchestra platforms – Gothenburg Opera House: 1 control panel; 2 synchronized cylinders; 3 hydraulically controlled shut-off valve; 4 lifting cylinder; 5 orchestra podium. Picture credits: Bosch Rexroth.

While the synchronization of several podiums with each other is controlled electronically, mechanical synchronization is preferred within a podium in order to enable safe movement of the individual podiums even in the event of failure of the control system in emergency operation. Different speeds of the two cylinders of a podium could result in jamming in the guides or even jumping out of the guides.

Figure 1.117 (left) explains the function of a *rope synchronization*. If several podiums are arranged in a row, the rope drums shown in Figure 1.117 (right) offer the possibility of a mechanical synchronization of several podiums by coupling the desired podiums to a continuous shaft. The old podium system in the *Vienna State Opera* was constructed in this way. Today, several podiums are synchronized with each other with the help of electronic control technology by giving the desired podiums the corresponding setpoints.

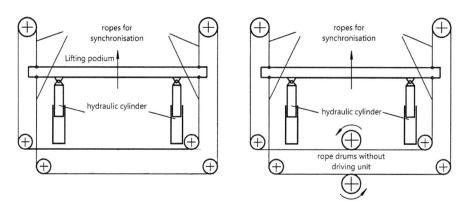

Figure 1.117: Rope synchronization – schematic drawing: (left) rope synchronization of the podium, (right) with rope drum for synchronization with other podiums.

Lifting platforms with rope synchronization that can be moved with hydraulic cylinders are installed, for example, in the *Hamburg State Opera* (see Figure 1.13).

Due to the elasticity of the ropes, a *shaft synchronization* as shown in Figure 1.118 should be preferred, as they were installed in the *Vienna State Opera* after the reno-

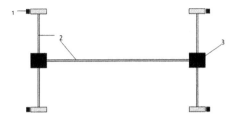

Figure 1.118: Shaft synchronization: 1: Rack and pinion, 2: Cardan shafts, 3: Transfer gearbox.

Figure 1.119: Podium drive with hydraulic cylinders in the Vienna State Opera, synchronization with pinions coupled by cardan shafts engaging in racks. Photo: Waagner-Biró.

vation (see also Figure 1.119). In the area of the four corner points of the podium with rectangular surface, racks are mounted parallel to the guides, in which pinions mounted on the podium and speed-coupled via cardan shafts and transfer gearboxes engage. This ensures precisely synchronized travel even under asymmetrical load conditions and irrespective of the clearance in the podium guides. These pinions and racks therefore do not serve as a lifting drive, but only as a mechanical synchronization device.

Regarding the geometric plant situation, it is often necessary to use cylinder drives in wells. With *telescopic cylinders* (see Figure 2.14(b)), the required well depth can be reduced or wells can be avoided. However, telescopic cylinders are very expensive and rarely used. Figure 1.120 shows podiums with telescopic cylinders.

Figure 1.120: Podium drive with telescopic cylinders. Photo: Bosch Rexroth.

Due to the compressibility of the hydraulic fluid, it is also necessary to provide mechanical locks for the platforms supported by hydraulic cylinders as for rope podiums. Mechanical fixation in any stroke position is possible by equipping the cylinders with clamping heads to fix the piston position by friction (see Section 2.3.1 and Figure 2.14(d)).

Scissor-type lifting platforms
All the podiums described so far have in common that they have to be guided in rails for vertical movement regardless of the drive type. In addition, depending on the type of drive selected, there is the requirement for the installation of wells, the installation of relatively space-consuming winches or the arrangement of cardan shafts to several drive elements to be operated synchronously.

Scissor-type lifting platforms as shown in Figure 1.121 (see also Figure 1.36) have a very low overall height when completely lowered, do not require any vertical guides, but due to their geometry they do not allow very large lifting heights unless coupled double scissors are arranged one above the other. If two shears are arranged next to each other on the same level to support a particularly long podium, they are referred to as tandem shears. Figures 1.122 and 1.123 show photos of scissor lifts, namely single scissors in the *Genoa Opera House* and double scissors in the *Kammerspiele Linz*.

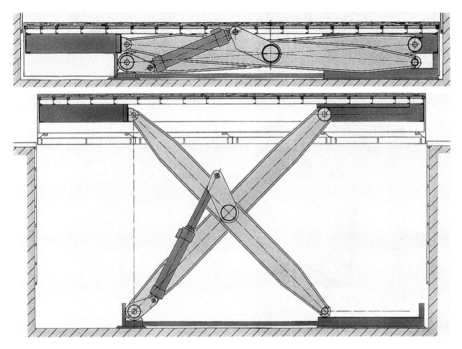

Figure 1.121: Scissor-type lifting platform – schematic drawing of a single shear. Picture credits: Bosch Rexroth.

In most cases, the drive is hydraulic with a hydraulic cylinder built into the shear. Figure 1.124 shows a hydraulically operated scissor lifting table, with the locking mechanism with bolts on the right. Instead of a hydraulic cylinder, a sliding or rolling screw drive (see Section 4.3.2) can also be used. Figure 1.125 shows a scissor-type lifting platform with a rolling screw drive in the *Lahti Theater in Finland*.

When the drive element is installed in the scissors, relatively large lifting forces are required as a result of the kinematics when the platform is in lowered position, which then become smaller and smaller in the course of the lifting movement. This can be avoided by installing folding spindles or push chains vertically between the lifting plane and the base plane; then the force required during the lifting motion is constantly equal to the lifting load. This arrangement can be seen in Figure 1.126 with folding spindles and in Figures 1.127, 1.49 and 1.50 with push chains (see also Figures 4.10–4.12).

If hydraulic cylinders are used, certain stroke positions must be locked, since even with leak-free barrier, slight height deviations occur when the temperature of the oil changes. The compressibility of the hydraulic fluid can also lead to slight changes in

Figure 1.122: Scissor-type lifting orchestra platform with a single shear – Genoa Opera House. Photo: Waagner-Biró.

Figure 1.123: Scissor-type orchestra lifting platform with double scissors – Kammerspiele Linz. Photo: Waagner-Biró.

position when the load changes or to vibrations (see Section 3.10). This positional fixation can be achieved in various ways. As described for podiums with hydraulic cylinder drive, clamping heads on the cylinders for stepless mechanical locking are also possible in this case, as well as bolt locks, e. g., as shown in Figure 1.124. Another possibility is, for example, to move an extendable "sword" along with the scissors, which is supported on precisely adjustable stops, if the sword is extended when a corresponding position is reached.

Figure 1.124: Scissor-type lifting platform: (left) double scissors, (right) detail of the locking device. Photos: Bayerischer BühnenBau.

Figure 1.125: Theater Lahti, Finland. Scissor-type lifting platform driven by ball screws. Photo: Kone (Finland).

Figure 1.126: Scissor lift table with folding spindle drive "Spiralift". Photo credit: Gala Theatrical Equipment.

In the *Schaubühne Berlin* (Figures 1.35 and 1.36), scissor lift tables are mounted throughout the floor area.

Scissor lift tables are especially commonly used as equalizing platforms and orchestra podiums. Sometimes they are also placed on primary podiums of the main stage as

Figure 1.127: BMW World. Scissor lift tables with push chain drive. Photo: Waagner-Biró.

secondary podiums, but then they prevent access through openings in the podium playing surface or the use of movable trap devices.

Scissor lift tables are used very frequently in materials handling technology and are therefore also offered as standard. However, special requirements must be observed for scissor lift tables in stage applications. For example, in general materials handling applications, in many cases much greater tolerances are permissible with regard to the movement accuracy of the platforms than in stage applications, both in terms of the exact parallel guidance of the platform surface and its exact vertical movement. Furthermore, scissor podiums suitable for stage operation must be sufficiently rigidly dimensioned in the vertical and horizontal directions so that they are not excited to vibrations during a performance.

It can also make sense to reduce the necessary lifting force of scissor lift platforms by using a planetary screw driven by an electric motor and an additional force generated by a hydraulic accumulator and hydraulic cylinder.

As an example, reference is made to the *Brussels opera house La Monnaie* (Figure 1.128), whose drive is similar to the podium drive in the *Stockholm Opera House* (Figure 1.110). However, another special feature of these scissor lift podiums is worth mentioning:

Figure 1.128: Brussels Opera House La Monnai – scissor-type lifting platform. Photo: Waagner-Biró.

It has already been pointed out that with drives according to Figure 1.121, particularly high forces are required when lifting from the lowest position, which decrease with the lifting height. This can be avoided – as already explained – by using drives according to Figures 1.126 and 1.127 with folding spindles or pressure chains.

Another way to avoid this is to have a pressure roller acting on a curved link attached to the scissor leg, which is anchored to a position-fixed pivot point, as shown in Figure 1.129. Through its shape, it can be achieved that a constant force is also required during the entire lifting process. This means that a constant lifting speed is also achieved at a constant drive speed. To simplify production, this link can also be constructed in an approximate circular shape, which means that the force varies somewhat and the stroke speed must be controlled. This design of a drive for scissor lift platforms was reportedly already used in earlier times, but had not become established for reasons of economy.

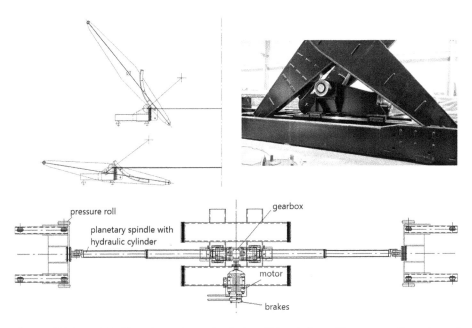

Figure 1.129: Brussels La Monnai opera house: (top) scissor lifting platform with pressure roller and curved link attached to the scissor leg, (bottom) drive unit of the scissor lift with planetary screw drive and hydraulic cylinder. Picture credits: Waagner-Biró, Novoscen.

The double drive acting to the left and right in Figure 1.129 drives two tandem shears, as can be seen in Figure 1.128.

It has already been pointed out at the beginning of this chapter that an essential characteristic of scissor lift platforms is that they do not require separate guides, since this task is performed by the scissors. However, in order to achieve the rigidity required in stage applications, the scissors must be of very massive design. This is mainly due to the fact that the scissor construction results in good stiffness ratios in the longitudinal direction of the scissors, but not in the transverse direction.

It can therefore be advantageous, especially for larger lifting heights, to also provide scissors in the transverse direction (in the direction of the narrow side of the podiums), as can be seen in Figure 1.130. In this case, the lifting or lowering movement is performed by means of pressure chains. In the application shown in the figure, it is a matter of a large number of podiums of small area to allow a high flexibility of the podium landscape on a total area of 135 m^2. The lifting height is 1.3 m.

Figure 1.130: Scissor lift podium with orthogonal scissors in the Leipzig Gewandhaus: (left) podium during factory testing, (right) concert hall at Gewandhaus Leipzig. Picture credits: SBS.

In addition to the scissor designs described so far, a scissor variant known as "*lambda scissors*" should also be mentioned (its design corresponds to the Greek letter λ). The podiums shown in Figure 1.131 were installed in the *Konzerthaus Berlin* for the variable design of the orchestra platforms. The primary podiums, equipped with four folding spindles, are guided in rails, while the secondary podiums, also equipped with four folding spindles, are guided with lambda shears.

Figure 1.131: Scissor-type lifting platform with "lambda shears" – Konzerthaus Berlin: (left) secondary podium with lambda shears, (right) primary and secondary platforms. Photo credit: SBS.

1.7.2 Stage wagons

Classification of stage wagons
Classification according to the orientation of the travel movement:
– Driven longitudinal: side stage ↔ main stage.
– Driven transversal: backstage ↔ main stage.
– Free riders: freely movable stage wagons.

Classification according to the support system:
– Multiroller stage wagon: moved on stage floor or as variant, supported on skids, can be moved on air cushion.
– Bridge car: travels like a crane bridge (without trolley) on lateral crane rails and bridges an open stage pit as a compensating trolley, when the lifting platforms are lowered.

Classification according to the type of drive:
– Manually pulled or pushed.
– With power-driven drive:
  – Pulled with ropes.
  – Pulled or pushed with chains.
  – Pushed with gears sunk into the floor (rack on the stage wagon) or friction wheels.
  – Self-propelled (drive on the stage wagon).

Classification of self-propelled machines with regard to the type of drive:
- Friction drive:
    - Travelling wheels are driven.
    - Stationary support wheels are driven.
    - Friction wheels (pressed on with spring) are driven.
- Gear drive:
    - Gears on the stage wagon, racks on the stage floor.
    - Gears on the stage floor, racks on the stage wagon.

Classification of self-propelled vehicles with regard to the energy supply:
- Electric drive with cable feed.
- Electric drive with batteries.
- Hydraulic drive with hose feed.
- Hydraulic drive with electric cable feed and hydro unit in the carriage.

More detailed description of stage wagons for various uses
As already described in Section 1.4 "Stage systems", stage wagons are used in the *sliding stage system*. These are platforms that can be moved horizontally on wheels. Depending on the orientation of the wheels, a distinction is made in relation to the rectangular shape of the stage trolley between *longitudinal drivers* for a travel movement between main and side stage and *transverse drivers* (Figure 1.132) for a travel movement between main and back stage.

There are also designs in which travel in both directions is made possible by swiveling the wheel axles or by alternative insertion of longitudinal or transverse wheels (raising or lowering the corresponding wheel sets), or the carriages are equipped with swivel rollers that allow travel in any direction. Recently, *swivel castors* – also called *"stage swivels"* and *"turtels"* – have been used for this purpose, which allow almost swivel-free travel in either direction and can also carry heavier loads (see Figure 1.142). Their running surface is made of polyethane elastomer, e. g., Pevodyn for medium and Vulkollan for larger load-bearing forces.

Depending on the load-bearing system of the platform wagon and the arrangement of the wheels, a distinction can also be made between *multiroller stage wagons* (Figure 1.132) and so-called *bridge wagons* (Figure 1.133).

Side stage wagons (longitudinal movement) of the usual design as shown in Figure 1.132(a) carry their dead load and payload via a large number of wheels onto the wooden stage floor below. The wheels therefore run on the surface of the stage platforms on podium stages in the main stage area and usually on compensating platforms on the side stage. Some wheels have a flange for positive rail guidance (e. g., as shown in Figure 1.132(c)).

In principle, *rear stage wagons* (transverse movement) can be built according to the same concept (Figure 1.132(b)), but the functional conditions of the main stage often

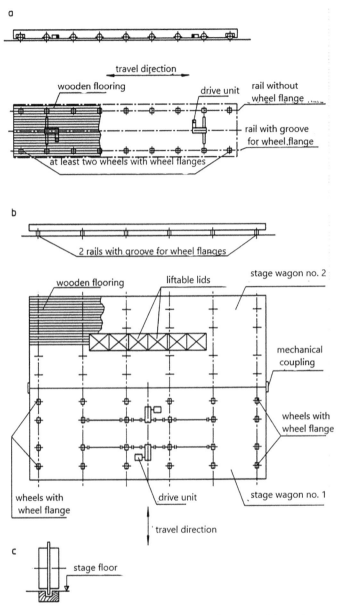

Figure 1.132: Stage wagon. Schematic drawing: (a) longitudinal driver, (b) transverse driver (two wagons in mechanically coupled condition), (c) wheel flange guide.

require a different design. If the stage pit is to be closed again to form a playing area when the main platforms are lowered into the lower stage, the stage wagons cannot be supported by a multiple roller system on the platforms below. In this case, a so-called *bridge trolley* similar to a crane bridge must span the entire width of the stage in a self-

Figure 1.133: Lifting platforms and compensating wagons (bridge wagons) in the revolving stage of the Vienna Burgtheater. Photo: Waagner-Biró.

supporting manner. The wheels roll on rails to the left and right of the platform pit. Such bridge carriages are also required as compensating carriages, as shown, for example, in the podium stage of the *Vienna State Opera* (Figure 1.14) and in the cylinder revolving stage of the *Vienna Burgtheater* in Figures 1.15 and 1.133.

Stage wagons with multiroller load transfer can be built with a very low height. The load on the rollers must be low enough not to exceed the permissible pressure on the wooden floor. Slightly higher pressures are permissible if the rollers run on hardwood rails or metal tracks embedded in the platform floor. Heavy bridge wagons require greater overall heights and steel running wheels rolling on steel rails, e. g., crane rails.

For scenic use, consider that bridge wagons running on steel rails can be moved with less vibration than stage wagons whose wheels roll on the stage floor, which can never provide the same evenness as a well-aligned rail track.

While bridge carriages are always rail-guided, other stage carriages are either guided by some track rollers in rails or by plug-in swords in slots, or they are freely movable in swivel castor design.

Often, a backstage wagon is equipped with a turntable, as already explained in Section 1.4 (Figure 1.12(b)). Figure 1.73 shows such a *turntable stage wagon* in the backstage of the *Semper Opera in Dresden*. When it is positioned in the main stage, a turntable is available. As a scenic effect, a turntable stage wagon can also be moved on an open stage with the turntable rotating.

Stage wagons with large diameter turntables can lead to space problems. Therefore, there are also technical solutions where the turntable is housed in parts in several stage wagons or partly in lifting platforms or fixed stage floor. In this case, the turntable can only be rotated by coupling the subsegments forming the turntable. Such a situation exists, for example, in the *Genoa Opera House* (see Figure 1.134).

The *Kammerspiele in Paderborn* serve as another example. There, the *backstage wagon* (9.0 m × 7.5 m × 0.333 m) designed as a *turntable stage wagon* does not contain the entire turntable with a diameter of 8.6 m, but a large turntable segment of the turntable

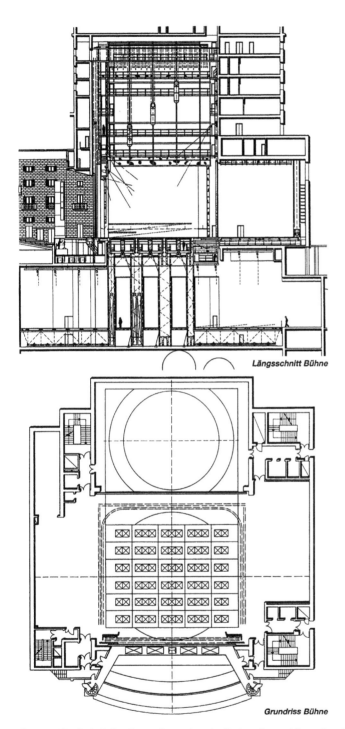

Längsschnitt Bühne

Grundriss Bühne

Figure 1.134: Carlo Felice Opera, Genoa. Longitudinal section and floor plan of the stage. Picture credits: BTR 4/1991, Waagner-Biró.

(see Figure 1.135). The small segment is located in the portal area and is coupled. Thus, the turntable in the carriage only forms a full circle with a circular segment embedded in the stage floor in front of the first stage platform. This allows a turntable diameter that utilizes the maximum width of the stage platforms, but also takes into account the storage possibilities on the backstage. Coupling takes place with the turntable carriage lowered. The turntable drive as a friction wheel drive is installed in the back stage wagon. The backstage wagon is driven by a roller chain attached to the underside of the carriage in the form of a pin gear and drives with fold-out pinions in the backstage floor and in a stage podium. The stage wagon together with the turntable segment can be lowered to stage level in the same way as the side stage wagons in the rear stage by means of a compensation podium. This *compensating podium* with a surface area of 9 m × 7.5 m has a stroke of 334 mm and is designed as a *wedge-type lifting podium* (see Figure 1.115).

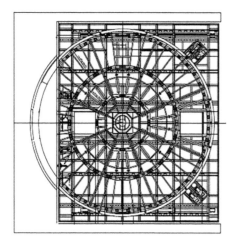

Figure 1.135: Turntable stage wagon in the Kammerspiele Paderborn. Photo credit: SBS.

The backstage trolley with cylinder revolving stage at *Oslo Opera* is also a special feature. The revolving stage carriage has its parking position in the backstage and can be moved to the proscenium with an electric rack and pinion drive. A cylinder revolving stage with a diameter of 15 m and a height of 3 m is installed in the carriage. This double-decker revolving stage thus also allows performances from below during the revolving movement (see Figure 1.136).

Stage trolleys can also be equipped with *tiltable covers*, similar to lifting platforms.

Modern *air-cushion technology* (cf. Section 4.6.2) opens up the possibility of moving stage wagons de facto frictionlessly on a thin film of air. The *Muziektheater Amsterdam*, for example, *was* equipped with such a stage wagon system. Each wagon of about 40 m^2 size is equipped with 8 air cushions of about 1 m diameter. Air is blown out through thousands of small holes, and it is possible to drive over gaps up to 2 cm wide as well as differences in height up to 1 cm. The drive is provided by push carriages. This technique was also used in the *Kuwait Conference Center* (see Figure 1.37).

Figure 1.136: Oslo Opera – turntable for installation in the backstage wagon. Picture credits: Bosch Rexroth.

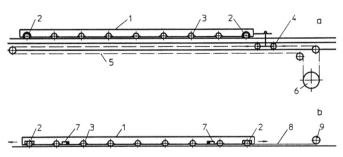

1	stage wagon	6	stationary drive unit
2	guide wheel	7	travelling drive
3	wheels	8	power supply via trailing cable
4	driving device – detachable	9	winch for electric cable
5	pull rope or chain		

Figure 1.137: Drive system for stage wagons: (a) with pull rope; (b) self-propelled unit with electrical power supply.

Stage wagon drive types

Smaller stage wagons can be moved manually by pulling or pushing and are usually referred to as *auxiliary stage wagons*. Larger stage wagons must be moved by motor. This can be done in several ways.

Drive with pull rope or pull chain

The stage wagon has no drive itself, but is moved by external drive elements by pulling the stage wagon with a *rope* or *chain*. Rope or chain can also transmit the tractive force in the manner of an underfloor drag conveyor via couplable carriers (Figure 1.137(a)).

Drive with pull-push chains

Figures 1.138 and 1.139 show new designs with *push chains* that have already proven their worth.

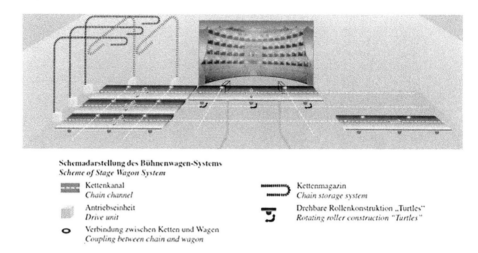

Schemadarstellung des Bühnenwagen-Systems
Scheme of Stage Wagon System

▦▦ Kettenkanal
     Chain channel

▨ Antriebseinheit
     Drive unit

○ Verbindung zwischen Ketten und Wagen
     Coupling between chain and wagon

⟿ Kettenmagazin
     Chain storage system

ᴣ Drehbare Rollenkonstruktion „Turtles"
     Rotating roller construction "Turtles"

Figure 1.138: Driving system of stage wagons with push chains. Arrangement of chains in main and side stages for longitudinal and transversal travel. Picture credits: Waagner-Biró.

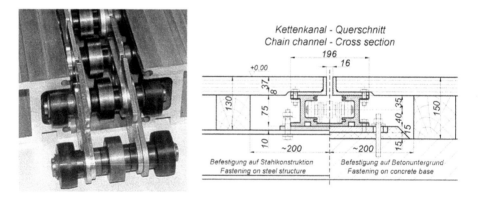

Figure 1.139: Drive of stage wagons with push chains, push chain guided in a chain channel. Picture credits, photo: Waagner-Biró, Serapid.

Chain channels are embedded in the stage floor and a push chain can be moved in each channel, which can absorb not only tensile but also compressive forces. If a stage trolley is connected to the chain by a driving pin, it can be moved in both directions by driving the chain via a sprocket.

As shown in Figure 1.138, the chain channels for the stage wagons are guided up the walls of the side platforms and deflected again at the ceiling. In this way, a channel length can be realized that allows the chain to be extended completely out of the travel range below the stage wagon. (Of course, the chain channels can also be deflected downward into the understage instead of upward.) In this way, chains for longitudinal and transverse travelers do not interfere with each other at the crossing points of the chain channels, as Figure 1.138 also illustrates.

Drive with gears meshing with a toothed rack

Another novel solution that has proven to be particularly low-noise is presented in Figures 1.140 and 1.141.

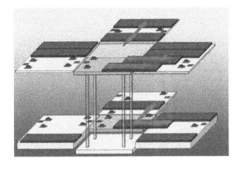

Figure 1.140: Stage wagon with stationary gear drives – schematic diagram. Photo credit: Waagner-Biró.

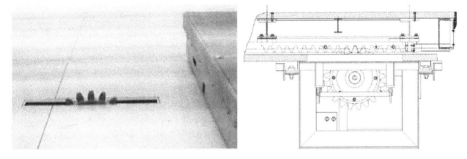

Figure 1.141: Stage wagon with stationary gear drives – details. Photo credit: Waagner-Biró.

Toothed racks are mounted on the underside of the stage wagon, and gear drives are installed stationary in the stage floor or, if necessary, in podiums (e. g., equalizing podiums). The pinions are lifted up from below through a mechanically opening slot and engage with the rack on the stage wagon. If racks for longitudinal and transverse travel are mounted on a stage wagon, they must be interrupted at the crossing points. The number and position of the drive units must be selected so that at least one unit is always engaged for the desired travel movement.

The exact positioning of the pinion and thus of the carriage is carried out by a laser system that determines the position of the first tooth of the rack mounted on the carriage. With this data, the control computer can calculate the travel distance for positioning respectively the rotational speed and position of the pinion, when it is swiveled in so that tooth cannot come to rest on tooth, resulting in a blocking of the drive.

If racks and drive units are present in the longitudinal and transverse directions of the stage (Figure 1.140), the wheels of the stage wagons must be mounted in swivels to allow different directions of travel – in *swivel castor sets* – also called "*stage swivels*" and "*turtels*" (see Figure 1.142).

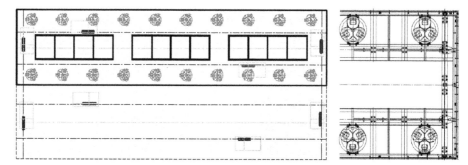

Figure 1.142: Stage wagon with swivel castor sets (also called "*Turtels*") as running wheels. Picture credits: Waagner-Biró.

Self-propelled unit with cable or hose feed

So-called *self-propelled stage wagons* are equipped with their own travel drive and must therefore be provided with a power supply. Cable feeds via cable drums for electrical energy or hose feeds via hose drums for hydrostatically driven stage wagons, if the power unit is not accommodated in the stage wagon, are possible options for this. One problem is usually that these power feeds have to be laid with some effort in such a way that they are not damaged when moving equalizing platforms.

Self-propelled with batteries

This problem of feeding can be avoided if *batteries* are installed in the stage wagons as an energy source. The accumulators are then recharged via busbars in the side or rear stage area or via electrical plug connections. Stage wagons equipped with batteries can be seen in Figure 1.143 (*Seoul Opera*) and Figure 1.144 (*Musiktheater Linz*).

The stage wagon shown in Figure 1.144 is guided in guide rails made of rolled steel profile; specially designed steel disk wheels serve as guide elements. By swiveling the wheels by 90°, both longitudinal and transverse travel is possible. Both stage wagons shown are radio-controlled.

Figure 1.143: Stage wagon with batteries – Seoul Opera. Photo: Waagner-Biró.

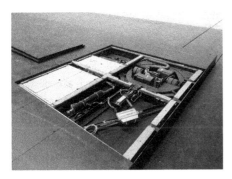

Figure 1.144: Stage wagon at the Musiktheater Linz. Photo: Waagner-Biró.

Self-propelled with adhesion drive or gear drive

The drive force is transmitted either by *adhesion* via the *load-bearing wheels* (see Section 3.4), or *gears* are driven which engage positively in toothed racks embedded in the platform floor or in pin chains serving as a drive shaft (see Section 4.4.1). In the case of multiroller stage wagons, this positive drive system should be given preference over a frictional system. Due to the statically indeterminate load distribution among a large number of rollers, it can hardly be ensured that the wheel loads of a few drive wheels are sufficient to actually transmit the required driving forces by adhesion.

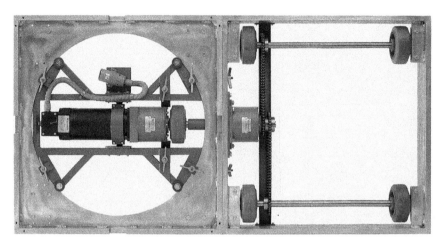

Figure 1.145: Friction wheel drive – "Zarga System" stage module. Picture credits: Max Eberhard AG - Bühnenbau (CH-Weesen), hereinafter referred to as "Eberhard" for short.

Figure 1.145 shows a modular *friction wheel drive* for a modular stage wagon system with Vulkollan running wheels.

Freely movable stage wagons with motor drive

Figure 1.146 shows two systems from the company Green Motion. The stage carriages can be assembled from $1\,\text{m} \times 1\,\text{m}$ modules. The overall height of the modules is $16\frac{2}{3}\,\text{cm}$ or 20 cm.

In the servo II system (Figure 1.146, top), steering during travel is achieved by applying two drives with different speeds. In the smallest version, these two drives are installed in one module; in larger units, two modules with one drive each are used.

In the vector III system (Figure 1.146, bottom), the module contains a 360° rotatable turntable with two drives that can be swiveled into the desired rotational positions both when stationary and during a travel movement. This makes it possible to change the direction of the axis from a stationary position without causing friction between the drive wheel and the ground.

The drives are controlled by radio communication via portable remote control panels or stand control panels with touch panel and computer control, but it is also possible to specify a certain travel path that the stage wagon must follow.

In the case of freely movable stage wagons, the navigation technology used can basically be the same as that used for automated guided vehicles (AGVs) in logistics. For this purpose, physical guidance systems can be used as inductive guidance systems (a sensor is used to follow an AC wire) or optical guidance systems (a sensor is used to follow a white or colored stripe). However, this requires conversions for different productions. Therefore, it is more practical to work with virtual guidance by reading in the driving

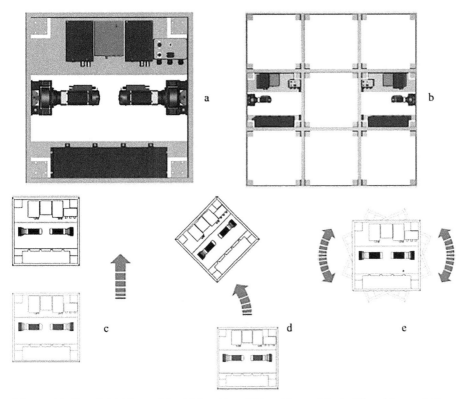

Drive module "Servo": (a) drive module with two drives, (b) two drive modules in different frames, (c) linear travel, (d) curve travel, (e) turning

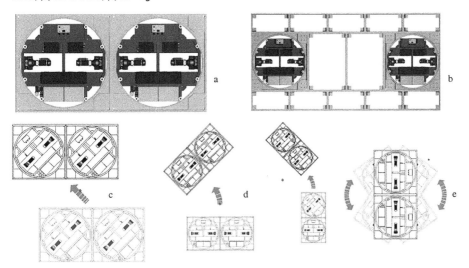

Drive module "Vector": (a) drive module with two drives, (b) two drive modules in different frames, (c) vector travel, (d) curve travel – longitudinal and transversal, (e) turning

Figure 1.146: Freely movable stage wagons. Photo credit: Green Motion.

curve through spatial coordinates and using laser navigation to measure the position in relation to installed reference points (cross-bearing).

1.7.3 Turntables and rotating platforms

Classification of revolving stage platforms
Structure according to the number of levels:
– Single turntable.
– Double-decker turntable = cylinder turntable:
    – Without built-in podiums.
    – With built-in podiums.

Classification according to the type of support:
– Circular rail with wheels rolling on it:
    – Stationary circle rail.
    – Circular rail on the rotating platform supported on stationary wheels.
– Central bearing with a ball bearing slewing ring.
– Circular rail and central bearing (statically indeterminate bearing).
– Column bearing (very rare older design).
– Hydrostatic bearing.

Classification by type of centering:
– With the wheel flanges on the rail head of the horizontal circular rail.
– With horizontal rollers on a circular rail.
– With centering pin.
– With a ball bearing slewing ring.
– With column bearing.

Classification according to the location of the drive:
– Drive stationary.
– Drive travels with the rotating platform.

Drive variants:
– Drive via the wheels running on the circular rail via adhesion.
– Drive via friction wheels (pressed on with spring force).
– Drive with gears.
– Rope drive via traction sheave.
– Drive with steel link chain.

More detailed description of revolving stage platforms
A revolving stage, resp. *turntable*, offers a very simple way to transform the stage set. It can accommodate two to three stage sets and face the audience depending on the rotational position. Many stages are equipped with fixed turntables in the main stage.

Also other types of mobile turntables can be used: For example, a *turntable* as shown in Figure 1.147 can be stored in the upper or lower stage. But there are also *folding turntables* that can be disassembled into small parts and stored like decorative material. Figure 1.148 shows a turntable that can be assembled from light metal elements from a set of standardized stage modules.

Figure 1.147: Folding turntable of the Vienna State Opera. Photo: Waagner-Biró.

Figure 1.148: Turntable, which can be assembled from components. Photo: Eberhard.

In the case of turntables that can be placed on the stage or built into stage wagons, designs with many small tangentially aligned wheels rolling on concentric circular rails are used. The drive for the rotary movement is provided, for example, by horizontally running friction wheels pressed on with sufficient spring force (see Figures 1.149 (right), 1.161(b) and 1.164) or positively via a horizontally lying chain pinion which engages in a toothed rim or a pin gear ring formed by a chain (Figure 1.161(d, e)).

As an alternative, it is possible to install a turntable in a stage wagon and move it to the main stage if required. Such a *turntable stage wagon* has already been mentioned in Sections 1.4 and 1.7.2 and is available, e. g., at the *Semper Opera in Dresden* (see Figure 1.73).

In Section 1.7.2 and Figure 1.136, respectively, a double-decker revolving stage was shown that was installed in the backstage wagon at the *Oslo Opera*.

Figure 1.149: Turntable stage wagon for the Kuwait Conference Center: (left) factory assembly, (right) friction wheel drive of turntable. Photo: Waagner-Biró.

Figure 1.150: Liftable core disc for a backstage wagon. Wuhan YueHu Culture and Art Center. Photo: SBS.

A complex backstage trolley was designed for the *Wuhan YueHu Culture and Art Center in China*. It incorporates a 17-m diameter circular ring disc and a 9-m diameter core disc that can be lifted 1.2 m by spiral lifts guided by scissor systems. Figure 1.150 shows the core disk in lifted condition.

As already mentioned in the list of stage systems (Section 1.4), large theaters are often equipped with so-called *cylinder revolving stages*. The revolving stage then consists not only of a turntable forming the stage floor (Figures 1.154 and 1.157(b)) but, analogous to a double-decker podium, of a revolving cylinder with two accessible levels. This design can therefore also be called a *double-decker turntable* (Figure 1.157(c)).

In most cases, cylinder revolving stages are also equipped with lifting platforms. The interaction between the rotation of the revolving stage and the vertical movement of the platforms provides opportunities for stage set transformations and scenic effects. Above all, a cylinder revolving stage with built-in lifting platforms, combined with a sliding stage system with stage trolleys, offers any possibilities for the rapid change of several prepared stage sets; see Section 1.4.

Figure 1.151 shows the model of a large revolving cylinder stage (see also Figure 1.83) in the *National Grand Theater Beijing* (diameter 16 m and height 18.4 m) with checkerboard podiums built into it, namely 13 square podiums guided inside and 2 rectangular podiums guided outside. In the photo on the right, you can see the nitrogen cylinders at the front and the piston accumulators at the back right.

Figure 1.151: Model of a cylinder revolving stage with built in podiums in the National Grand Theatre Beijing. Photos: SBS-Rexroth consortium.

The *Musiktheater Linz* has already been mentioned in Section 1.4. Figure 1.152 shows the large *transport revolving stage* and the *performance revolving stage* installed in it. Both revolving stages are – as already explained in more detail in Section 1.4 – supported on a circular rail as well as on a slewing ball bearing. This statically indeterminate bearing was necessary due to the very low overall height available. In both cases, the slewing ball bearing takes over the centering function.

The heavy transport turntable runs on the circular rail with steel running wheels in balancers guided by horizontal pressure rollers (see Figure 1.153). (Balancers are two running wheels in an articulated wheel housing for equal load distribution on both wheels.) The drive is provided by eight friction wheels on the circular rail supported on the concrete wall of the foundation.

The much lighter performance turntable is quietly running on wheels with a Vulkollan running surface on the circular rail. Eight wheels are driven.

For permanently installed turntables, a wide range of bearing and drive options is possible, which will be systematically discussed in the next section.

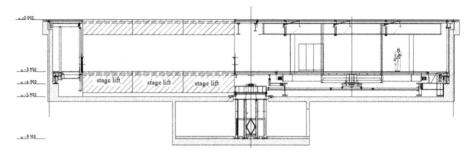

Figure 1.152: Musiktheater Linz – Large transport revolving stage and integrated performance revolving stage. Picture credit: Waagner-Biró.

Figure 1.153: Musiktheater Linz – wheels on the circular rail of the transport revolving stage. Photo: Waagner-Biró.

Various design of support methods of revolving stages

The same developments can be traced in the designs for the bearing arrangements of turntables and slewing platforms as in crane construction for the bearing arrangements of slewing cranes.

Bearing arrangement with circular rail

The oldest design was the *bearing arrangement on a circular rail*; it is shown in different variants in Figure 1.154. Depending on the design, the wheel is centered by the flanges of the vertical running wheels (Figures 1.154(a) and 1.155) or by horizontal rollers on the circular rail (Figure 1.154(b)), by a centering pin in the axis of rotation (Figure 1.154(c)) or by a centering pin which is also capable of absorbing vertical forces (Figure 1.154(d)).

Figure 1.155 shows the bearing of a heavy cylinder revolving stage in the *Sydney Opera House* on a horizontally laid circular rail. Heavy crane wheels mounted in balancers support the loads resulting from the high dead weight of the stage and the live loads. The wheel flanges provide the centering.

Whereas steel wheels have to be used for this large cylinder revolving stage, the much lighter revolving stage of the *Theater in der Josefstadt in Vienna* according to Figure 1.156 can use very quietly running wheels with plastic bandages.

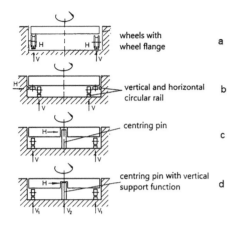

Figure 1.154: Bearing of turntables on a circular rail. Schematics of different variants. Centering: (a) with the flanges of the wheels, (b) on the circular rail with horizontal wheels, (c) on a centering pin, (d) on a centering pin with vertical support function.

Figure 1.155: Cylinder revolving stage of the Sydney Opera – factory photo during assembly. Photo: Waagner-Biró.

Figure 1.156: Turntable in the Theater in der Josefstadt (Vienna) – bearing on a circular rail. Photo: Waagner-Biró.

This figure shows both the vertical rollers mounted in balancers to support the vertical loads and horizontal rollers to center the turntable. In addition, as described in Section 1.7.1, sliding traps in the turntable floor and a transportable trap mechanism for persons are also shown.

In order to reduce noise even with heavy revolving stages, a solution can serve in analogy to running gears with movable grandstands (see Section 1.7.4), as it was realized in the *Oslo Opera House* (see also Section 1.7.2). When the revolving stage is at rest, it is lowered on spring-loaded friction surfaces. If the stage is to be set in rotation, a total of 24 wheel sets are pressed hydraulically against the running surface of the turntable and 6 wheel sets are driven electrically. Since the heavy revolving platform is not supported by the wheels when at rest, plastic treads can be used on the wheels to minimize running noise.

In the circular rail solutions mentioned so far, the wheels were located on the rotating part and the rail on the foundation. However, it is also possible, in reversal of the principle, to mount the rollers in the fixed foundation and to mount the turntable with one rail or even several concentric rails on circularly arranged and tangentially aligned rollers.

Column bearing

Slewing cranes for large load moments were built with so-called *column bearings*. In this design, the vertical forces are absorbed centrally by a roller bearing at the tip of the column. Tilting moments from the eccentricity of the vertical loads and from horizontal loads are absorbed, on the one hand, by a roller bearing at the top of the column and, on the other hand, by horizontal pressure rollers supported on a circular rail lying in the horizontal plane. In one construction variant, the conical column is fixed with the tip pointing upwards and the rotating framework is placed over it in the shape of a bell; in another variant, which is also used in stage engineering, the column rotates with the tip pointing downwards, as can be seen in Figures 1.157(a) and 1.158 (*Stadttheater St. Pölten*). In stage engineering, the column bearing has been realized only very rarely.

Bearing arrangement with ball bearing slewing ring

Another bearing arrangement which has become widely accepted in crane construction is also being used more and more frequently in stage engineering, namely the bearing arrangement with a *ball bearing slewing ring*. This is a special type of slewing bearing as shown in Figure 1.159, which is capable of supporting not only radial forces but also very large axial forces in both directions, i. e., upward and downward. This bearing is thus suitable for absorbing both axial forces from the sum of vertical loads and tilting moments from their eccentric position. These bearings are also built equipped with an external or internal ring gear, as can be clearly seen in Figure 1.160, so that simple concepts also result for the rotary drive. Bearings of this type are characterized by particularly smooth running, since the rotary motion is achieved by rolling balls in a precisely

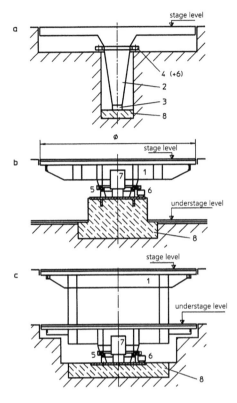

1	turntable	5	ball bearing slewing ring
2	column with circular rail for horizontal pressure rollers	6	drive
3	axial/radial bearing	7	slip ring assembly
4	horizontal pressure rollers	8	foundation

Figure 1.157: Central bearing systems for revolving stages: (a) column bearing of a turntable; (b) turntable on a slewing ball bearing; (c) cylinder turntable on a slewing ball bearing.

machined ball race. It goes without saying that the rotational resistance resulting from friction in this rolling bearing arrangement is much lower than the rotational resistance on wheels in the other bearing systems. Therefore, very low drive powers are also sufficient.

However, such *central bearing constructions* must also be considered in the construction concept of the stage building, since the entire load, concentrated on a small area, must be absorbed by this central bearing and its foundation. In the case of support on a circular rail, these loads are distributed over a much larger foundation area.

Figure 1.157(b) shows the central bearing of a plain turntable with one platform and Figure 1.157(c) that of a cylinder turntable with two platforms in schematic representation. Figure 1.160 shows the centrally mounted turntable of a theater in Linz (Austria) and its drive.

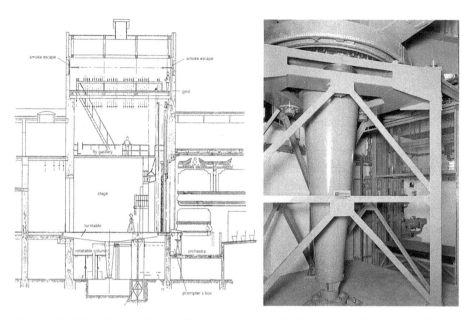

Figure 1.158: Column bearing of a turntable in the Stadttheater St. Pölten. Longitudinal section and photograph of the stage. Picture and photo credit: Waagner-Biró.

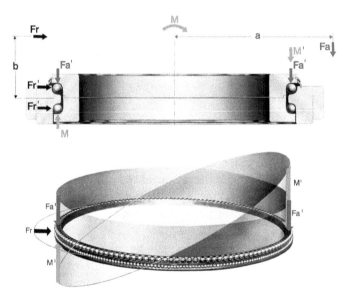

Figure 1.159: Double-row ball bearing slewing ring – sectional drawing and explanation of load-bearing function: (top) the axial forces (F_a), radial forces (F_r) and moments (M) acting on the ball bearing slewing ring $M = F_a \cdot a + F_r \cdot b$; (bottom) the axial forces (F_a) and moments (M) acting in a cylindrical surface and radial force (F_r) acting in a plane. Picture credits: Hoesch/Rothe Erde - Schmiedag AG (D-Dortmund).

Figure 1.160: Cylinder revolving stage of the Linzer Landestheater. General view during factory assembly and detail of the drive. Photos: Waagner-Biró.

Hydrostatic bearing

Another, but very rarely used, variant of the bearing of revolving stages results from the application of a *hydrostatic bearing concept*. The turntables installed in the main and proscenium stages of the *Friedrichstadtpalast in Berlin* in Figure 1.46 are mounted in this way. Below a circular slideway are plates with a diameter of about 500 mm through which oil is forced so that the disc floats on a film of oil about 0.3 to 0.4 mm thick.

Construction methods for the drive of turntables respectively rotating stages

There are many drive options, which are described below and summarized in Figure 1.161. Basically, electric or hydrostatic drives are again possible.

Drive via frictional locking (see also Section 3.4)

If there is sufficient frictional contact at certain running wheels, the drive can be effected via these running wheels as *friction wheel drive*. This type of drive can therefore be used,

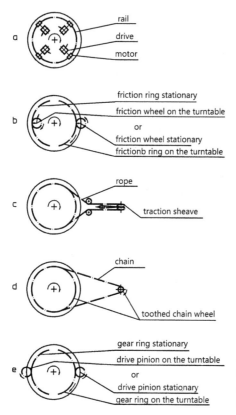

Figure 1.161: Drive systems of revolving platforms – schematic representation of different variants: (a) adhesion drive via carrying wheels, (b) adhesion drive via friction wheels, (c) rope drive via traction sheave, (d) drive via chain wheels, (e) gear drive.

for example, when a revolving platform is mounted on vertically positioned wheels that roll on a horizontal circular rail (Figure 1.155). If there is insufficient frictional contact due to dead and live loads, the drive can be provided by *spring-loaded friction wheels*. The friction wheels can also be installed in a horizontal position and roll on a vertically positioned friction rim as shown in Figure 1.149.

Rope drive

Predominantly in older installations, the *drive* was designed as a *traction sheave drive* by looping a rope with sufficient pretension around the turntable and a usually double-grooved drive sheave. Figure 1.161(c) shows this principle, while Figure 1.162 presents a photo of the *Tiroler Landestherater*.

Drive via positive locking (see also Sections 4.2 and 4.4)

In a *gear drive*, one or more drive pinions mesh with a large ring gear concentric to the revolving stage. In large cylinder revolving stages with circular rail bearings, a pin gearing is often used as the drive ring (Figure 1.163, *Burgtheater in Vienna*). At a pin

Figure 1.162: Rope drive of the revolving stage in the Tyrolean Landestheater. Photo: Waagner-Biró.

Figure 1.163: Revolving stage driven by pin gearing – Vienna Burgtheater. Photo: Waagner-Biró.

gearing, the teeth are formed by bolts with a circular cross-section arranged next to each other at the pitch distance of the gear teeth (see Figure 4.21(d)). The shape of the pinion should be adapted to these pins as cycloids according to the geometrical conditions of the gearing theory.

If the bearing arrangement is made with a slewing ball bearing, these slewing bearings can be procured already equipped with a gear rim in involute toothing as standard. Gears with standardized involute flank are then used as drive pinions (see Figure 1.160 or 4.21(a)).

However, positive-locking gear drives can have the disadvantage that backflash caused by the tooth flanks striking during acceleration or deceleration can impair the fundamentally smooth running provided by the bearing arrangement in a slewing ball bearing. The clearance between the teeth can have a negative effect because revolving stages mounted in this way have a particularly low frictional resistance during rotary motion. This can lead to alternating contact between the tooth flanks when the speed is adjusted to an exact setpoint. For this reason, a friction wheel drive is sometimes preferred for this type of central bearing as in the *Graz Schauspielhaus*, for example, with spring-loaded friction wheels acting on both sides of a steel ring – see Figure 1.164.

Figure 1.164: Revolving stage driven by friction wheel in the Schauspielhaus Graz. Photo: Waagner-Biró.

Chain drive

As a special form of a gear drive, it is also possible to realize the pin gearing ring by means of a link chain looped around the turntable. This solution is sometimes used for lay-on turntables. Standardized sprockets are often chosen as drive sprockets, which are mounted in a swing arm and held in the chain tooth engagement under spring load. It should be noted that the tooth flanks of sprockets do not correspond exactly to the geometry of the drive pinion toothing (see Section 4.4.1).

As an alternative to the gear drive now described, another kind of *chain drive* should also be mentioned. In this case, the drive is provided by a chain transmission consisting of a stationary small-diameter drive sprocket and a concentric large-diameter sprocket mounted on the turntable as well as a chain wrapped around both tooth gears (see Figure 1.161(d)). Often it is also sufficient to simply loop the chain around a cylindrical wall, e. g., covered with wood, on the turntable and the friction is sufficient to move the turntable by driving the chain via a small chain wheel.

1.7.4 Mobile podiums and stands

Pedestrals (practicables)

If lifting platforms cannot be used for this purpose, or in addition to lifting platforms, the stage set can be designed with so-called *"practicables."* This designation, which has been in use for a very long time, was used for walk-on stage scaffolds to distinguish them from only imitation scaffolds. In Figure 1.165, practicables are shown as simple immovable stackable "wooden boxes."

Figure 1.165: Pedestals designed as "wooden boxes". Picture credits: Eberhard.

DIN 15920 standardizes a modular system of construction elements. It covers stage pedestals ("practicables") of various shapes and sizes (cuboids, triangular prisms, etc.), steps, stairs, freely movable small stage wagons and stage railings. This standardization is intended to facilitate the realization of coproductions and guest performances.

Furthermore, there are many different plug-in and screw systems for platforms and scaffold superstructures. Figures 1.166 and 1.167 show light metal platforms developed according to a modular principle and adjustable in height and inclination.

Figure 1.166: Adjustable stage platforms in aluminum construction "Scebomat" system. Picture credits: Eberhard.

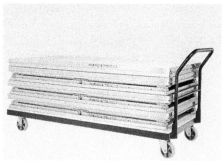

Figure 1.167: (left) Lifting table with scissor construction adjustable from above, (right) transport trolley. Picture credits: Bühnenbau Schnakenberg.

Movable concert hall podiums

In concert halls, there is sometimes a need to arrange the orchestra podium differently. Podiums that can be moved like drawers, but also partially lifted, can serve this purpose. Figure 1.168 shows an example of the podiums in the *Mozart Hall of the Vienna Konzerthaus*.

Mobile grandstands

However, special mention should also be made of *telescopic stands* for the variable design of the spectator area.

These stands can be deposited in a wall niche by sliding them together as in Figures 1.169(a) and 1.170, lowered below the auditorium level on lifting platforms as in Figure 1.169(b), or moved in stacked form as in Figure 1.169(c) by means of a lift truck.

A problem with mobile stands is often that the running wheels cause pressure marks on the floor during long periods of standstill as a result of the relatively high pressure on the wheel contact surfaces or the plastic running surfaces of the wheels

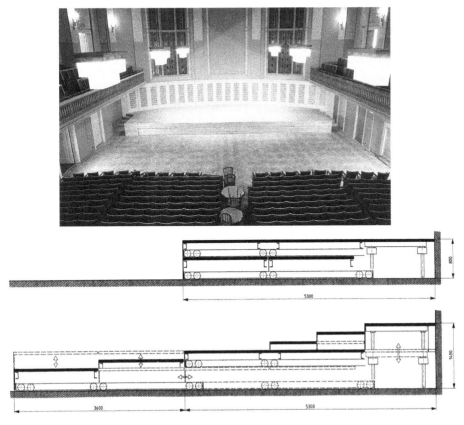

Figure 1.168: Extendable podiums in drawer design combined with a lifting podium in the Mozart Hall of the Vienna Konzerthaus – photo and schematic drawing. Photo, image credit: Waagner-Biró.

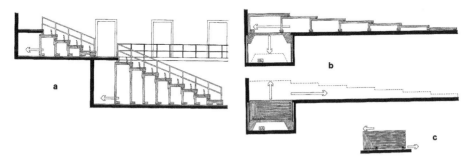

Figure 1.169: Telescopic bleachers: (a) parking in wall niches, fixed seats with folding backrests, (b) retractable below auditorium level with lifting platform, (c) retractable on extendable rollers. Picture credits: Bayerischer BühnenBau.

Figure 1.170: Telescopic grandstand. Photo: Bayerischer BühnenBau.

become out-of-round due to flattening. For this reason, there are systems in which the stationary podium is supported on skids over a large area and the running wheels take over the load transfer only for transport. In the solution shown in Figure 1.171, this is done by filling a pneumatic hose.

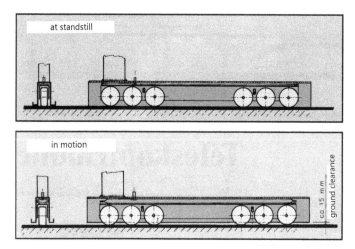

Figure 1.171: Mobile grandstands with a pneumatic hose device for lifting the parking skids and placing the load on the wheels. Picture credits: Bayerischer BühnenBau.

Grandstand supports on air cushions are a particularly elegant solution. When stationary, a sufficiently large ring surface is available for load transfer, and when moving, the plate-shaped air cushion elements are supplied with air and allow the stands to float on a thin film of air so that they can be moved with very little effort. Further information on air cushion technology is given in Section 4.6.2. A grandstand construction with air cushions can be seen in Figures 1.172 and 1.173.

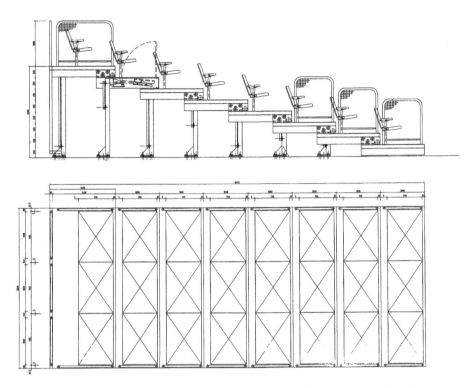

Figure 1.172: Telescopic grandstand with air cushion as traversing device. Picture credits: Bayerischer BühnenBau.

Figure 1.173: Telescopic grandstand with air cushion as traversing device in extended and stacked position. Photos: Bayerischer BühnenBau.

If there is a need to equip a hall with a level floor either with or without rows of seats, this can be done using a seating system shown in Figure 1.174, for example: The seating is mounted on seat row supports equipped with casters that can be moved on rails laid in floor slots. The seat rows are coupled by spring steel carrier belts. As shown in the illustration, when the steel belt is extended, it forms the cover of the floor slot for

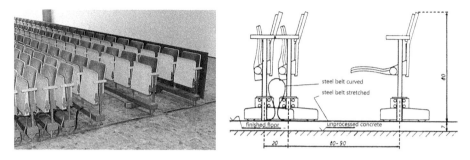

Figure 1.174: Seating system with rows of seats moving on rails, drive with rope winch – "Elochair" system. Picture credits: Vogel.

Figure 1.175: Grandstand transport: (left) steerable transport trolley (Arena Madrid), (right) air-cushion transport trolley (Madrid Sports Palace). Photo: Waagner-Biró.

the rails. If the rows of chairs are pushed together to form a package in the park position on one side of the hall, these gaps must be covered manually with cover profiles. The last seat row carrier is moved by means of a rope winch drive and carrier. Depending on the direction in which the rope drum is turned, one or the other end of a rope loop laid in the floor is wound up and the seat row carrier is pulled into stacking or seating position on the right- and left-hand sides of the hall.

Figure 1.175 shows transport systems for grandstands. The left picture shows a transport unit for the *Arena Madrid*, the right picture for the *Sports Palace Madrid*. The stands in the arena are moved on steerable transport carts with battery-powered electric drives, while the stands in the Madrid Sports Palace are moved "floating" with air cushions to other locations in the hall, so that very different configurations can

Figure 1.176: Telescopic grand stands in various positions. Photo: Waagner-Biró.

be created. Different telescopic positions can be seen in Figure 1.176 with a different number of elements extended (see also Section 1.6, Figures 1.53 and 1.54).

1.8 Technical equipment of the upper stage – upper stage equipment

This section deals with those technical systems that are to be assigned to the stage area above the stage level.

Fixed installations in the ceiling and wall area

In large stages, the stage space in the main stage area is more than twice as high as the proscenium portal opening in order to be able to move the decorations upwards out of view of the audience. On the ceiling, the *fly grid* is a walk-on girder grid construction. At small stages such a high free space above, the stage is not existing.

The side walls and the back wall of the stage house are equipped with *work galleries*, from which it is possible to perform various manipulations on the stage.

Portal construction – proscenium

The portal construction of the *proscenium* opening belongs also to the equipment of the upper stage. Here, too, mechanical systems can be integrated to vary the size of the portal opening; in special cases, the entire proscenium can be moved in the direction of the longitudinal axis of the stage to change the depth of the stage space. In the case of multipurpose use of a venue, it may also be envisaged to remove the entire portal structure and lift it into the stage tower, for example (see also Figure 1.183(d)).

In the proscenium there are also curtains for opening or closing the proscenium opening. These are *curtains* made of fabric as curtain lift or curtain pull systems and the *iron curtain* as a fire protection device. *Gate constructions* with fire and sound protection functions can also be used between the main stage and the side stage or backstage.

Mechanical equipment on the fly grid

The ropes of the hoisting equipment hang down from the fly grid in large numbers. In the past, natural fiber ropes were used, today wire ropes. These facilities include the following devices (Figure 1.177):

– *Bar hoists* with load bars (*load bar hoists, drop hoists*) oriented normal to the longitudinal axis of the stage and spanning the entire width of the stage.
– *Point hoists* with – as the name already expresses – a single load suspension point.

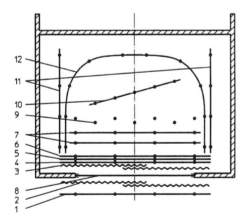

1	forestage hoist	7	load bar hoist, drop hoist
2	decorative curtain	8	fire curtain
3	play curtain	9	point hoist
4	gauze curtain	10	freely positioned bar hoist
5	festoon (cloud type curtain)	11	panorama curtain
6	sound curtain	12	cyclorama

Figure 1.177: Different kinds of flies in the upper stage (based on DIN 56 920, sheet 3).

- Mechanical devices for delimiting the playing area from the working galleries, such as *panorama hoists* and *circular bar hoists*, also called *cyclorama hoists*, and *horizontal dragged cyclorama* (*wrap around curtain*).
- Mechanical equipment for lighting technology, such as *lighting bridge, batten light hoists* and *sidelight racks*.
- Special equipment, e. g., *flying machines, flybar systems*.

1.8.1 Fixed installations in the upper stage

Fly grid

In the simplest case, the *fly grid* consists of a walkable ceiling in the uppermost area of the stage tower. The grate-like design of the ceiling construction allows ropes to be passed through for the suspension of decorative elements. The ropes of hoists usually have to be redirected toward a stage side wall to the counterweight slides of manual counter hoists or the winches of machine hoists. Figures 1.178(a)–(d) show various arrangements for load bar hoists.

In Figure 1.178(a), the ropes are guided via single-groove rope pulleys from the grid to a multigroove head pulley on the side wall; in Figure 1.178(b), the ropes are guided horizontally to the side wall supported in partly multigroove pulleys.

For better accessibility of the fly grid, the wire ropes of the hoists can also be guided below the fly grid by installing the rope sheaves on the underside of the grid (Figure 1.178(c)). The disadvantage of this arrangement, however, is the inconvenient maintenance possibility.

If sufficient space is available, the most suitable grid design is that two levels are created, namely the *walkable grid* through which the ropes are passed vertically, and a *rope pulley grid* located approx. 2 to 3 meters above it, in which the deflection pulleys for passing the ropes on are situated (Figure 1.178(d)).

The walkable grid floor can be equipped with a *slatted grid* or a *light grid*. The slatted grid, made of sheets bent into a U-shape, offers long lead-through slots for the ropes, so that transverse displacement of the ropes is possible. A light grid allows only a single point through which the rope must be rethreaded each time its position is changed; however, the light grid can, of course, be designed to provide rectangular rope slots at specific locations. A walkable grid in a slatted grid design with a rope pulley grid above it can be seen in Figure 1.179 (*Großes Festspielhaus Salzburg*), and a design with a light grid can be seen in Figure 1.180 (*Stadttheater Duisburg*). A rope guide below the grating as shown in Figure 1.178(c) as it is provided, for example, in the *Stockholm Municipal Theater* (Figure 1.181).

A rarely realized solution can be found in the *Frankfurt Opera House* and can be seen in Figure 1.182. A solid ceiling is installed above the walkable fly grid as a foundation for the electric winches. The hoisting ropes leave the rope drums vertically upwards and are drawn to the required points via rope pulleys in a roller grid on the ceiling and

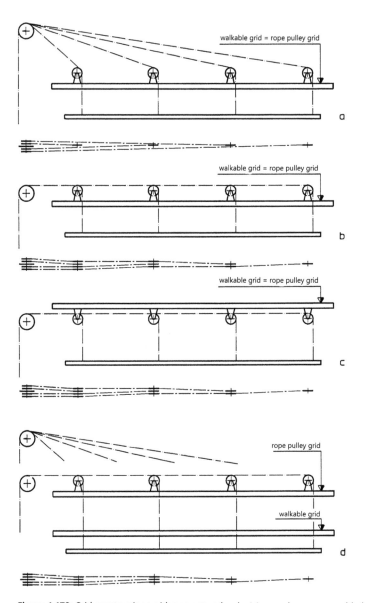

Figure 1.178: Grid constructions with one or two levels: (a) rope sheaves assembled on the grid, inclined rope guidance to the wall; (b) like (a), but horizontal guidance of the ropes; (c) rope sheaves below the grid; (d) rope pulley grid above the walkable grid.

guided vertically through the ceiling by small openings. There are thus three levels in total, with a walkable grid at the bottom, a solid ceiling of fire protection and sound-proofing effect above, and the rope pulley grid above.

Figure 1.179: Walkable grid with slatted frame and rope pulley grid above at the Großes Festspielhaus in Salzburg. Photo: Waagner-Biró.

Figure 1.180: Walkable grid as light grid and rope pulley grid above at the Stadttheater Duisburg. Photo: Krupp.

Load capacity of the grid

The grid offers the possibility of installing a large number of hoists. Each of these hoists has a maximum load-bearing capacity that results from the dimensioning of the structural elements of the respective hoist. This results in enormous total loads for the supporting structure of the grid, which ultimately have to be transferred to the building foundation via the building structure.

When stage equipment is replaced or major alterations are made to older buildings, this often results in considerable problems, whether the roof trusses, the masonry or the foundations of the building cannot support these loads. If necessary, this can also mean that the overall load has to be restricted.

Two specific situations can be described in relation to this problem:

When the *Vienna Volkstheater* was renovated a few decades ago, the entire support structure of the grid was placed on a steel portal structure built into the building, so that steel columns transferred the load directly into foundations without loading the building itself.

Figure 1.181: Rope-guide with pulleys below the grid at the Stadttheater Stockholm. Photos: SBS.

During the reconstruction of the *Berlin Opera House "Unter den Linden"*, it had to be determined that, for geological reasons, the building subsoil would not at all be able to bear the total load that would result if all existing hoists were loaded with the maximum life load capacity. Therefore, it is necessary to limit the size of the total load and to support the grid construction on load cells to exclude overloading.

Work galleries

The walls of the stage – the side walls and the rear stage wall – are equipped with *work galleries*. These are several accessible areas arranged one above the other and connected by stairs – often also by an elevator system – from which a wide variety of manipulations can be carried out with good visibility to the stage.

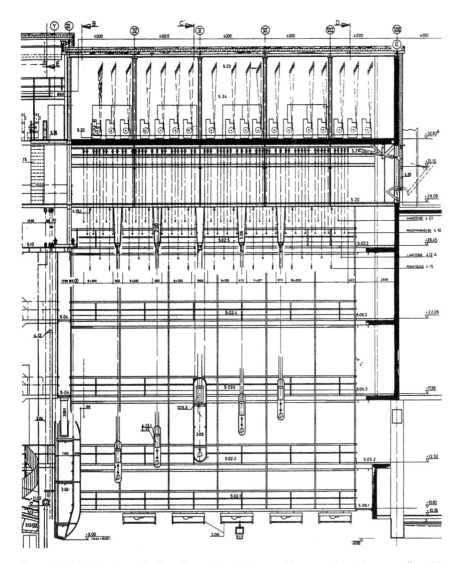

Figure 1.182: Grid-system at the Oper Frankfurt am Main – machine room above the rope pulley grid. Picture credits: BTR 2/1992.

Of particular importance are the working gallery for the operation of the manual counterweight hoists, the working gallery for the loading and unloading of the counterweight carriages, and those working galleries on which control panels for the control of mechanical drives of the lower and upper platforms are located.

The handrails of the working galleries are designed to be more massive than those of other walkways, since the handrail is also intended to be used for attaching spotlights and for suspending simple *manual operated natural fiber manual string hoists* (see Section 1.8.3).

1.8.2 Proscenium facilities

Design of the stage portal

Proscenium is the architecturally designed framing of the stage opening. The stage side of the proscenium wall is provided with platforms and catwalks on larger stages, primarily to allow lighting equipment to be installed. If the size of the portal opening is to be variable, the lateral boundary can be made adjustable by movable *portal towers* or, in the case of fixed portal towers, by sliding panels. The upper boundary can be varied by changing the lifting position of the movable *portal bridge* or if a fixed portal bridge is present by moving an aperture vertically.

The portal towers can be moved either manually or by motor. With an appropriate rope guidance of a winch drive, a positively guided symmetrical movement of both towers can also be achieved.

For vertical movement of the portal bridge, a manual drive is out of the question because of the usually very large moving masses, despite partial counterweight compensation on larger platforms.

In its function as a *lighting bridge*, a movable portal bridge can generally be lowered to stage level so that it can be easily equipped with spotlights. Fixed guides for the portal bridge are provided above the highest portal opening. The movable portal towers installed below must usually be moved to the position of the smallest gantry opening to provide guidance for the portal bridge in order to lower it to stage level. If the portal bridge can be built sufficiently high, the guidance in the movable portal towers is not necessary and portal towers and portal bridge can be adjusted completely independently of each other.

Large portal bridges are built with several levels. Access to the catwalks on the portal bridge from the working galleries is made possible, if necessary, by means of movable connecting elements to bridge level differences.

Figure 1.183 schematically summarizes various possibilities of forming the portal zone. Figure 1.184 shows the movable portal towers and the height-adjustable portal bridge in the *Festspielhaus Bregenz*, Figure 1.185 those in the *Großes Festspielhaus Salzburg* and Figure 1.186 in the *Genoa Opera*.

Curtain constructions

In the proscenium area, larger stages usually have several curtains installed, whose functions and construction methods are briefly described below, namely:
- Show curtain.
- Veil curtain, veil fly.
- Ornamental curtain.
- Sound curtain.
- Iron curtain.

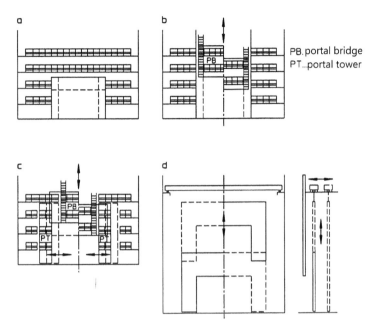

Figure 1.183: Different kinds of design of the portal zone: variants regarding portal towers and portal bridge: (a) fixed portal construction, modification of the portal opening possible by movable panels, (b) movable portal bridge, fixed portal towers, (c) movable portal bridge and movable portal towers, (d) complete portal construction liftable and movable in the longitudinal direction of the stage.

Figure 1.184: Movable portal towers and movable portal bridge in the Festspielhaus Bregenz. Photo credit: Waagner-Biró.

Figure 1.185: Portalzone in the Großes Festspielhaus Salzburg. Photo credit: Waagner-Biró.

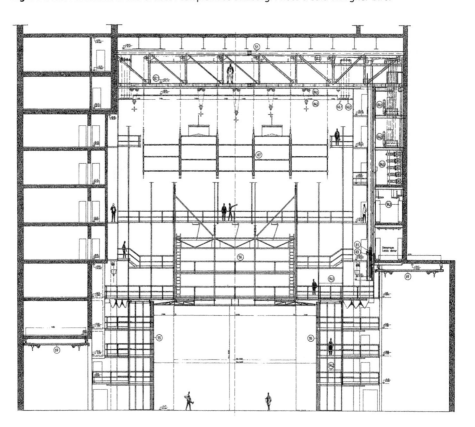

Figure 1.186: Movable portal towers and movable portal bridge at the Genoa Opera House. Picture credit: Waagner-Biró.

Show curtain

The *show curtain* is used to open and close the proscenium opening. The following three possibilities of the movement sequence are offered:

– *Lifting* the curtain.
– *Dividing* the curtain by warping the halves of the curtain to the left and right.
– *Diagonal lifting* of the curtain halves in the direction of the lateral upper corners.

In the *lifting curtain* according to Figure 1.187(a) – also called *German curtain* – a load bar with the curtain mounted on it is moved vertically like a bar hoist. The drive can be manual or motorized.

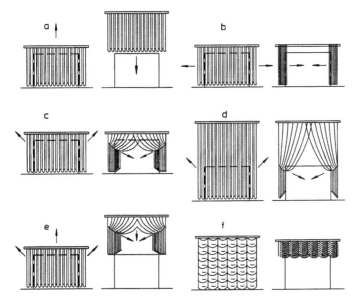

Figure 1.187: Show curtain, constructions and functions according to DIN 56 920, sheet 3: (a) lifting curtain (German curtain), (b) partial travers curtain (Greek curtain), (c) diagonally opening curtain (Italian curtain), (d) diagonally opening curtain (Wagner curtain), (e) liftable and diagonally opening curtain (French curtain), (f) cloud curtain, festoon.

With the *partial traverse curtain* according to Figure 1.187(b) – also called *Greek curtain* – two curtain halves are moved sideways on a curtain track. The curtain is driven manually or electrically by ropes. If each curtain half is pulled in the closing or opening direction at the end facing the center, an uneven pleat division results on the curtain during the movement. Figure 1.188(a) shows the suspension of the curtain on glides (*glide pull device*), Figure 1.188(b) the suspension on rollers (*roller pull device*).

In most cases, the curtain sliders or rollers are moved with ropes or chains. Figure 1.189 shows a system in which a movable toothed belt is used:

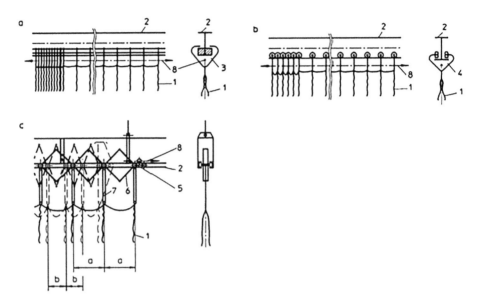

Figure 1.188: Traverse curtain: (a) with sliding suspension, (b) with wheel suspension, (c) with scissor pull device for a pantograph-type curtain (1 curtain, 2 curtain track, 3 curtain glider, 4 curtain roller, 5 roller, 6 scissors, 7 suspension, 8 pull rope).

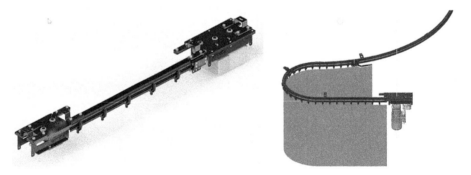

Figure 1.189: BELT-TRACK: (left) running rail with toothed belt and detail of toothed belt, (right) example of an application. Picture credits: Gerriets GmbH.

BELT-TRACK is a motorized curtain track for medium to heavy fabrics, for screens as well as for backdrop parts. With this track, single or double track, straight as well as individually curved track runs can be realized. The curtain parts can be individually and positively coupled to the toothed belt and moved in the same direction or in opposite directions. The slip-free, positive drive via toothed belt offers the possibility of exact and reproducible positioning by encoder and corresponding control and guarantees absolutely precise and reproducible motion sequences, as they must be guaranteed especially on cruise ships and at shows.

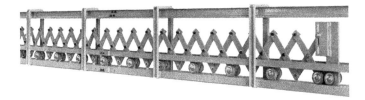

Figure 1.190: Scissor-type – pantograph-type curtain. Photo: Schneider Bühnentechnik (D-Kiel).

In the so-called *scissor curtain* (*pantograph-type curtain*) as shown in Figure 1.188(c) or 1.190, the distance between the suspension points of the curtain is uniformly increased or decreased, thus maintaining a uniform drape at each curtain position.

A large scissor curtain was installed in the *Tashkent Congress Center* built in 2009 (see Figure 1.192). The scissor curtain consists of two scissors with the end position at the left and right ends. Both scissors are installed laterally offset and overlap in the middle in the closed state. The total length of the curtain is approx. 36.7 m, the overlap normally 1.0 m, in special cases maximum 3.0 m.

The support-beam of the scissors can be suspended in ropes (13 ropes at a distance of approx. 2.9 m) and thus raised or lowered as a lifting curtain via a rope winch. The drive of the scissor traverse-curtain (Greek train) is located on the scissor beam and is a traction sheave drive with a three-grooved traction sheave (3 times 180° wrap). There is also a hemp rope emergency drive available.

If necessary, this curtain design could also be equipped with a drive for diagonal opening curtain as an Italian curtain.

A characteristic feature of a *diagonal opening curtain* is that two halves of the curtain are drawn diagonally to the upper corner points of the proscenium opening. Depending on the height of the portal opening, the starting point and the position of the diagonal pull, it is also referred to as an *Italian curtain* (Figure 1.187(c)) or a *Wagner curtain* (Figure 1.187(d)). When the movement is combined with a lifting movement according to Figure 1.187(e), the term *French curtain is* commonly used. The pull on the ropes for the opening movement must take place at an ever decreasing speed during opening. In the past, this was achieved by using a conical rope drum driven at constant speed (Figure 1.191); today, the speed of a cylindrical rope drum is changed. For the closing process, the drum was disconnected from the drive to close the curtain.

In the proscenium, curtains of different modes of operation can be installed one after the other in the portal zone. But it is also possible to build combined curtain systems, where the same curtain fabric can be moved in two or three different ways. In this sense, there are combined lifting curtain and partial travers curtain, partial travers and diagonally opening curtain, lifting and diagonally opening curtain or, when all three movement possibilities are combined, there is a lifting and travers and diagonally opening curtain.

Figure 1.191: Historic picture of a conical winch for a diagonally opening curtain. Photo: Waagner-Biró.

Figure 1.192: Pantograph-type curtain for Tashkent Congress Center. Image credit: Gerriets.

Example of lifting/traverse/diagonally opening curtain system

Figure 1.193 shows the new curtain system of the *Vienna State Opera*. It allows three variants of operation, as a lifting/traverse/diagonally opening curtain. Figure 1.193(b)–(d) shows the schematic of the overall system, Figure 1.193 the drives and rope guides in three specific situations:

- Opening or closing as a traverse curtain is performed with a rope winch, whereby two ropes are wound on a drum and two counter ropes are unwound in each case,

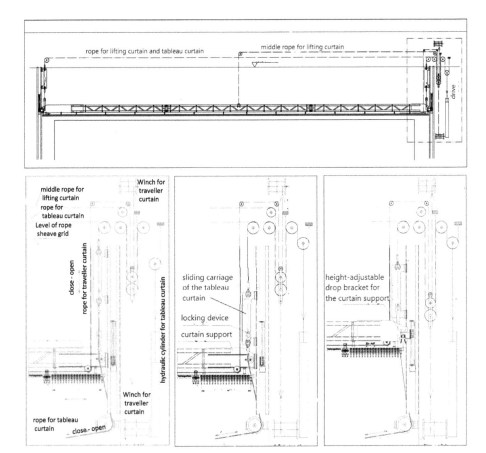

Figure 1.193: Combined Greek, Italian, and Wagner curtain of the Vienna State Opera House: (a) overview; (b) function as diagonally opening and traverse curtain – curtain support and raffle slide for the tableau curtain not interlocked; (c) function as lift curtain – curtain support and raffle slide for the diagonally opening curtain locked – travel position; (d) function as lift curtain – curtain support placed on a height-adjustable bracket – curtain in open position. Picture credits: Klik Bühnensysteme GmbH.

so that both directions of movement are made possible via tensile forces. (Only one rope is shown schematically in the illustration.) Since the ropes are guided over rollers of the lifting carriage, which is used to operate as a lifting curtain, a traverse curtain can be used at different heights of the curtain support.

- When opening and closing as a diagonally opening curtain, this curtain support is set down on support pins. The two ropes coming from the curtain are each attached to a sliding carriage of the tableau curtain (tableau curtain carriage) which is adjusted in its height position by a rope pulley block with the ropes of the lifting and tableau curtain for opening or closing by extending or retracting the piston of a hydraulic cylinder.

- When the curtain is raised or lowered as a lifting curtain, the tableau curtain carriage is interlocked with the curtain support so that moving this carriage also raises or lowers the curtain support. Since the ropes of the lifting and tableau curtain are also guided over the lifting carriage as a pulley block, the tableau curtain carriage is moved by a lifting winch via a pull rope engaging the lifting carriage. When the lifting curtain is in the raised position, the two ropes of the lifting and tableau curtain are relieved by setting the curtain support down on a height-adjustable set-down bracket.
- Since the bending stiffness of the curtain beam as a beam on two supports is not sufficient, the hoist rope center is attached in the center of the beam, which is moved along with the ropes of the lifting and tableau curtain. However, since the curtain beam must not change its height position during the movement of the curtain, this rope is not guided to the cylinder via the pulley block and is suspended after the lifting carriage.

Besides these three most common curtain systems, other variants are sometimes used. As examples may be mentioned:
- The *winding curtain*, where a textile is wound on a horizontal drum.
- The *cloud curtain*, in which the fabric is pushed together upwards by several lifting ropes guided in rings (Figure 1.187(f)).

Other curtains used in the proscenium area:

Veil curtain
Veil curtain is a curtain made of light translucent textile. It can be moved as a lifting curtain at a particularly high speed and is used to achieve scenic effects. As a proscenium device, it is also called a *portal veil*. However, load bar hoists installed in the stage area behind the proscenium can also be used for curtain functions and in particular for veil curtains.

Ornamental curtain
A decorative curtain can be installed in front of the iron curtain if the curtain leaf of the iron curtain on the side facing the audience is not decorative. For safety reasons, the proscenium opening, which is closed with the curtain, is usually opened in the presence of the audience only a few minutes before the performance begins and is immediately closed again after the performance ends.

Sound curtain
A *sound curtain* is a curtain that is arranged behind the play curtain and can be lowered after the play curtain is closed in order to have a sound-insulating effect between the stage and the auditorium. This is intended to prevent sounds from being perceived too loudly in the auditorium during stage alterations.

In the portal zone in the proscenium, one or two load bar hoists may also be installed for variable use.

Iron curtain
Although only textile curtains have been mentioned so far, the iron curtain should also be mentioned in this list as a *fire protection curtain*. It is described in more detail in Section 1.9.1.

1.8.3 Hoists with ropes

Classification
Classification by construction method and intended use:
- Bar hoists:
  - Drop hoist.
  - Panorama hoist.
  - Round bar hoists, cyclorama.
  - Freely positioned bar hoist.
- Point hoists.
- Manual operated natural fiber rope hoist ("handline hoist").
- Batten light hoist.
- Performer flying.

Classification according to the type of drive:
- Manual:
  - Load bar hoist as manual counterweight hoist (ordinary or double purchased).
  - Winch hoist.
- Electric motor drive.
- Hydraulic drive:
  - With hydraulic cylinder.
  - With hydraulic motor.

Classification by type of load-bearing equipment:
- Wire rope.
- Natural fiber rope (as hand pull).
- Chain.
- Steel belt.
- Synthetic fiber rope in special cases (see Section 1.8.8 or Figure 1.258).

Terms and basic construction methods
Suspended decorative elements are carried by hoists and can be lowered into or raised out of view of the audience in the stage house. For this purpose, the *grid* described in Section 1.8.1 is located in the roof area of the stage house.

Irrespective of the type of drive, a distinction is made between *bar hoists* and *point hoists* depending on the intended use and the type of *load* suspension.

With bar hoists loads can be mounted on a vertically movable load bar, i. e., uniformly distributed loads or concentrated loads (point loads).

As a load bar are used, among others:

- Tubes, e. g., outer diameter 63.5 mm, wall thickness 4 mm, meter mass 5.9 kg/m, area moment of inertia $I = 33.2\,\text{cm}^4$, axial section modulus $W = 10.5\,\text{cm}^3$.
- Rectangular hollow profile, e. g., external dimensions 100 mm × 50 mm, meter mass 6.6 kg/m, axial geometric moment of area $I_y = 106\,\text{cm}^4$, axial moment of resistance $W_y = 21.3\,\text{cm}^3$.
- Special profiles made of aluminum alloy (e. g., according to Figure 1.196).

The load capacity of a load bar hoist is determined by three criteria:

- By the maximum total sum of all loads suspended from the load bar;
- By the maximum permissible load of each hoisting rope on which the load bar is suspended; and
- By the maximum permissible bending stress on the load bar.

Therefore, the following data are provided as characteristic values of the load-carrying capacity (see Figure 1.194):

- The maximum effective mass in kg or the maximum load in N or kN or, if this load is distributed as a uniform load over the length of the bar-section, possibly also in N/m or kN/m.
- The maximum point load under a hoist rope.
- The maximum point load in the center of the bar section between two hoisting ropes or at the end of the bar.

Figure 1.195 shows possible designs of adjustable rope attachments. Since a load bar hangs on about 4 to 6 ropes depending on its length, it must be possible to adjust the length of the ropes. In the designs shown in Figure 1.195(a, c), turnbuckles are used for this purpose, in the design shown in Figure 1.195(b) a sliding adjustment lug. Fixing lugs welded to the center of the tube as shown in Figure 1.195(c) make it possible for carriages whose rollers run on the top left and right of the tube at an angle of about 45° to pass. Such load bars are then also suitable for use as rail for performer flies (see Section 1.8.8). Figure 1.196 shows special profiles made of light metal which offer universal application possibilities.

Load bar hoists are mounted in large numbers oriented normal to the stage axis and are also called *drop hoists* in this application. The length of the load bars corresponds approximately to the width of the stage, the distance between the load bars is approx. 180 to 300 mm. If the spacing is very close, there is a risk of snagging on neighboring bars already fitted with loads.

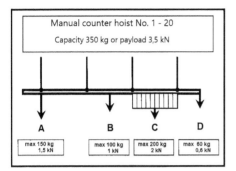

The load specifications A through D mean:

A   maximum point load below the rope suspension – depending on the rope dimensioning

B   maximum point load in the center of the field between two rope suspensions – depending on the dimensioning of the load bar and the maximum load A

C   maximum uniformly distributed load between two rope suspensions – depending on the dimensioning of the load bar and the maximum load A

D   maximum point load at the end of the bar – depending on the dimensioning of the load bar and the maximum load A

Figure 1.194: Payload table for load bar hoists.

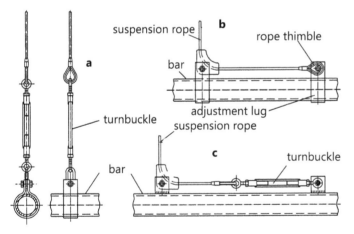

Figure 1.195: Adjustment of the rope suspension with (a) turnbuckle vertical, (b) sliding adjustment lug, (c) turnbuckle horizontal, bar useable as rail (see Figure 1.256).

If hoists are located in the front or rear stage area, they are also referred to as *front* or *rear stage hoists* according to their spatial assignment.

If load bar hoists are oriented along the stage side walls in the longitudinal axis of the stage in order to cover these walls decoratively with their hangers, they are called *panorama hoists*. In addition to load bar hoists with a straight *bar*, there are also so-called *round bar hoists*, or *cyclorama hoists*, with a curved load bar. These hoists are

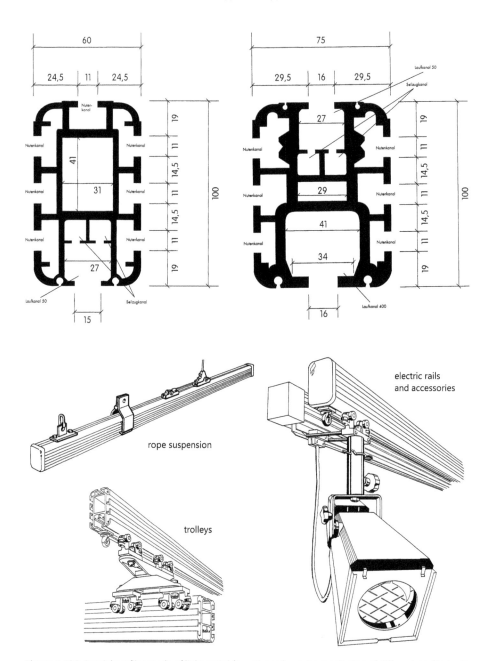

Figure 1.196: Special profiles made of light metal for universal use: (top) selection of different profile types, (bottom) rope suspension, trolleys, electric rails and accessories. Picture credits: Bühnenbau Schnakenberg GmbH & Co. KG (D-Wuppertal).

often used in combination with panorama hoists to delimit the stage area. Movable bar hoists of small load capacity with movable rope pulleys on the grid and in any spatial position are also called *freely positionable hoists*.

In modern stages, it is usually important to install many *point hoists*. Point hoists are not intended for line loads but for point loads and offer the possibility of suspending a single load with one hoisting rope or, in the case of a block and tackle system, with two rope lines and a bottom pulley.

Usually, several point hoists work together in a synchronous group, e. g., to lift a ceiling. Of course, a load bar can also be hooked into several point hoists. Therefore, their use is generally only possible in systems that can be networked in terms of circuitry and where it is possible to regulate the speed of the point hoists concerned with a feedback control.

The stage-specific lifting devices listed so far usually have not very high load capacities. Load bar hoists are usually designed for a load capacity of approx. 300 to 1000 kg or a load capacity of 3 to 10 kN, point hoists for 100 to 300 kg or 1 to 3 kN. If the installed hoists have a low load capacity, as it is mainly the case with manual counterweight hoists, individual *heavy-duty hoists* are often also installed in the upper platform in order to be able to manipulate higher loads if necessary (see Figure 1.235).

In most cases wire ropes are used for hoists, but also chains or steel belts are available (see Sections 1.8.4 and 1.8.5).

Construction types of bar hoists

In the past theaters used to be equipped almost exclusively with manually operated hoists. Motor-driven hoists were used relatively rarely. Bar hoists operated with hydraulic cylinders via simple manual control valves had also proved their worth in old theaters. Electric drives with suitable control characteristics could hardly be realized in earlier years, or only at economically unacceptable expense. As a result, hydraulic drives were initially most popular for motorization, because they could be controlled sensitively. With the possibilities of modern electrical engineering, electric drives have now become established again, as will be explained in more detail in Section 2.4.

Manual drive

Manual hoists are generally built for a load capacity of up to 300 kg (3 kN) payload. Such loads could not be moved quickly enough with pulley blocks or winches in scenic applications. For this reason, stage technology uses so-called *manual counterweight hoists* in which the lifting load is balanced by counterweights. The design is shown in Figure 1.197(a).

The load bar hangs from several wire ropes which are guided vertically to the fly grid respectively to the roller grid and via deflection pulleys to the counterweight wall. At the *counterweight wall*, they are tied into a *counterweight cage* which is loaded with

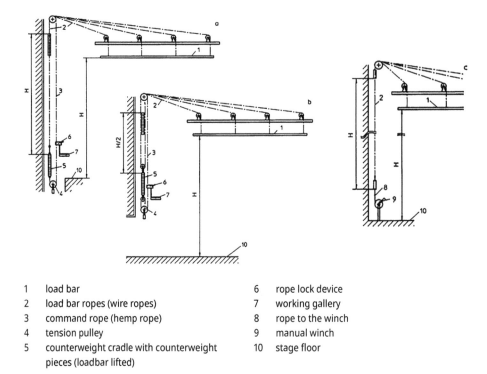

1	load bar	6	rope lock device
2	load bar ropes (wire ropes)	7	working gallery
3	command rope (hemp rope)	8	rope to the winch
4	tension pulley	9	manual winch
5	counterweight cradle with counterweight	10	stage floor
	pieces (loadbar lifted)		

Figure 1.197: Manually operated bar hoist: (a) (ordinary) manual counter hoist, (b) doubled purchased manual counter hoist, (c) manual winch hoist.

manually applied weight elements to compensate the load of the bar and the load hanging on the bar. These counterweight cages are loaded and unloaded from a working gallery provided for this purpose; the individual weights should not weigh more than 15 kg. To raise or lower the load bar or counterweight cage, a *command rope* is pulled manually from a working gallery. This is a natural fiber rope (e. g., hemp rope) that is tied into the counterweight cage in a vertical rope loop. In this arrangement, the counterweight cage covers the same lifting distance as the load bar, i. e., a maximum lifting distance roughly equal to the difference in level between the fly grid and the stage floor. Therefore, the counterweight wall must also reach down from the grid to the stage floor or even lower. Figure 1.199 (*Festspielhaus Bregenz*) shows a counterweight wall for manual counterweight hoists.

If a side stage adjoins the main stage and the counterweight wall can therefore not reach down to stage level, *doubled purchased manual counterweight hoists* must be used instead of the *ordinary manual counterweight hoists* described above, as shown in Figure 1.197(b). In the case of doubling, a 1:2 pulley block is interposed between the load bar and the counterweight. Therefore, a counterweight mass twice as large must balance the load, which then, however, only covers half the lifting distance of the load bar (cf. Section 4.1.2). The command rope can also be doubled or directly tied.

To fix a load bar at a certain height, the command rope is clamped with a *rope lock device*. There are a large number of rope lock designs of varying quality; above all, the hemp rope should not be subject to too much wear at the clamping point. Figure 1.198 shows a rope lock for a command rope. Sometimes, however, only simple *rope clamps*, shown in the same figure, are used.

Loading the counterweight cage with counterweights of 15 kg requires a great deal of work and puts strain on the spine. For this reason, efforts are being made not only

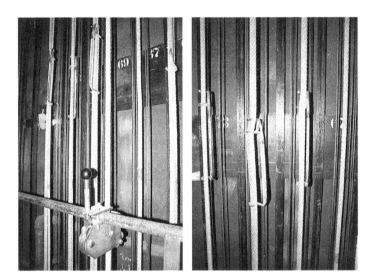

Figure 1.198: Rope lock devices for manual counterweighted hoist: (left) rope lock, (right) rope clamp. Photos: Waagner-Biró.

Figure 1.199: Counterweight wall for ordinary manual counterweight hoists in the Festspielhaus Bregenz. Photo: Waagner-Biró.

as a rationalization measure but also in the interests of safety at work to replace manual counterweight hoists with machine hoists or to limit the load capacity of manual counterweight hoists to very small payloads.

In the Netherlands, for example, a regulation has been passed according to which only a maximum of 20 manual counterweight hoists may be installed in a stage, the load-bearing capacity of which is limited by the fact that – regardless of whether the hoists are simple or doubled – the counterweight cage may only have a maximum mass of 75 kg; the individual counterweight pieces may not weigh more than 6 kg. As a motivation for this regulation, it should be noted that setup work is particularly frequent as a result of the "guest play" system common in the Netherlands.

No scenic use of hoists is required in the backstage, under stage and side stage areas, so simple *manual winch hoists* as shown in Figure 1.197(c) can also be used. The load bar suspension ropes are combined into one rope that is wound onto the drum of a hand winch. This avoids the need for manipulation with counterweights; however, only low speeds are achievable. A simple hand winch is shown in Figure 1.200. Figure 1.201 shows manual winch and manual counterweight hoists.

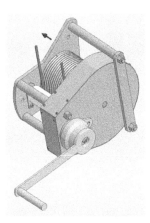

Figure 1.200: Manual winch hoist. Photo: Köstner.

Figure 1.201: Hand winch and hand counter-weight hoists. Picture credits: Eberhard.

Bar hoists with hydraulic cylinder drive
Bar hoists driven by hydraulic cylinders had been used for many years as an alternative to manual counterweight hoists, using water–oil emulsions as hydraulic fluid. For
example, in the Vienna State Opera hydraulic linear hoists were installed beside a large
number of manual counterweight hoists.

Later, in the course of converting manual counterweight hoists to motorized hoists,
it was often common practice to integrate hydraulic cylinders into the counterweight
wall.

In the case of hydraulic linear hoists, the suspension cables of the load bar are
guided over the roller grid to the side wall of the stage and moved with a hydraulic piston. In order to reduce the required cylinder stroke, a block and tackle system is used
so that the hydraulic piston has to apply i times the force, but only covers the ith part of
the distance (see Section 4.1.2 and Figure 1.202(a)).

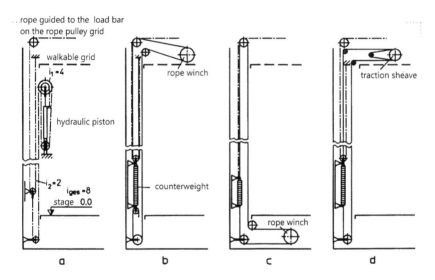

Figure 1.202: Motor-driven hoists: (a) hydraulic linear hoist, (b) winch hoist with counterweight doubled
(1:2) suspended, winch situated on the grid, (c) winch with counterweight, winch situated in the understage, (d) traction sheave drive.

Since there are relatively large quantities of oil in a hydraulic cylinder for actuating
a load bar, temperature changes in the hydraulic fluid result in shifts in the piston position, which then change the height position of the load bar by i times the amount. This
can mean that prospect bars adjusted when the oil is warm are no longer in the correct
positions after the hydraulic fluid has cooled. For this reason, hydrostatic winches with
hydraulic motors are generally preferred for modern hydraulic upper stage systems,
since the rotary position of the rope drum is fixed by a mechanical brake.

However, this possibly disturbing temperature drift on hydraulic cylinders can be
compensated by special measures, such as those taken for the linear drives of the upper

machinery in the *Schauspielhaus Kassel*. If the load bar of a bar hoist has left its starting position due to a drop in temperature of the hydraulic fluid after a longer standstill, a special control circuit is activated and fluid is supplied from a hydraulic accumulator until the load bar returns to the correct position.

Winch hoists driven by rotating motors

In a winch hoist, the load bar ropes are wound up or unwound on a drum. The drum holds the suspension ropes of the load bars in spiral grooves. Regardless of whether the hoist is driven by an electric or hydraulic motor, it can be constructed with or without a counterweight. In a normal case, motor-driven winch hoists are built without counterweights, because the required power is not great because of the relatively small payloads.

Figure 1.202(b)–(d) shows design variants which are mainly used when manual hoists are or have been converted to motor hoists and the existing counterweight cage continue to be used, but with an unchangeable counterweight size, advantageously the size of the dead weight of the load bar plus half the maximum payload. In this way, a minimum value is achieved for the drive power (see Section 3.6).

The hemp rope for manual operation is replaced by a wire rope whose ends are wound up or unwound on the rope drum. Such bar hoisting winches from the *Düsseldorf Opera House* can be seen in Figure 1.203.

Figure 1.203: Electrically driven winches for bar hoists with counterweight compensation of dead-weight plus half payload, winding or unwinding only one rope to move the counterweight cage (system according Figure 1.202(c)) – Düsseldorf Opera House. Photo: Krupp.

As an alternative, point hoist winches can be used to move the counterweight carriage (e. g., the "Fly" point hoist system according to Figure 1.232).

Nowadays, almost only winch hoists driven by electric motors are installed (see Section 2.2.2).

Depending on the room situation and space conditions, the rope drums can be installed with a horizontal or vertical axis.

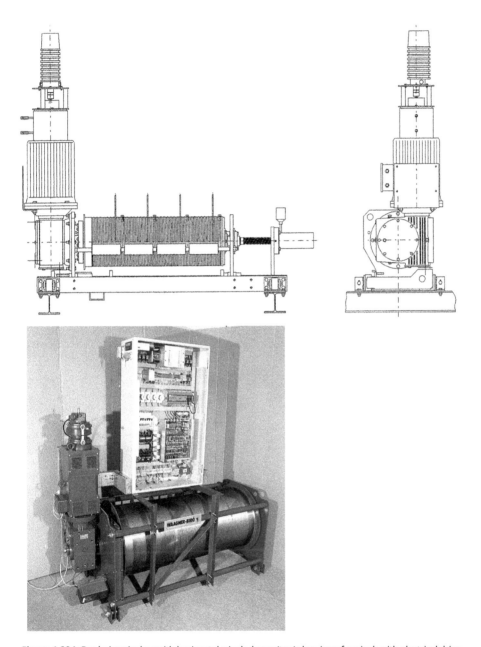

Figure 1.204: Bar hoist winches with horizontal winch drum: (top) drawing of a winch with electrical drive, (bottom) photo of an electric winch. Photo credit: Waagner-Biró.

Figure 1.204 shows a winch with horizontal drum. The deflection pulleys for further rope guidance are far enough away from the rope drum so that permissible deflection angles at the drum can be observed (see Section 4.1.1).

In the winch with vertical drum shown in Figure 1.205, the hoist ropes are deflected so close to the winch that the permissible deflection angles would be exceeded. The rope drum or the pulley support is therefore shifted when the rope drum is rotated in the direction of the drum axis in such a way that an equal deflection angle is always maintained.

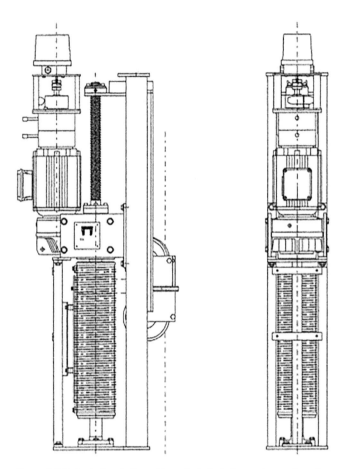

Figure 1.205: Bar hoist winches with vertical winch drum – Lucent Danstheater Den Haag. Photo credit: SBS.

For installation on the counterweight wall when replacing manual counterweight hoists with motor-driven winches, narrow-diameter bar hoist winches with vertical drums have been specially developed (Figure 1.206). In the case shown, when the drum rotates, it is not the drum but the roller carrier that is displaced vertically.

Figure 1.207 shows the three variants for proper winding of the rope drums.

Electromechanical winch drives are shown in Figures 1.204–1.206. The rope drum is driven by an electric motor with the interposition of a gearbox. With modern power elec-

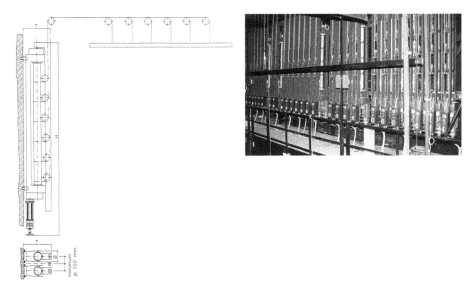

Figure 1.206: Bar hoist winches at the Theater an der Wien. Photo: Waagner-Biró.

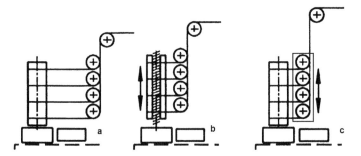

Figure 1.207: Variants for avoiding excessive deflection angles at the rope drum: (a) distance of the deflection battery to the rope drum sufficiently large not to exceed permissible deflection angles, (b) distance too small, therefore the rope drum is axially displaced when the rope drum rotates, (c) distance too small, therefore the deflection battery is axially displaced when the rope drum rotates.

tronics, the speed of electric motors can be well controlled. Whereas externally excited DC motors were used almost exclusively in the past, modern three-phase technology now offers far better application possibilities. Modern *servo motors* offer particularly favorable control behavior. With them, a particularly high control ratio between maximum and minimum operating speed can be realized; indeed, the load could even be held by the motor at standstill (see Section 2.2.2). Control ratios of 1:1000 are considered to be generally common in modern venues.

It is, of course, possible to keep noise emissions very low by careful design of the gearing, the backlash of the tooth gearing, etc., and the motor design. Figure 1.208 shows a bar hoist winch with the proprietary name "whisper winch."

Figure 1.208: Whisper winch. Photo: Statec.

In hydraulic winches (Figures 1.209 and 1.210), the drum is driven by a hydraulic motor. The hydraulic motor is usually a low-speed motor, so that there is no need for an intermediate gearbox or only a small gear transmission ratio is required. Noise emissions from fast-running electric motors and gears in transmission gears can thus be avoided. By combining hydraulics and electronics, very sensitive and precise control systems can be realized with the aid of modern servo valve technology. However, slow-running hydromotors do not permit large transmission ratios, as they run nonuniform at very low speeds due to their design.

Figure 1.209: Hydraulically driven bar hoists – Gothenburg Opera House. Photo: Bosch Rexroth.

If very large transmission ratios are to be achieved, it is also necessary to use faster-running hydromotors for hydraulic drives, since hydromotors no longer rotate absolutely evenly at too low speeds. In this case, however, high noise emissions result, which must be countered by appropriate sound insulation measures, as has been done in the Grand Thèâtre de Genève and Hanover Opera House (Figure 1.211).

Figure 1.210: Hydraulically driven winch hoist with rope drum in vertical position with movable deflection pulley battery. Photo: Bosch Rexroth.

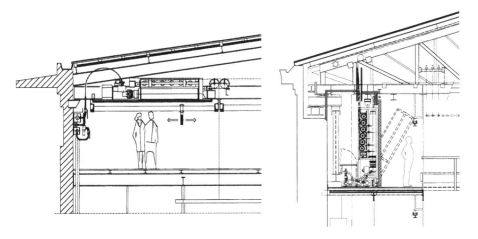

Figure 1.211: Hydraulic bar hoists winches in: (left) Grand Thèâtre de Genève, (right) Hanover Opera House. Picture credits: Bosch Rexroth.

(The technology of hydraulic drives is discussed in more detail in Section 2.3.)

Figure 1.209 shows a hydraulic motor-driven bar hoist winch as used in the *Gothenburg Opera House*. A special feature of the winch shown in this image is also the slightly inclined position of the rope drums, which makes it possible to design the rope run-off to the grid in a favorable way.

The vertical drum of bar hoist winches shown in Figure 1.210 are equipped with vertically displaced pulley batteries.

Electric bobbin winch

Normally, the hoist ropes are wound in a single layer on a rope drum provided with grooves. In special cases, the rope winding can also be in the form of a bobbin. This makes it possible to design small compact units for small lifting heights and low lifting speeds, since long rope drums are avoided. The bobbin winch shown in Figure 1.212 is designed for 4 ropes for a load capacity of 3 kN, a hoisting speed of 0.1 m/s and a hoisting height of 6 m and is equipped with an emergency hand control with attachable crank handle.

Figure 1.212: Bar hoist with a bobbin winch. Photo: Waagner-Biró.

Motorized traction sheave hoists

In a winch drive, the tractive force is transmitted via positive locking. Alternatively, the force can also be transmitted to the rope by frictional locking (see Section 4.1.4). This is the case with load bar hoists with counterweight driven by *traction sheaves* (Figure 1.202(d)). However, it is problematic to ensure sufficient frictional locking (see Section 4.1.3) in stage operation under all possible load situations. For this reason, traction sheave hoists are rarely used. Furthermore, even if no sliding slip occurs as a result of overloading, so-called expansion slip (see Section 4.1.4) is unavoidable with traction sheaves.

The HCWA winch of the company ASM-Steuerungstechnik GmbH

"HCWA" stands for "*hand counterweight automation.*" It is an extremely narrow designed bar hoist drive (180 mm) that can be installed in place of a manual operated counterweight hoist without major modification of the steel structure because the hoist is attached under the existing pulley structure.

The rope drum is driven by a steel belt (plus a second safety belt) and enables lifting speeds of up to 1.8 m/s. The maximum load capacity is 1000 kg.

Four to six ropes carrying the load bar are wound on top of each other as a bobbin on a large diameter plastic drum (800–1000 mm). In addition to the ropes, two stainless steel bands are wound up in the opposite direction to the ropes. The stainless steel belts have the function of the drive. One of the belts is the main carrier and the other is a safety belt. (However, there is also a design with only one belt available.) A torque transmission via the hub of the drum is therefore not necessary due to this belt drive which acts on the circumference of the drum. The two stainless steel belts are wound on a smaller drum, which is situated upstream the winch of the large drum with electric motor drive. When two belt drives are used, a compensation mechanism is also included, for the different possible belt lengths of the main belt and the safety belt. For larger load capacities, two drive motors of smaller size are installed to maintain the narrow design.

The construction method is shown in Figure 1.213. Figure 1.214 shows the design for the *City Theater in Constance* and for the *National Theater in Mannheim*.

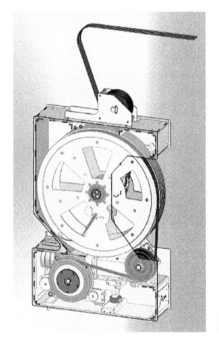

Figure 1.213: Construction of the HCWA winch. Picture credits: ASM-Steuerungstechnik GmbH.

Convertible bar hoists regarding drive system

Manual counter hoists convertible to machine hoists
Since the manual counterweight hoist is considerably less expensive than the mechanical hoist, for cost reasons theaters sometimes install mixed systems in which the stages are equipped with manual counterweight hoists and a certain number of mobile mechanical winch units are also available. By mechanically coupling one of these mobile

Figure 1.214: HCWA winch in the Mannheim National Theater. Picture credits: ASM-Steuerungstechnik GmbH.

winch units, a manual counterweight hoist can be converted to a machine hoist, so that it can possibly also be used to move larger loads.

In the *Bregenz Festival Theater*, only manual hoists were available in the upper stage. In order to meet the requirements of the many operating conditions as an internationally renowned venue but also as an exhibition and congress center, the upper machinery in the Festspielhaus was modernized. Whereas previously only 60 manual hoists with a load capacity of 350 kg were available, 15 (expandable to 20) motor driven units were now to be installed, allowing each of these manual hoists to be converted into a mechanical hoist with a load capacity of 500 kg and a lifting speed of 0.001 to 1.20 m/s, depending on the requirements (see Figures 1.215 and 1.216).

Drive units were developed for this purpose, which are arranged in the understage in front of the counterweight tracks, and can be moved on rails. For the use of a manual hoist as a machine-driven bar hoist, the counterweight slide including the basic weights and the command rope are dismantled. For the machine hoist, only the upper guide with the rope rake is used, which has to be combined with the machine rope.

On the rails, the complete drive units can be moved over the entire stage and locked at any point with a spacing of 500 mm. Each manual hoist can thus be converted into a fully-fledged machine hoist, especially for festival operations. Various soundproofing measures were taken to prevent sound transmission or propagation to the stage.

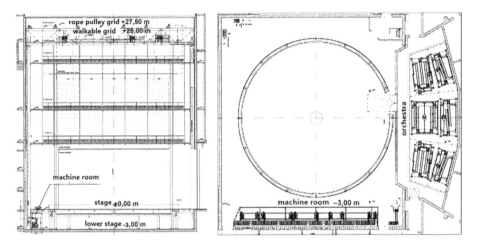

Figure 1.215: Location of the moveable bar hoist drives in the under stage of the Festspielhaus Bregenz. Picture credits: Bühnenplanung Walter Kottke Ingenieure GmbH.

Machine hoists convertible to manual counter hoists

In contrast to this basic idea, which has already been implemented frequently, "all flies manual – some optionally machinable," there was also the contrary model "all flies machinable – some optionally manualizable," which, however, hardly makes sense any more, since the modern control and regulation technology of motor-driven trains covers all the requirements. Nevertheless, this concept is explained with Figure 1.217.

If a stage is completely equipped with mechanical bar hoists, specially developed moveable manual counterweight hoists enable the user to use each of the machine hoists manually as a manual counterweight hoist. In the case of several moveable manual counterweight hoist units available, several of the machine hoists can thus also be used simultaneously as manual hoists.

To make this possible, a rope loop is formed in the horizontal rope run in the area of the stage house side wall for each mechanical pull. A vertically movable rope sheave is located in this loop. In machine operation, this loose sheave is locked by a bolt.

Coupling the hand counterweight hoist to a machine hoist requires only a few manual operations: The lock is released and when the command cable of the manual counterweight hoist is moved, the loose sheave and therefore the load bar moves.

The simple reeving caused by the loose sheave is cancelled out with the moveable manual counterweight unit, so that the load bar stroke and the travel of the counterweight or the natural fiber command rope correspond 1:1. Thus, it is an ordinary non-doubled manual counterweight hoist.

Motorization of manual hoists with friction wheel drive

Finally, one drive variant should be mentioned. The idea was to replace the operation of the command rope on the manual counterweight hoist with a friction wheel drive

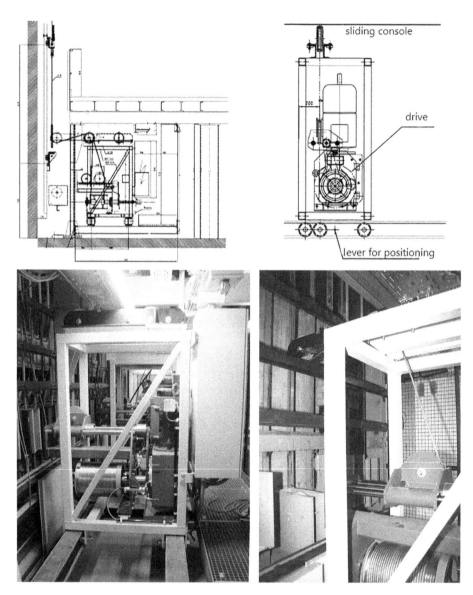

Figure 1.216: Movable bar hoist drives at the understage of the Festspielhauses Bregenz. Picture credits: Bühnenplanung Walter Kottke Ingenieure GmbH.

acting on the command rope. However, this was based on the assumption that the load and counterweight cage had to be balanced in the same way as with manual drive. Since this means that the strenuous loading of the counterweight cages cannot be dispensed with, this system shown in Figure 1.218 has not become established.

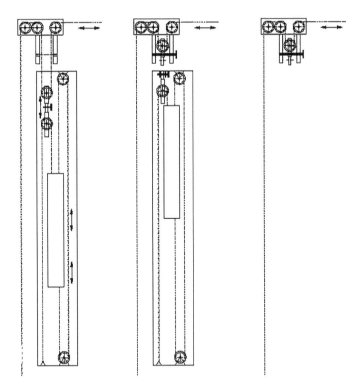

Figure 1.217: Convertible load bar hoist: (left and center) manual operation, rope sheave can be moved, (right) machine operation, rope sheave locked. Picture credits: Theatertechnische Systeme GmbH.

Figure 1.218: Auxiliary drive for a manual counterweight hoist. Photo: Bader Maschinenbau.

Tubular shaft hoist

In smaller theaters and for subordinate purposes, so-called *tubular shaft hoists* are also used (Figure 1.219).

In this case, the hoisting ropes of the load bar are not guided via deflection pulleys to the stage side wall to be wound up together on a drum there, but a tubular shaft is

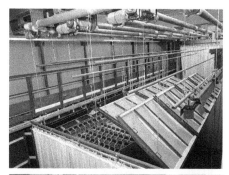

Figure 1.219: Tubular shaft hoist at the Hartberg Theater. Photos: Waagner-Biró.

laid over the entire width of the stage, which can be regarded as a long rope drum on which the hoisting ropes of the load bar are wound up or unwound directly. As can be seen in the picture, the tubular shaft is supported in support rollers and driven by a motor. If individual drums are coupled instead of a continuous tubular shaft, the term *transmission shaft hoist is* commonly used.

Waagner-Biró offers a *modular tubular shaft hoist* ("MRZ") (see Figures 1.219 and 1.220). It has an extremely flexible design and consists of the modules drive, rope drum, tubular shaft and angular gear. It is designed in such a way that standing, suspended or lateral mounting is possible. Since the tubular shaft for torque transmission does not have to be machined, the modules can still be interchanged on site. In the case of fixed rope passages in the ceilings, the MRZ can also be equipped with a shifting device for the rope drums – to align the rope run. The shifting of the drum modules and the drive module is done with a rotating spindle and a fixed nut on the carriage module. The spindle thread corresponds to the thread of the rope drum.

In the standard program, 2 types with payloads of 3 and 5 kN and a nominal speed of 0.15 and 0.3 m/s are available.

Construction types of point hoists

In addition to the equipment of the fly grid with load bar hoists, point hoist winches are often installed to move individual loads. However, multiple use of point hoists did not

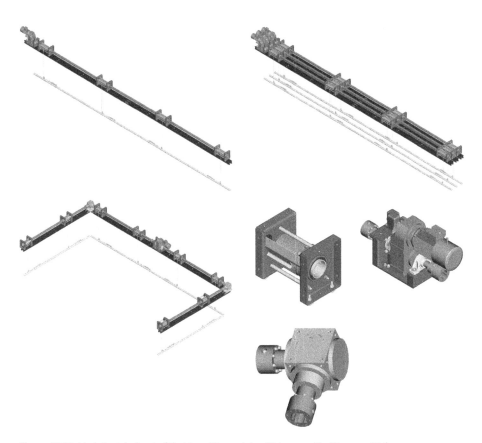

Figure 1.220: Modular tubular shaft hoist and its modules. Picture credits: Waagner-Biró.

make sense until modern control technology made it possible to precisely specify the lifting and lowering speeds.

Load bar hoists are usually built as rope drives. In point hoists, the rope can also be replaced by chains or steel bands. Therefore, chain and belt drives are also dealt with in Sections 1.8.4 and 1.8.5.

In principle, it should be noted that by mechanically coupling of a load bar with point hoists and synchronizing the drives of the point hoists, a load bar hoist can also be realized with several point hoists, regardless of whether a rope, a chain or a steel belt is used as the means of traction.

Wire rope hoists with manual drive
Of course, point loads can be lifted with simple hand winches. In the main stage area, the use of such *manual winch hoists* is hardly an option, but on small stages it is.

If the winch is omitted for very small loads and a natural fiber (hemp) rope is pulled directly via one or more pulleys, this manual operated rope hoist is called *manual string hoist* (a manual operated natural fiber rope hoist). However, this is only permissible for very small loads (e. g., for loads of maximum 30 kg).

Motor driven single point hoists

By pulling the hoisting rope over *offset pulleys* (*misalignment sheaves*) on the walkable grid, the operating position on the grid can be varied despite the fixed installation location of the winch. Figure 1.221 illustrates the type of installation. Like the winches for load bar hoists, these winches can also be driven electrically or hydraulically. Figure 1.222 shows a hydrostatically driven point hoist winch.

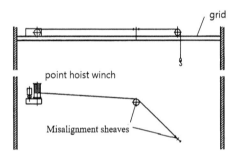

Figure 1.221: Point hoists: Determination of the point of use with misalignment sheaves (shown in ground plan and elevation).

Figure 1.222: Point-hoist winch with hydrostatic drive. Photo: Bosch Rexroth.

Figure 1.223 shows an offset pulley design which can be mounted on light grids or slatted grids and which allows any spatial position of the deflection pulley by swiveling about two axes. This means that horizontal and vertical rope deflections, but also rope deflections oriented anywhere in space, can be realized.

As the hoisting rope on the drum changes the point of take-up or take-off during winding and unwinding, the first offset roller after the drum must be set in such a distance that no excessive rope deflection angles occur, unless the rope drum is axially displaced during rotation – as can be seen in Figure 1.224.

Figure 1.223: Offset roller (misalignment sheave). Photo: Waagner-Biró.

Figure 1.224: Electrically driven point hoists with shifting drive unit – left for vertical, right for horizontal rope pull-off. Photo: Köstner.

In the winches shown in Figure 1.224, the rope drum moves in axial direction according to the groove pitch during rotation, so that the position of the rope take-off remains unchanged. Of course, if the rope is further guided via offset pulleys, it must again be ensured that the angle from the offset pulley plane is not too large (see Section 4.1.2).

Point hoists movable parallel to the proscenium
A frequently realized construction method of point hoists is a *trolley-system* as shown in Figures 1.225 and 1.226. Instead of a bar hoist, a rail is laid across the width of the stage at the fly grid, on which small trolleys can travel. The hoisting rope is led from a fixed point on one side of the stage to this small trolley, then down to a hook block, from there back up to the trolley and then to the opposite stage wall and on to the rope drum winch. With this drum winch, lifting and lowering of the load can be made. The position of the crab on the rail determines the position of the load point in the transverse direction of the platform. By moving the crab on the rail, the position of the load point

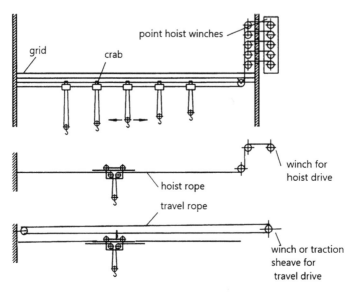

Figure 1.225: Point hoist system – point hoist with on a rail movable crabs.

Figure 1.226: Point hoist crabs at the Genoa Opera House. Photo: Waagner-Biró.

can be changed without affecting the height position of the load. The crab can be moved by hand or motorized via a rope system. In a very common design variant, the change of the crab position is performed from the stage floor by pulling one of the two ropes leading to the hook block. Usually, five to six point hoists are arranged on one rail with a separate winch drive for each point hoist.

There are also systems in which the crab can be moved by means of a wire rope system, as is common in tower slewing cranes. This means that the point hoist can be used as performer flying which can be moved in the plane (see Section 1.8.8).

A variant of the system just described, which is also implemented but not very practical, is that all five or six load ropes are operated by a single winch. With this solution, unneeded suspension points must be pulled to the outside and let them run along out of sight without function.

These movable point hoists offer greater variability in use, but also the possibility of hooking in a load bar if required. A disadvantage compared to offset rollers would be that the same freedom of movement in the choice of the pay-off point is not given if only a few point hoist tracks are available, since the crabs can only be moved in the transverse direction of the stage. It also has to be mentioned that point hoists with such movable crabs require a double-strand suspension with bottom block.

As already mentioned in connection with load bar hoists, electric and hydrostatic drives can again be used as hoist drives for the point hoists.

In the variant shown in the rope schematic in Figure 1.227, both rope ends are wound on the rope drum in this case so that the rope only has to be moved with the speed v to gain the hoisting speed v. If only one rope end is wound, the rope running speed $2v$ is required for the hoisting speed v, which results in higher rope running noise. The installation shown in Figures 1.227 and 1.228 is located in the *Schauspielhaus Frankfurt*. The point hoist system provides six separately driven suspension points each with

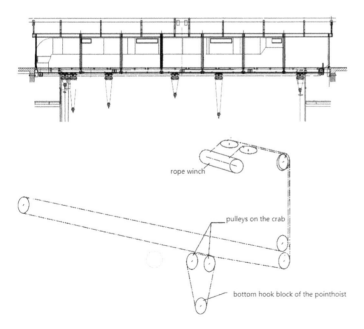

rope winch

pulleys on the crab

bottom hook block of the pointhoist

Figure 1.227: Point hoist system at the Schauspielhaus Frankfurt. Picture credits: SBS.

Figure 1.228: Point hoist system at the Schauspiel-haus Frankfurt. Photo credit: SBS.

250 kg load capacity in 15 aisles. The six drives are arranged one above the other as a block.

Displaceable point hoist winches

If the walkable fly grid is kept as free as possible, *displaceable point hoist winches* can also be used. The higher their load capacity, the greater their dead weight, of course. Winches with a low load capacity of up to about 150 kg can be designed as portable point-hoist winches; winches with a higher load capacity are built to be movable, e. g., on wheels or rollers (see Figures 1.229 and 1.230).

Figure 1.229: Displaceable point hoist. Picture credits: Statec.

Displaceable winches can also be driven electrically or hydraulically. However, it should be borne in mind that hydraulic drives require hose lines and quick-connect couplings, since the winch is intended to be used at any location on the grid. Therefore, de facto only electrically driven winches are used.

Figure 1.231 shows a view of the upper stage of the *opera house in Muscat, Oman.* The accessible fly grid is provided with a light grid, and the pulleys for the load bar hoists

Figure 1.230: Displaceable point hoist. Photo credit: Waagner-Biró.

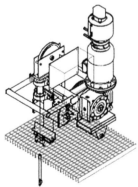

Figure 1.231: Movable point hoist winches at the Royal Opera House Muscat. Photo credit: SBS.

can be seen in the roller grid above. On the left and right of the light grid, two of the six movable point hoists can be seen. The hoist rope is guided vertically downwards from the drum via a swivel roller. A plastic guide prevents contact between the cable and the grating. The many cable drums are also noticeable. They offer the possibility of using each machine hoist as a lighting hoist.

The *point hoist system "Fly"* from Waagner-Biró can be used very flexibly. Depending on the installation situation, the FLY can be used standing on the grid or suspended mounted on a fixed cantilever or on trolleys. When used suspended, the FLY is moved into position via a manual or motorized bottom flange crab and via a manual or motorized crane bridge system.

Figure 1.232 shows the installation in the *"Ronacher" stage in Vienna*. The point hoists are suspended in traverses, which in turn can be moved manually between the rope aisles on the fly grid. A redundant axis computer and the complete electronics are integrated in the rope winch. Nevertheless, the FLY weighs only about 175 kg with a maximum payload of 400 kg.

Figure 1.232: Suspended moveable point hoists "Fly". Photo: Waagner-Biró.

The FLY can also be used to convert manual counterweight hoists to machine hoists by mounting the FLY in the counterweight shaft and attaching the rope to the counterweight cage.

Figure 1.233 shows a new standardized point hoist winch with a distanced deflection pulley from Waagner-Biró.

The rope winch and the rope's guide pulley form a unit connected by a spacer rod. This ensures that the grid at the contact patch of the winch is only loaded by the weight of the payload and the pulley. This can prevent the permissible load per square meter of the grid from being exceeded. In addition, the distance between winch and pulley ensures that the permissible deflection angles to the groove pitch on the drum winch are not exceeded.

Figure 1.233: Point-hoist winch with distanced rope pulley: (left) in working position, (right) with spacer rod folded up. Credit: Waagner-Biró.

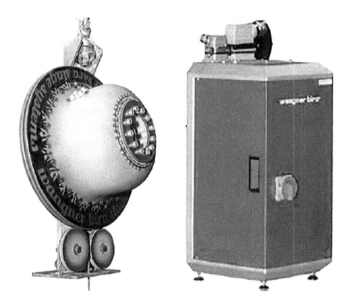

Figure 1.234: Slim Hoist. Photo: Waagner-Biró.

Figure 1.234 shows the so-called *"Slim Hoist"* from Waagner-Biró with the associated control cabinet. The hoist has a dead weight of only 130 kg and can lift a maximum payload of 300 kg at 1.5 m/s.

Bobbin winch

Figure 1.212 shows a bar hoist bobbin winch. This can of course also be used as a point hoist with a single-groove bobbin pulley.

Heavy duty hoists

In a statically suitable place of the grid, hoists are often installed with higher load capacity, which are also a kind of point hoists. This was common especially when the upper stage was equipped only with manual counter weight hoists of small load capacity. However, in accordance with generally required safety regulations for winches used in stage applications, the installation of a second brake is required (see Chapter 5). For this reason, commercially available standard electric rope hoist winches were not available. But an electric wire rope winch of special design for stage use, as shown in Figure 1.235, was offered already many years ago with the name "Adler Winde."

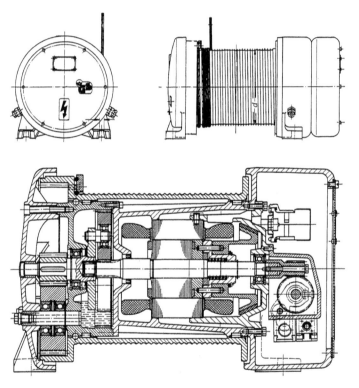

Figure 1.235: Heavy-duty hoist "Adler winch": (top) winch with additional brake, (bottom) sectional view of winch without additional brake. Picture credits: Friedrich Köster GmbH & Co. KG (D-Heide).

Tirak clamp hoist

Finally, a point hoist variant should be mentioned that was purchased by some theaters but is hardly ever used anymore. These are hoists whose drive technology is similar to that of the *clamping drives* described in Section 4.1.5. The rope runs through a fittingly-sized groove in the traction sheave and is also pressed into place by means of pressure rollers with compression springs.

Whereas clamp hoists – of a different design – have proven themselves in many cases, especially as erection equipment, these clamp hoists are not really suitable for stage applications.

Figure 1.236 shows the Tirak clamp hoist. The left picture shows the complete unit consisting of drive motor, gear, clamping disk and rope storage drum; the right picture explains the clamping principle. The loose rope end is pushed into the rope storage drum and pulled out again in the countermovement.

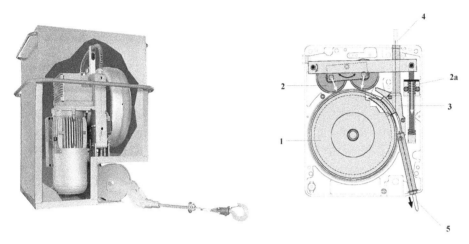

Figure 1.236: Tirak clamp hoist (X series): 1 traction sheave, 2 pressure rollers, 2a pressure spring, 3 rope guide, 4 loaded rope (inlet), 5 loose rope (outlet). Picture credits: Greifzug Hebezeugbau GmbH (D-Bergisch Gladbach).

Interlinked point hoist systems

Mechanical linkage

Point hoists can only be used universally if several point hoists can be connected together in groups and moved synchronously together. Synchronization is possible mechanically by means of synchronization shafts in combination with suitable coupling mechanisms. Figure 1.237 on the left shows a mechanical point hoist system in which the point hoist drums can be mechanically coupled to a central motor and thus combined to form a synchronously operating group. This achieves exact synchronization without electronic synchronization control. As an alternative, Figure 1.237 on the right shows a solution with two central motors, whereby the point-hoist units can then be coupled to one or the other motor, so that two point-hoist groups can be formed which are independent of each other in their movement. Mechanically coupled point hoist systems are no longer installed today (sometimes in the forestage area), and only electronically linked point hoist systems are built.

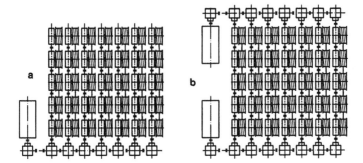

Figure 1.237: Point hoist system with mechanical linkage: (left) with one drive motor, (right) with two drive motors.

Electrical interlinking

Mechanical shafts can be replaced by "electrical shafts." The classic three-phase technology offered the possibility of synchronizing motors. However, this type of speed coupling had not become established in stage technology and is no longer realized.

Electronic interlinking

By using travel measuring systems and electronic feedback control of the operating speed according to predefined desired values, point hoist winches can now be variably linked to form complex point hoist systems. In principle, point hoists and load bar hoists can be combined, regardless of whether they are electrically or hydrostatically driven.

Winch modular design

A trend in technology is to offer compact units by integrating the electrics and electronics belonging to a drive into a separate control cabinet. This makes it easy to expand or retrofit stage technology equipment; this modularization is proving advantageous (see also Sections 2.5.6 to 2.5.8).

As an example, Figure 1.238 shows several "Unicorn" control cabinets from Waagner-Biró together with the winch drive, in which components such as frequency converter, line filter, braking resistor (mounted on the rear panel), maintenance switch, main contactor, braking contactors, power supply and axis controller are installed for each winch.

Low speed electric motors

In principle, it is possible to use multipole motors with a low rated speed instead of three-phase motors with a high rated speed. As a result, transmission stages in the gearbox can be omitted or reduced, and the emitted sound power can be reduced. (Motor and brakes can also be integrated into the drum).

Figure 1.238: Compact winch with "Unicorn" control cabinets. Photo: Waagner-Biró.

As explained above, the use of low-speed rotors for drives with hydraulic motors is a matter of course for reasons of noise protection. However, if you then want to achieve control ratios of 1:1000 – as is common today – hydromotors generally run unevenly at such low speeds, while electric motors with the state-of-the-art control technology can still be controlled down to standstill in extreme cases. So far, however, the use of low-speed electric motors has not become widespread.

That the use of electric low-speed motors is possible is shown by the fact that these are used as standard in traction sheave elevators in the USA, while in Europe motor-gearbox combinations with high-speed motors are common.

Figure 1.239 shows an electric winch for hoists developed according to this concept.

Figure 1.239: Drum winch with direct electric drive (without reduction gear). Photo: Waagner-Biró.

1.8.4 Hoists with chains

Chain hoists are mainly used as point hoists, especially for rigging purposes. According to safety guideline VPLT: SR2.0 – "Provision and use of electric chain hoists," there are three basic chain hoist design variants on the market.

The *"D 8 hoist"* is an electric chain hoist for lifting loads as a simple erection and assembly device. This hoist does not meet the requirements of stage technology, as it is only equipped with one brake. Therefore, it must not be used for assembly/disassembly and setup work or for holding a load when persons are present under the load. The load must be secured by other means after it has been lifted.

The *"D 8 Plus hoist"* is an electric chain hoist for lifting loads with the special feature that loads may be held above persons without secondary protection when at rest. This is because this type of hoist complies mechanically – especially with regard to the redundant braking technology (two redundant brakes) – with the regulations for use on stages. Therefore, a load can be suspended from it even though persons are below the load. However, it should be noted that it lacks crucial control technology components. Thus, in group operation, e. g., when moving a ceiling due to small differences in the travel speed of the individual hoists (due to the load dependency of the rotation speed – asynchronous travel), incorrect load distributions could occur, which could lead to damage to the load or the load handling device.

A *"C 1 hoist"* is an electric chain hoist whose travel speed can be controlled by a feedback control system, and is therefore permissible for holding and moving loads and people, in stage engineering.

It should also be mentioned that, according to the European Machinery Directive, hoists with a load capacity of 1000 kg or more must in principle be equipped with an overload cut-off. This provision is therefore also complied with in the case of the D 8 hoist. Both D 8 Plus and C 1 chain hoists are always equipped with an overload cut-off, regardless of their load capacity.

Figure 1.240 shows chain hoists in different versions.

Figure 1.240: Chain hoists: (a) fixed suspension – single strand, (b) climbing hoist – single strand, (c) fixed suspension – double strand, (d) climbing hoist – double strand. Photo: Chainmaster.

1.8.5 Hoists with steel belts

Advances in materials technology have made it possible to produce high-strength spring steels which can be wrapped like a bobbin as wear-free tensile elements and used in so-called *belt hoists*. The stainless steel strips have a cross-section of 40 mm × 0.5 mm, for example, and have a tensile strength of approx. 1700 N/mm^2, i. e., of a similar order of magnitude to steel wires used in ropes. However, the elongation under load is lower because the modulus of elasticity of a steel strip is higher than that of a wire rope as a result of the stranding of the wires.

Whereas ropes have to be specially designed or require special manufacturing in order to operate without rotation carrying a load in single strand arrangement, this problem does not exist with belts. Due to their smooth surface, they allow very quiet running. For this reason, belt hoists are being used more and more frequently (Figure 1.241).

Figure 1.241: Belt winches: (a) bobbin-type wound belt element, (b) single-strand belt hoist, (c) double-strand belt hoist. Photo: ASM – Lightpower.

Figure 1.242 shows the drive of belt hoists on which load bars are suspended. A special feature of this system is also that the load bars are provided with hinged joints at the suspension points, thus forming a statically determined system. A load between two suspensions thus does not lead to unintentional unloading of an uninvolved suspension means.

During the last renovation of the *Vienna Volkstheater*, ASM belt hoists with a load capacity of 250 kg were installed as point hoists on the fly grid; see Figure 1.243.

1.8.6 Special installations for limitation of the stage space

For the design of the stage set, it is usually necessary to use hanging decorations to separate the playing area from the stage walls equipped with technical equipment or from the side and back stages.

Backdrop, convertible backdrop
The back wall of the stage can be covered in a simple way by a load bar hoist as *backdrop hoist*.

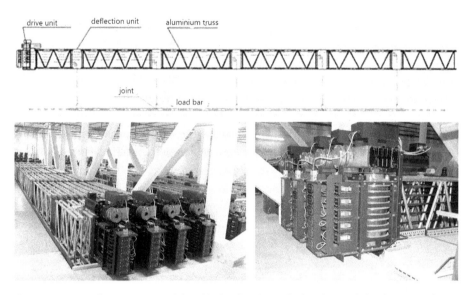

Figure 1.242: Overall arrangement and drive blocks of belt hoists for load bars – Thalia Theatre Hamburg. Photo: ASM.

Figure 1.243: On rails movable belt hoist – Volkstheater Vienna. Photo credit: Waagner-Biró.

As described in Section 1.1, so-called *"convertible backdrops"* were also used for this purpose, especially in earlier times. Their original purpose was to arrange several images (landscapes, houses, etc.) next to each other on a screen in order to be able to create different stage sets.

As a further development, e. g., in the *Branch theater of the Bolshoi Theater in Moscow*, a fabric track of approx. 120 m in length and approx. 11.5 m in height has recently been installed as a convertible stage limitation, which also allows the effect of a passing landscape to be simulated with two winding cones according to the system of the horizontal dragged cyclorama, e. g., when someone is sailing along a river on a boat. The speed of the fabric web is infinitely variable.

Panorama bar hoist

Panorama hoists are load bar hoists, which are not positioned at right angles to the stage axis like the usual drop hoists, but parallel to the longitudinal axis of the stage along the stage side walls (see Figures 1.177 and 1.244). Panorama hoists in combination with a back drop hoist can completely enclose the playing area. The designation *side* and *rear panorama hoist is* also sometimes common.

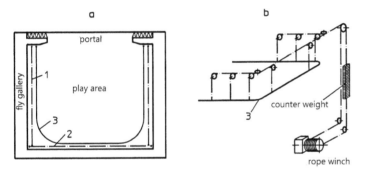

Figure 1.244: Panorama and cyclorama hoist: (a) arrangement on the stage, (b) drive of a cyclorama bar hoist with winch and counterweight. 1 panorama hoist (side panorama hoist), 2 prospect hoist (rear panorama hoist), 3 cyclorama hoist.

Round bar hoist

If one wants to make the demarcation in the rear left and right stage corner area by a curtain guided in an arc, the drop hoist must be given a bar with rounding arcs. Sometimes the three wall areas are also covered by a complete round bar hoist, which is also called a *cyclorama hoist*.

Figure 1.244 (left) shows the arrangement on the stage in plan view and Figure 1.244 (right) a winch with counterweight compensation. Sometimes doubled hoisting is also provided.

Cyclorama horizontal dragged

In the cases described so far, the textile curtain is raised into the stage tower or lowered from there by hoists to limit the playing space. Another possibility is to move a curtain in a rail that is guided horizontally around the playing area. In the simplest case, a circular curtain rail is used to fling manually the curtain.

In large stages, a system also known as a *cyclorama* (shown in Figure 1.245) consisting of curtain storage drums and curtain track is sometimes installed. Usually, storage drums are mounted on both sides of the proscenium in the area of the portal towers. The storage drum, in the form of a winding cone, is simultaneously lowered when rotating in the winding direction and simultaneously raised when rotating in the unwinding direction, so that the height of the fabric web relative to the curtain rail is maintained during the winding process. The conical shape of the drum ensures free hanging of the fabric in the coiled condition. Usually, one proscenium side accommodates a storage drum with white fabric and the other proscenium side accommodates one with black fabric (night sky). In Figure 1.245(b), the special design of the rail can be seen. The rope, which is sewed into the upper flap of the fabric track, runs in a groove with a slit open at the bottom. Thus, on the one hand, the rope serves as a supporting element for absorbing downward weight forces and, on the other hand, as a traction rope for moving the curtain. Figure 1.246 shows the winding cone and rail in the photo.

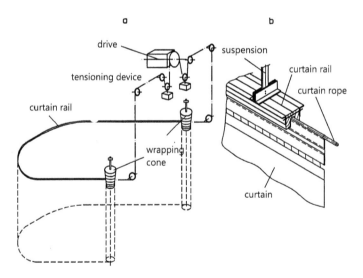

Figure 1.245: Cyclorama built as wrap around curtain (horizontal dragged cyclorama): (a) schematic diagram of the system, (b) rail guidance.

Figure 1.247 shows a cyclorama design in which the entire rail and winding cone system can be lifted into the stage tower. A special rope guide must be used to ensure that the ropes remain tensioned regardless of the lifting position.

Cyclorama of the classical design according to Figure 1.245 are becoming increasingly rare. In the *Vienna State Opera*, the circular horizontal dragged cyclorama was

Figure 1.246: Cyclorama horizontal dragged – built as wrap around curtain in the Großes Festspielhaus Salzburg (no longer exists). Photo: Waagner-Biró.

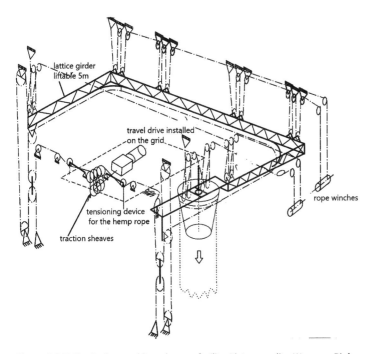

Figure 1.247: Vertically movable cyclorama facility. Picture credits: Waagner-Biró.

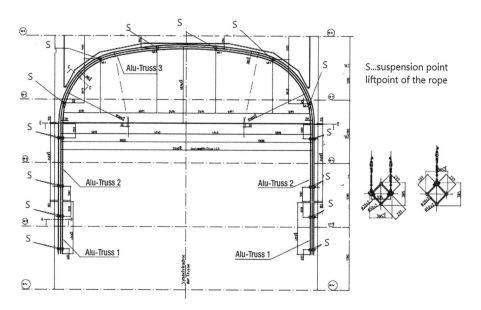

S...suspension point
liftpoint of the rope

Figure 1.248: New cyclorama facility at the Vienna State Opera. Photo credit: Waagner-Biró.

replaced by a system of independently or coupled movable load bar hoists as shown in Figure 1.248.

The installation consists of a total of five trusses with four hoists, twice Alu-Truss 1 and 2 as panorama hoists and Alu-Truss 3 as curved cyclorama hoist. In addition to the electronic synchronous control of the five elements, the individual segments in the main tubes of the trusses can also be mechanically coupled by conical coupling pieces. Trusses 1 and 2 have a lifting capacity of 360 kg, truss 3 of 1590 kg, the maximum lifting speed is 0.3 m/s. The circular curtain consists of a single continuous sheet of white cotton material 22 m high and 54 m long. The curtain is attached to the lower aluminum truss tubes and can also be stored there rolled up or stored rolled up like a backdrop. The fabric is very light with 175 g/m² and therefore weighs only a little over 200 kg in total.

Play space limitation by decoration elements parallel to the proscenium
Side panorama hoists and round bar hoists and cyclorama devices have the major disadvantage that no aisles remain for the performers to enter and leave. Side lighting is also obstructed or has to be omitted.

If the view of the side walls is to be covered by decoration elements or hangers oriented parallel to the portal plane, their covering effect must be checked by a sightline construction according to Figure 1.249. Using the same principle, soffits can be used to cover the view to the grid without inserting a plafond-like element.

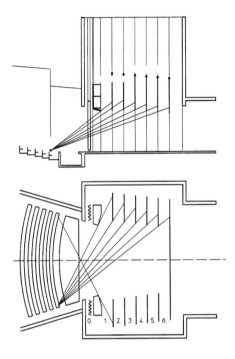

Figure 1.249: Sightline construction in top view and elevation.

1.8.7 Mechanical equipment for lighting technology

The use of lighting systems in the stage area also requires equipment of mechanical stage technology.

Fixed or movable *lighting bridges* can be provided across the width of the stage to mount spotlights, projectors, etc., on them. A lighting bridge on the proscenium wall is called a *portal lighting bridge* or *portal bridge* (Figure 1.250).

The *portal bridge* is thus an element of the proscenium and possibly together with sliding panels on the audience side, forms the upper boundary of the proscenium opening. As described in Section 1.8.2, the portal bridge can usually be moved in height and, guided in rails, can be lowered to stage floor level to facilitate the installation of lighting fixtures.

When designing height-adjustable portal bridges, care should be taken to ensure that the portal bridge is mounted stable and free of clearance with regard to the guide and suspension on the lifting device. Even small movements of the portal bridge caused by a clearance also move the spotlights mounted on it and their light cones and the projection images emitted by the portal bridge. For example, a clearly backlash-free contact of the guide rollers with the guide rails can be achieved by eccentric suspension of the portal bridge in the transverse direction. The lift drive for the platform can be provided by rope winches or hydraulic cylinders. Figures 1.250, 1.251 and 1.252 show the portal bridge and portal towers of two stages: in the *Semperoper Dresden* and in the *Grosses*

Figure 1.250: Portal bridge at the Semper Opera House in Dresden. Photo: SBS.

Figure 1.251: Proscenium zone in the Großes Festspielhaus Salzburg. Photo: Waagner-Biró.

Figure 1.252: Lighting bridge in the Bregenz Festival Theater. Photo: Waagner-Biró.

Figure 1.253: Lighting fly hoist in the Großes Festspielhaus Salzburg. Photo: Waagner-Biró.

Festpielhaus Salzburg. Figure 1.252 also shows the crane-like movable lighting bridge in the *Festpielhaus Bregenz.*

So-called *batten light hoists*, or *lighting hoists*, are installed distributed over the depth of the stage, often at about the third points. These are usually winch hoists with a higher load capacity similar to load bar hoists, but with load carriers or frames that can be raised and lowered to accommodate spotlights. Figure 1.253 shows light-ing hoists with pleated belts for electrical feed. However, cable drums can also be used for this purpose, although only motorized cable drums can be considered for greater lifting heights because spring cable drums would result in a too great tension-ing force.

Slow hoisting speeds are sufficient for these devices. They are usually driven by electrically operated rope winches. If large masses have to be moved, as in the case of a lighting bridge, counterbalancing is useful. However, hydrostatic cylinder drives such as those shown in Figure 1.254 can also be considered.

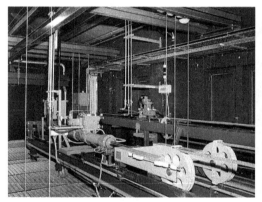 

Figure 1.254: Hydraulic linear drives for movable lighting scaffolds: (left) drive of a portal bridge, (right) drive of a lighting hoist. Photos: Bosch Rexroth.

A flexible use of light hoists is also offered by a system in which two adjacent load bars of motor-driven bar hoists are coupled and allow the installation of spotlights.

Sometimes it is also common to mount mobile and/or liftable *sidelight racks* on the side working galleries. These can be seen in Figure 1.255 (*Branch theater of the Bolshoi Theater in Moscow*). In this picture, among other things, the drop store elevators can also be seen (see also Section 1.7.1).

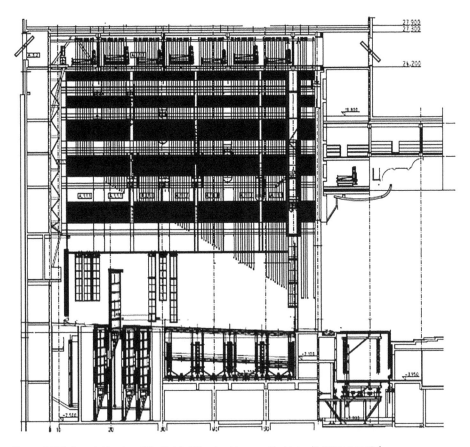

Figure 1.255: Branch theater of the Bolshoi Theater Moscow. Photo credit: Waagner-Biró.

1.8.8 Performer flying devices

Load bar hoists used as fly bars for flight movements in the plane

Performer flying devices are special devices in the upper stage that can be used to lift or lower and move persons or decorative parts in scenic use. With regard to the construction method, there are various possibilities. Two older systems are described here as examples:

– Variant A (Figure 1.256(a)). The running rail is fixed in its height position. A rope driven trolley (see Figure 1.225) can be moved on it. If this lifting device is actuated, the lifting corset is raised or lowered; if the travel rope drive is actuated, the trolley and together with it the lifting corset are moved.

– Variant B (Figure 1.256(b)). The height of the rail and the travelling trolley running on it can be changed. The lifting corset is thus raised and lowered by raising or lowering the bar used as running rail. The trolley is moved by means of a rope pull. In the system shown, the rope guide must be designed in such a way that the travel

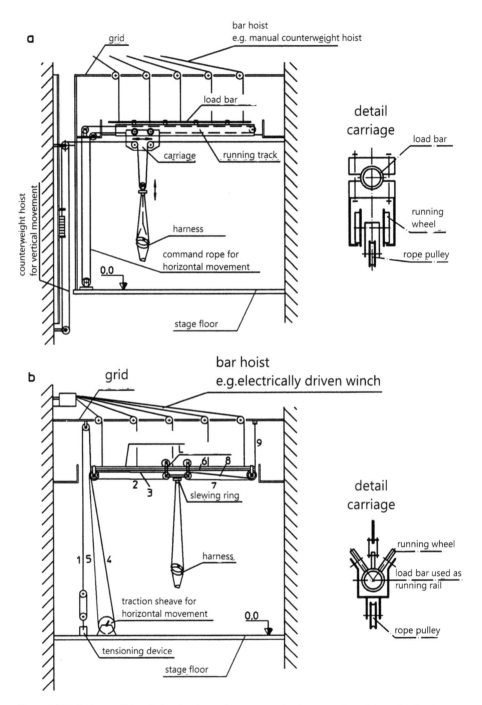

Figure 1.256: Performer flying devices in schematic representation for two variants A (a) and B (b). (Segments 1–9 show rope guidance for horizontal travel.)

rope loop remains the same length and thus always taut, regardless of the height position of the running rail.

Regardless of the system selection just described, a manual counterweight hoist or a machine hoist – an electrically or hydrostatically operated winch hoist or a hydrostatic cylinder hoist – can be used as the lift drive.

For the travel movement, a rope loop can also be operated manually or the rope pull can be motorized by winding and unwinding the right and left travel ropes on a rope drum or by driving them frictionally with a traction sheave under sufficient pretension. In Figure 1.256(b), the travel motion is performed with a traction sheave.

However, the two figures also show another detail with two design variants. In Figure 1.256(a), a separate rail is clamped to the load bar to accommodate the travelling trolley. In Figure 1.256(b), the load bar is designed to serve directly as a rail for the trolley.

Reference is therefore made once again to the design of the load bars shown in Figure 1.195(c) (Section 1.8.3), which makes it possible to install a fly on each load bar hoist. The special continuous cast profile shown in Figure 1.196 has a helmet profile at the bottom to accommodate a trolley.

Movable crane bridge as performer flying system for spatial flights
A fly system with more movement options was installed at the *Schauspielhaus Dresden*. It is a crane bridge that can be moved in the longitudinal direction of the stage and spans the entire width of the stage, arranged a few meters below the grid. A travelling trolley can be moved hanging from the lower chords of the bridge as a rope-driven crab. From the trolley, the bottom block can be raised or lowered with the load carrying device, e. g., with a person suspended in a corset, but a slewing gear installed in the bottom block also allows rotation. Figure 1.257 shows this fly device during testing in the factory.

Flying system generated by spatially arranged rope winches
Spatial motion sequences can also be represented by suspending the flying object or the flying corset for a person from at least three-point hoist ropes and programming the three-point hoist winches to produce straight-line, circular, spiral or any three-dimensional trajectories. Such a system was implemented by Bosch Rexroth in the *National Theater in Budapest*, for example. In this way, modern computerized control technology enables freely configurable flight movements in space with three or more suitably placed winches.

Special airframe for artists
Figure 1.258 shows an *acrobatic hoist* developed by Waagner-Biró for the Far East. The hoist is dimensioned for loads of 200 kg per hoist with rope speeds of up to 4 m/s. For a three-dimensional flight, four computer-controlled winches are used.

Figure 1.257: Flight equipment for the Schauspiel-haus Dresden. Photo: SBS.

Figure 1.258: Artist winch for a three-dimensional fly system. Photo: Waagner-Biró.

Whereas according to EN 17206 only steel ropes (and to a very limited extent also natural fiber ropes) may be used for ropes as suspension means, synthetic (plastic) ropes are also used in this case because it is intended to utilize their special elasticity, as in the case of mountaineering ropes, in order to keep deceleration and force effect as low as possible in the event of an emergency stop (category 0) (see also Section 4.8). In addition, this results in a potential separation between winch and load.

However, it is not a plastic rope made of polyamide (nylon, perlon, etc.), as is the case with mountaineering ropes, which already loses its strength at relatively low temperatures, but a *"Liquid Crystal Polymer"* ("LCP"), which is characterized by extremely high tensile strength (similar to steel wires), a high modulus of elasticity and by its high melting point, and therefore also allows applications in the high-temperature range. In this specific case, the rope in question is a rope with the brand name "Vectrus" from the company Ticona (Texas) with a diameter of 6 mm and a breaking load of approx. 31 kN.

To avoid slack rope on the drum winch despite the high speeds during lowering, the outgoing pulley is also driven and runs at a slight overspeed compared to the winch. The built-up slip keeps the rope under tension. An earlier wear out of the rope is accepted with this technology. Special attention also had to be paid to the fixing of a synthetic rope into the rope drum.

It should also be noted that special attention had to be paid to the deceleration and force effects in the event of an emergency stop (category 0).

Figure 1.258 also shows that the rope drum is also moved axially during its rotation so that the payoff point of the rope remains unchanged. The rope is guided at the bottom of the grid through an eye with a smooth surface (see Figure 1.259).

Figure 1.259: Eyelet for rope feed-through ("Donut"). Photo: Waagner-Biró.

1.9 Safety equipment for fire protection

Before the invention of electric lighting systems, work was done with pine chip, oil lamps and candles, later with gas light. Therefore, there was a particular fire hazard. Devastating theater fires with many human casualties are known from history. In Europe, this was particularly the case with the fire at the Ringtheater in Vienna in 1881. This fire catastrophe in particular resulted in strict building regulations being issued to ensure the highest possible level of fire protection and to be able to evacuate people as quickly and easily as possible in the event of a fire breaking out; these regulations set an example for many other countries.

Even though modern lighting technology has reduced the potential dangers in theaters and opera houses, strict fire safety regulations still exist in most countries.

Here, only the safety devices of the mechanical stage equipment will be discussed. These are primarily *mobile partition walls* between different fire compartments in the stage area, in particular the *fire curtain* – also known as *iron curtain* – between the stage and the auditorium, as well as the *smoke escape systems*, the opening of which is intended to dissipate smoke gases and overpressure on the stage because of heat. The interaction of these two devices is intended to prevent flames from spreading from the stage to the auditorium, on the one hand, and to prevent lethal smoke gases from entering the auditorium, on the other.

Opening the smoke escapes prevents overpressure on the fire side in the stage area due to the heating of the air, which could result in deformation of the safety curtain and the flow of smoke gases into the auditorium. Due to the chimney effect, fire with smoldering smoke is fanned and an air flow is created, which hardly allows the gases to enter the auditorium. Although this measure promotes the destruction of the stage, it protects the auditorium and the visitors.

1.9.1 Fire protection curtains

In classical theater construction, depending on local building regulations, a separation of the auditorium and stage area into different fire compartments is prescribed above a certain theater size. Therefore, the stage house must be structurally separated from the building section for the audience by a fire wall. The "unavoidable" proscenium opening must be able to be closed by an appropriate sufficiently fire-retardant material. In most cases, this is a so-called *iron curtain* in the form of a *lifting curtain* (Figure 1.260). Normally, the iron curtain is raised only a few minutes before the start of the performance to unclose the proscenium opening; it is lowered immediately after the end of the performance.

Figure 1.260: Großes Festspielhaus Salzburg – iron curtain, orchestra platforms. Photo: Waagner-Biró.

In the case of large stage systems with side and/or back stages, the stage area may also be divided into several fire compartments by installing side or back stage curtains. These usually also have soundproofing tasks to perform. When closed, they should allow decoration work to be carried out in the side or back stage during rehearsals and performances (Figure 1.261). In most cases, these are also lift-up doors, sometimes also other closing elements such as rolling (shutter) gates (see Figure 1.8 – Backstage, longitudinal section).

Figure 1.261: Side stage and rear stage shutters at the Festspielhaus Bregenz. Photo: Waagner-Biró.

In addition to the standard design as a door leaf to be lifted away upwards, there are also special solutions for structural reasons. If there is not enough space above the proscenium, the curtain can be lowered down instead of up, as is the case at the *Kammerspiele in Vienna*, for example.

The curtain blade can also consist of two parts; the parts are telescoped up or down, or one part is moved up and the other down.

Sometimes also *sliding curtain leafs* are installed, where the door blades are moved sideways. A sliding safety curtain is installed, for example, in the *Schönbrunner Schlosstheater in Vienna*; Figure 1.263(c) explains how it works.

The iron curtain leaf
The door leaf should be smoke-tight and fire-retardant for at least 30 min. However, depending on local regulations, a longer fire resistance period may be required.

With regard to the dimensioning of the door leaf, it should be noted that in the event of a fire, high thermal loads and relatively large pressure loads can occur as a result of an air pressure difference between the audience area and the stage area. Therefore, the door leaf is usually designed as a steel grating construction with fire-retardant plate coating on the stage side. In many countries, a fire-retardant coating must be provided on the stage side in any case. In the past, asbestos was used for this purpose; since it has become known that asbestos fibers are harmful to health, other mineral panels are used. Instead of this plate coating adequate cooling of the door leaf on the stage side with water is ensured by installing a *deluge spray system*.

Loading the door leaf with forces from an air pressure difference of approx. $400\,\text{N/m}^2$ (value depending on the specification) causes large bending stresses. If the curtain blade is to be considered as a bending element supported only in the lateral guides, i. e., as a grating hinged at two edges, very high bending stresses result in the case of wide platform openings, resulting in large thickness dimensions and an exceedingly high blade weight.

A considerable weight reduction is possible by considering the door leaf not only as a surface element supported on two sides in the guides, but also as an element supported on four sides, i. e., also at the top and bottom. For this purpose, the bottom of the door leaf is provided with conical bolts which, when closed, engage in corresponding holes in the stage floor. This is based on the assumption that during the closing process, the door leaf is not yet deformed to such an extent that these pins would no longer be able to enter the holes.

Figure 1.260 shows the curtain wall of the *Großes Festspielhaus in Salzburg* with a 30 m wide portal opening. This curtain leaf is about 1 m thick in the central area.

However, the smoke-tightness of the protective curtain can only be achieved to a limited extent. At the contact surface of the door leaf with the wooden floor of the stage, for example, elastic material or sealing elements which swell at high temperature can be used; sand channels as shown in Figure 1.262 are often installed at the upper edge of the curtain leaf. However, other proofs are also possible, e. g., sealing beads made of mineral wool wrapped in heat-resistant materials (e. g., Zetex® from Newtex Industries, USA) can be applied at the top, bottom and in the side guides.

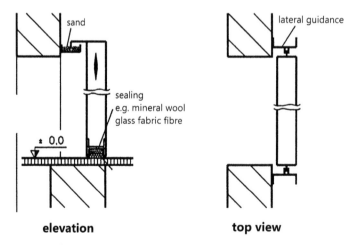

Figure 1.262: Guide of the curtain blade and sealing: (left) elevation, (right) plan view.

The drive
In normal operation, the door leaf will be moved with a relatively low nominal speed in the sense of opening or closing. In the case of emergency closing, however, the closing process must take place at a greater speed. Regulations, for example, require a maximum closing time of 30 seconds. The emergency closing operation must be triggered by an authorized person. Emergency closing must be possible even if the electrical power supply fails. The driving force for the movement of the curtain blade therefore usually results from the gravitational effect of deadweight or counterweight masses acting in the sense of the closing movement. However, the closing process must also be controlled by automatically acting braking devices so that the door blade does not reach the closing position at too high speed.

For a lifting safety curtain (Figures 1.263(a) and 1.263(b)) with the door leaf raised during a performance, this results in the following construction methods, for example: The dead weight of the curtain blade is compensated by counterweights except for a

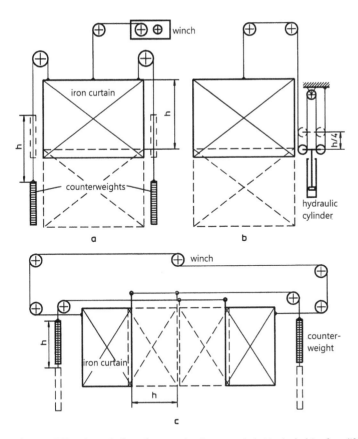

Figure 1.263: Drive units for safety curtains (iron curtains): (a) winch drive for a lifting curtain, (b) cylinder drive for a lifting curtain, (c) winch drive for sliding gates (sliding curtain).

residual mass of about 1–2 t. The weight of the curtain blade is then reduced by the weight of the counterweights. Thus, a lifting drive for a lifting force of approx. 10 to 20 kN is sufficient to carry out the operational opening or closing process.

In the event of an emergency closure, the release by opening the hoist brakes causes an automatic falling process of the curtain blade as a result of the excess weight of the curtain blade. However, this lowering movement must be braked automatically so that the lowering speed does not increase steadily according to the laws of free fall and the curtain blade hits the platform floor with a too high speed. In older systems, the braking was performed by an automatically acting centrifugal brake, i. e., via mechanical friction. At the last few centimeters the gate blade impact on hydraulic buffers and slowly moves into its final position.

In modern plants, the centrifugal brake has been replaced, e. g., by a hydraulic pump. This hydraulic pump connected, for example, to the motor shaft or another fast-running gear shaft is driven during the falling process and works against a throttle. When using throttles as hydrostatic brakes, an emergency closing process can be very well controlled in the motion sequence by path-dependent variation of the throttle effect (initially a throttle with a larger opening, near to the final position a throttle with a smaller opening) and the installation of hydraulic buffers may be omitted.

However, it is also possible to achieve an analog braking effect purely electrically without using hydraulics by driving a synchronous motor generatively during emergency closing and using braking resistors instead of throttles. Its rotor has a permanent magnet since for safety reasons it must be assumed that no power supply is available. This type of motor also requires very little maintenance, as it does not require brushes.

In addition to the winch drive just described for the lifting movement of the curtain blade, a hydrostatic linear drive is also possible as an alternative, in which the lifting or lowering process of the curtain blade is carried out with a hydraulic cylinder. By using an "inverted" pulley block, a short-stroke cylinder can be used despite a large travel distance of the door leaf corresponding to the height of the proscenium opening (see Figure 1.263(b)). In this case, too, a very precisely controllable emergency closing process is possible by means of travel-dependent throttling controlled by the position of the leaf. When using hydraulic cylinders, counterweights can usually be dispensed with.

In the case of special solutions, such as the variant already mentioned, in which the curtain leaf is lowered into the basement, the counterweights must be overweight in relation to the curtain, so that the closing process can also take place without energy supply in this case. Also in the case of sliding curtains, weights have to provide the propulsive forces for the horizontal movement of the gate elements via rope pull (see Figure 1.263(c)).

Fabric fire curtains

In some countries, the responsible authorities have recently also approved rollable fire protection curtains made of fiberglass fabric for smaller stage openings to separate the fire sections of the stage house and auditorium. To withstand higher pressure loads, stainless steel wires are woven into the fiberglass fabric. The pressure difference between the stage and auditorium permitted for the curtain material depends on the size of the curtain area and is generally far lower than the values prescribed.

The curtain is guided laterally in rails as shown in Figure 1.264 and pulled over upright stainless round bars when closed. In the event of fire, the flue gas tightness is created by the horizontal tensile forces acting in the fabric.

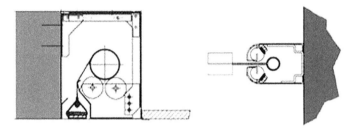

Figure 1.264: Safety curtain system "Fibershield" installation example on top and side guidance. Picture credits: Stöbich.

1.9.2 Smoke escape systems

Adequately sized *smoke escapes* must be provided in the area of the stage and auditorium. Their minimum opening cross-section is precisely defined in regulations and depends on the floor area of the stage area resp. the floor area of the auditorium. Similar to the protective curtain, their opening process must be possible without energy supply in case of fire. In some countries, there is also a requirement that the smoke vents open automatically at a certain overpressure or temperature. But it seems to be more useful, if the fire brigade decides whether the smoke escapes shall be opened or should stay in closed position. However, since the smoke escapes are also used for ventilation purposes, normal operation control must also be provided.

A very common construction method are so-called *smoke flaps*. These are approximately vertical wall elements located in the roof area that can be opened via a horizontal axis, with weights ensuring automatic opening. Winch drives or hydraulic cylinders can again be used for operational opening and closing. Figure 1.265 shows smoke flaps operated by ropes, namely vertical outer and horizontal inner flaps connected in series, and Figure 1.266 shows hydraulically operated simple smoke flaps. Where icing of the smoke flaps can occur due to climatic conditions, special care must be taken to ensure that sufficiently large forces are applied to push the flaps open. This is possible, for example, hydrostatically using accumulators. Wind forces must also be taken into account.

Figure 1.265: Smoke escape system of the Großes Festspielhaus in Salzburg, operated by rope pull (smoke escape open). Photos: Waagner-Biró.

Figure 1.266: Smoke flaps in Vienna's Raimundtheater with hydraulic actuation (smoke flaps closed). Photo: Waagner-Biró.

In addition to the smoke flaps just described, there are also construction variants in which roof elements – so-called *smoke lantern* – are raised so that the prescribed smoke escape areas are released (Figure 1.267). Especially when snow loads have to be taken into account, relatively large lifting loads can result, which also have to be moved again by stored energy in the event of a fire. Counterweights or pressurized hydraulic accumulator systems can be used for this purpose.

However, smoke escape units have also been designed that are actuated pneumatically with compressed air compressed in a compressor during operation and moved with CO_2 from pressure cylinders during emergency shutdown (e. g., in the *Musical Theater Basel*).

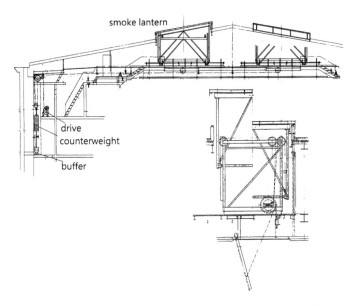

Figure 1.267: Smoke lanterns and smoke flaps at the Semperoper Dresden – rope winch drive, emergency closing by counterweight.

1.9.3 Water extinguishing systems

For firefighting by the fire brigade, *extinguishing water plugs* are available in every larger venue. In some cases, however, it is also common practice to install *water spray extinguishing systems* in order to be able to immediately combat the rapid spread of fire. This involves a network of pipes with extinguishing nozzles, connected to an efficient water supply. If the amount of water from the general water supply is not sufficient, water can also be stored in water tanks, especially in pressurized water tanks to ensure a sufficient pressure level. The release of the extinguishing system is not triggered automatically by fire sensors, but by an authorized person.

In a water spray extinguishing system, as just described, all extinguishing nozzles are open and spray extinguishing water as soon as they are supplied with pressurized water in the piping system. Of course, several separately pressurizable piping systems can also be provided.

In contrast, *sprinklers* are automatically heat-responsive spray nozzles that are always under water pressure and kept closed by glass ampoules containing a special liquid with an air bubble. In the event of a fire, the liquid in the glass ampoules heats up, expands, and the ampoules burst, opening the nozzles and allowing water to escape from the sprinkler pipe network. In a sprinkler system, therefore, all nozzles are pressurized with water, and only those spray extinguishing water whose closure is burst open.

Notes for supplementary literature to Chapter 1

Bühnentechnische Rundschau
Der Theaterverlag - Friedrich Berlin GmbH.
Formerly: Erhard Friedrich Verlag GmbH & Co. KG resp.
Orell Füssli + Friedrich Publishers Ltd.
In particular, the stages listed in the book are described in more detail in the following editions or special reference is made to them:

Jg. 50 (1956) Nr. 1	*Vienna State Opera*
Jg. 71 (1977) Nr. 3	*National Theatre London*
Jg. 73 (1979) Nr. 4	*Semper Opera House Dresden*
Jg. 76 (1982) Nr. 2	*Schaubühne Berlin*
Jg. 78 (1984) Nr. 5	*Friedrichstadtpalast Berlin*
Jg. 79 (1985) Nr. 3	*Zurich Opera House*
Jg. 79 (1985) Nr. 4	*Festspielhaus Salzburg*
Jg. 79 (1985) Nr. 6	*Friedrichstadtpalast Berlin*
Jg. 80 (1986) Nr. 3	*Decoration magazine Munich*
Jg. 80 (1986) Nr. 3	*Graf Zeppelin House Friedrichshafen*
Jg. 81 (1987) Nr. 3	*Muziektheater Amsterdam*
Jg. 82 (1988) Nr. 3	*Hamburg State Opera*
Jg. 82 (1988) Nr. 5	*Alvar Aalto Theater Essen*
Jg. 83 (1989) Nr. 2	*Kuwait Conference Center*
Jg. 83 (1989) Nr. 3	*State Theater Stuttgart*
Jg. 83 (1989) Nr. 4	*National Theater Munich*
Jg. 83 (1989) Nr. 4	*Prospect store, National Theater Munich*
Jg. 83 (1989) Nr. 5	*Opera de la Bastille Paris*
Jg. 84 (1990) Nr. 4	*State Opera Munich*
Jg. 85 (1991) Nr. 1	*Theater New Flora*
Jg. 85 (1991) Nr. 4	*Teatro Felice – Genoa Opera*
Jg. 86 (1992) Nr. 1	*Residenztheater Munich*
Jg. 86 (1992) Nr. 2	*Deutsche Oper Berlin*
Jg. 86 (1992) Nr. 2	*Opera Frankfurt on the Main*
Jg. 87 (1993) Nr. 4	*Megaro Musikis Athinon*
Jg. 87 (1993) Nr. 6	*National Theater Maribor*
Jg. 98 (1995) Nr. 1	*National Theater Maribor*
Jg. 98 (1995) Nr. S	*(special volume) Theater am Goetheplatz in Baden-Baden*
Jg. 98 (1995) Nr. S	*(special volume) Vienna State Opera*
Jg. 99 (1996) Nr. 1	*Burgtheater Vienna*
Jg. 99 (1996) Nr. 4	*Musical Theater Basel*
Jg. 100 (1997) Nr. 2	*Semper Opera House Dresden*
Jg. 100 (1997) Nr. 5	*Congress Center Frankfurt*
Jg. 101 (1998) Nr. 5	*Hanover Opera House*
Jg. 102 (1999) Nr. 1	*Geneva Opera*
2001 Nr. 2	*Berlin ensemble*
Nr. 3	*Als die Matrosen auf den Schnürboden zogen (When the sailors moved to the grid)*
	Theater an der Wien – 200 years old (Part 1)
	Restauration of the Bremerhaven Municipal Theater

| | Nr. 4 | *Ein festes Haus bis in die fernste Zukunft hinaus (A permanent home far into the future)* |

Nr. 4 *Ein festes Haus bis in die fernste Zukunft hinaus (A permanent home far into the future)*
       The Maxim Gorki Theater in Berlin-Mitte
       Theater an der Wien – 200 years old (Part 2)
       Stage technology in the last decades
2002   Nr. 2 *The Tempodrom Berlin*
       New theater building in Basel
       The reconstruction of the Berlin Volksbühne
       Nr. 4 *Regensburg Municipal Theater – General renovation of stage technology*
       Nr. 5 *The Hungarian National Theater in Budapest (Part 1)*
2003   Nr. 1 *The Hungarian National Theater in Budapest (Part 2)*
       Modernization of the upper machinery at the Saarbrücken State Theater
       Nr. 2 *The renovation of the large Thalia Theater in Halle*
       Nr. 3 *Konzerthaus Berlin – Reconstruction of the orchestra rehearsal hall*
       Nr. 6 *Grand Théâtre Luxembourg*
2004   Nr. 1 *E. T. A. Hoffmann Theater in Bamberg*
       Nr. 2 *The new theater building in Erfurt*
       Nr. S *(special volume) The New Philharmonic Hall Essen*
2005   Nr. 1 *Muziektheater Amsterdam*
       Nr. 2 *La Fenice in Venice – Where she was, how she was*
       Thalia-Theater Hamburg – Renewal of the Upper Machinery
       Nr. 3 *Palace of Arts Budapest*
       Nr. 4 *The renovation of the lower machinery in the Schauspielhaus Frankfurt*
       Nr. 5 *The restoration of the Bolshoi Theater*
       Nr. 6 *State Theater Wiesbaden*
2006   Nr. 2 *Circus theater "Care" in Amsterdam*
2008   Nr. 5 *Stage technology "Made in Europe" for Oslo*
       Nr. 6 *New stage technology for Düsseldorf Schauspielhaus*
2011   Nr. 6 *The Bolshoi Theater in new splendor – The renewal of the stage technical equipment – Part 1*
2012   Nr. 1 *The Bolshoi Theater – a Rebirth – The Renewal of the Stage Technical Equipment – Part 2*
       Nr. 2 *Opera construction in the Gulf States – The new Royal Opera House Muscat(Oman)*
       Nr. 3 *Shanghai Culture Square – A new musical and show stage in Shanghai*
       Nr. 4 *Move and store decorations – The new container and prospectus warehouse of the Bolshoi Theater*
2013   Nr. 3 *The stage machinery at the Musiktheater Linz*
2017   Nr. 5 *New work in the renovated house – The renovation of the Staatsoper unter den Linden – Part 3 Stage Technology*
2020   Nr. 2 *New Radiance for the 21st Century – The State Opera in Prague*

(In the figure legends, the figure reference for 87 (1993) No. 6, for example, is in the form BTR 6/1993).

Prospect
Magazine of the OETHG for Stages & Event Technology
Publisher and media owner: Österreichische Theatertechnische Gesellschaft - OETHG
In particular, the stages listed in the book are described in more detail in the following booklets:

1996	No. 1	*The new stage machinery for the Vienna Burgtheater*
	No. 3	*High-tech on the high seas*
		The Vienna Volksoper through the ages
2000	No. 1	*Vom Zug zum Flug (From a hoist to a flying system)*
2001	May	*175 Years Grazer Schauspielhaus*
		200 Years Theater an der Wien (Part 1)
	Sept.	*200 Years Theater an der Wien (Part 2)*
2003	March	*The new "Helmut List Hall" in Graz*
	June	*A new stage floor for the Seebühne in Mörbisch*
	Dec.	*200 Years Landestheater Linz 1803–2003*
		The Black Box of the Burgtheater in the Casino
2004	March	*The reopening of "La Fenice" in Venice*
	June	*New intelligent stage wagon – Ideen-Schmiede Werfing*
	Dec.	*Stable, lowerable new orchestra podiums for the Vienna Burgtheater*
2005	March	*Tiroler Landestheater: Etappenweise Umrüstung des Schnürbodens auf Elektroantriebe*
		(Conversion of the equipment on the grid to electric drives by stages)
	June	*Arena Rockódromo Madrid*
	Oct.	*Fascination of the stage: Baroque stage technology*
	Dec.	*The Palacio de las Artes – Valencia, Spain, has a new opera house*
		Volksoper Wien – Reconstruction of the drive for the retractable core disc of the revolving stage
2006	No. 1	*Spectacular reconstruction of the Viennese "Ronacher"*
		House for Mozart
	No. 3	*The University – Mozarteum in Salzburg shines in new splendor*
		Festspielhaus Bregenz – Refurbishment in only 319 days
		The sliding prospectus hoists in the renovated Festspielhaus
		Unique new stage wagon system by Waagner-Biró for the new opera in Copenhagen
	No. 4	*Renewal of the main curtain beam at the Vienna State Opera House*
2007	No. 2	*The new opera house in Oslo is taking shape*
	No. 3	*Renewal of the CATV4 understage control system for the Vienna State Opera House*
2008	No. 1	*"China National Gran Theatre" – a pearl for Beijing*
	No. 2	*Reopening of the Vienna Ronacher after functional renovation*
	No. 4	*The Oslo National Opera – "Cultural Building of the World 2008"*
2009	No. 1	*Opera reopens at Seoul Ars Center*
	No. 2	*The Slovak National Theater in Bratislava*
		New music theater in Linz – Realization has begun
	No. 4	*37 m long curtain scissor lift for Tashkent – by Gerriets*
2010	No. 3	*The stage technology for the new Musiktheater Linz*
		Multiversum – New multifunctional event center for Schwechat
		Vienna State Opera – New circular horizon system
	No. 4	*Arena "Reyno de Navarra" in Pamplona – The hidden treasure chest in the arena*
2011	No. 1	*Neues Musiktheater Linz – Europe's most modern opera house with fully automated decoration store*
	No. 4	*The Bolshoi Theater shines in new splendor*
		Graz Opera – Renovation of the stage technology
2012	No. 1	*Curtain systems with heavy duty rails*
	No. 2	*The stage technology of the newly opened Bolshoi Theater in Moscow*
		New Music Theater Linz – The Final Spurt
	No. 3	*Neues Musiktheater Linz – the multifunctional transport revolving stage*

2013	No. 1	*The new music theater in Linz*
		Tyrol – The Erl Festival Theatre is open
		The new concert hall in Stavanger – Stage technology by Waagner-Biró
		Mariinsky Theater – Extension with 2000 seats
	No. 3	*Musiktheater Linz – Airport technology in the opera: The automated scenery warehouse is the secret star of this new theater*
	No. 4	*Wiener Kammerspiele – In new splendor after general renovation*
2014	No. 1	*Vienna Opera Ball – Transformation of the opera house into a huge ballroom*
		The brake test in stage technology – A stepchild in terms of safety considerations
		Standardization in stage technology and dimensioning of brakes
	No. 3	*Burgtheater Wien – Start of the 2014/15 season with new stage control technology*
	No. 4	*The "Globe Vienna" in the Marx-Halle*
2015	No. 4	*Royal Opera Stockholm*
2017	No. 3	*Staatsoper Berlin – A tour – A lot of technology under the Linden*
2019	No. 1	*150th Anniversary of the Vienna State Opera – Technical History of a World Stage*
		Globe Vienna – Renovation and upgrade – The end of a compulsory brake
	No. 3	*Sydney Opera House Renovation – The Digital Twin of the Opera House*
2020	No. 2	*Prague State Opera – New technology in old splendor*
2021	No. 1	*Wiener Volkstheater – The game can begin*
	No. 3	*Vienna State Opera – Virtual Stage Planning at the Haus am Ring*

Merkblätter über sachgemäße Stahlverwendung

Issue 289 – *Steel structures in theater construction*

Consulting office for steel utilization Düsseldorf DK 725.82; 624.94/95

In the *figure legends*, the image reference is made with the indication "Merkblatt 289."

Kranich, Friedrich: *Bühnentechnik der Gegenwart*, 2 Bde., Munich-Berlin: Oldenbourg, 1929 (reprint 1992).

Unruh, Walter: *Theatertechnik – Fachkunde- und Vorschriftensammlung*, Berlin-Bielefeld: Klasing, 1969.

2 Drives for stage equipment

Motor drives – apart from special cases – ultimately always use electrical energy that is taken from the general power grid. Only when an emergency generator is used, electrical energy is generated directly in a generator driven by combustion engine. In stage equipment, this electrical energy is then converted into kinetic energy in two ways:

– The conversion takes place by means of electric drives by generating the working movements with the help of electric motors. The rotating motion of the motor is converted via a gearbox into rotating motion of a different speed, into linear motion or into kinematically complex motion sequences. Electric linear motors are rarely used.

– Working movements are generated hydrostatically by causing with pressurized fluid linear movements by hydraulic cylinders or rotary movements by hydraulic motors. The pressurized fluid is provided by hydraulic pumps driven by electric motors or is taken from accumulators, which, however, must also be filled in most cases by electrically driven hydraulic pumps.

In the following sections, both drive variants will be explained in more detail. This is preceded by some comments on manual drives.

2.1 Manual drives

Until the invention of motor drives, people had to rely on *muscle power*. However, muscle power can be replaced, or at least the amount of power required from humans can be greatly reduced, for example, by using *gravity*. One possibility is the use of counterweights to partially or completely balance the weight of the payload. In very old hydraulic systems, the *gravity pressure* of water stored in the roof level of the building may also have been used to perform lifting work with simple plunger cylinders. Later, to achieve higher pressures, the *hydrostatic pressure* was increased by *compressed air* pumped into a closed water tank.

For the application of greater forces, *block and tackle systems, levers* or *gears* can also be used.

Today, technical equipment on the understage is hardly ever moved manually, apart from smaller stage trolleys and lowering devices for people. Only in very old theaters can one possibly still find revolving stages and larger lowering devices operated by muscle power.

However, manual drives are still often found in the upper stage, for proscenium hoists and curtain drives. Firstly, it is much cheaper to equip the upper stage with manual hoists, and secondly, in the past some operators were of the opinion that sensitive movements adapted to the play could only be realized with a manual hoist. This view originates primarily from the time when only poorly controllable or adjustable machine

https://doi.org/10.1515/9783111366968-002

hoists were actually installed in stages, either because the technology was not yet mature enough, or because technically inferior solutions were chosen for cost reasons. An electrically or hydraulically operated bar hoist using modern drive technology is superior to the manual hoist in many respects, but, of course, more expensive. Manual counterweight hoists will be used less and less in the future, also for reasons of operational organization. Manipulating the counterweights is heavy work that puts a lot of strain on the spine, and using the flies requires a great deal of time and personnel.

Manual drives are also used in stage technology as *emergency drives* in order to be able to maintain at least emergency operation with restrictions in the event of technical faults. In such emergency situations, slower motion sequences can then also be accepted.

Ergonomic aspects should be taken into account in the design with regard to the arrangement of handwheels and cranks. It should also be ensured that excessive hand forces are not imposed. At the circumference of a handwheel with a smooth gripping surface, the hand force can be about 200 N; at a hand crank, somewhat higher forces are possible for short periods. A pull on a rope or chain should preferably be in a vertical direction and the tensile force should not exceed 300 to 400 N.

Taking into account the data for a hand crank drive, a torque of about 100 Nm can be generated at the drive and, assuming a rotational speed of about 30 rpm, a power of about 300 W can be produced for a short time. Over a longer period of time, the power that can be applied by muscle power is much lower.

The torque and corresponding force effects can be increased by gear ratios, but the speed of the movement is reduced to the same extent. The power, which is very small for manual drives and which is proportional to the product of the acting force and the speed of movement according to Eq. (3.26), can, of course, not be increased, unless several persons are used at the same time. The limitation of the power therefore means that either large forces or torques can be overcome with only small speed or large speeds or revolutions can be realized only for small forces.

If the operating speed of a manual emergency drive is too low, an emergency drive can also be designed so that a hand drill or similar device is temporarily coupled to a shaft provided for this purpose in order to achieve higher drive powers.

2.2 Electric drives

2.2.1 Direct current drives and three-phase drives of classic design

Three-phase motors can be fed directly from the *three-phase supply*, unless you want to change the feed frequency of the drive motor. The DC current required to operate *DC motors* must first be generated via rectifiers. Only motors of low power can be operated from *batteries* or accumulators. However, the battery charger is again supplied with mains current.

Which motor type and circuits to choose depends on the operating conditions and requirements. Essential criteria are:

- Whether the motor is powered from the mains or from a battery.
- Whether the motor must be operated only at its rated speed.
- Whether an additional slow speed is desired.
- Whether several speeds or a stepless adjustment of the speed is required.
- How large the ratio between maximum and minimum speed should be.
- Furthermore, it must be considered whether a speed adjustment is sufficient as a *control*, accepting a certain load dependency, or
- Whether *load-independent speed feedback control* is required for a single drive or *control* for *synchronous operation* of several drives.

In a closed-loop drive with a feedback control, the measured actual value of an operating variable, e. g., the operating speed, is compared with the specified setpoint and the speed is corrected accordingly.

Only one nominal speed is sufficient, for example, for pump drives in the hydraulic pressure control center or for drives that are not used for scenic purposes and do not require high operating speeds. This is usually the case with orchestra platforms, compensating platforms, drives for changing the inclination of the decks of platforms, safety curtains and heavy-duty assembly hoists.

In most cases, however, very high demands are placed on the drives to be used in the lower and upper stage, namely stepless adjustment of the operating speed, no load dependency, good controllability in the sense of high control dynamics and a high control ratio of maximum to minimum speed. For this reason, only electric drives with feedback control are generally used in modern stage systems for hoists, stage wagons and revolving stages.

DC motor

The DC shunt-wound motor and the *separately excited DC motor* with the winding circuit in stator and rotor according to Figure 2.1(b, c) have a low load dependence of the speed, because a magnetic flux Φ is generated in the excitation winding (stator winding) which is independent of the load. For comparison, Figure 2.1(a) also shows the circuit of the series-wound motor, which, however, is not used in stage technology because of its extreme load dependence.

The externally excited DC motor also offers very good possibilities of infinitely variable speed control and regulation, and was and is therefore very often used in stage technology. The following relationships (proportionalities) apply to speed and torque:

$$n \approx \frac{U - I_A \cdot (R_I - R_V)}{\Phi}, \tag{2.1}$$

$$M \approx I_A \cdot \Phi, \tag{2.2}$$

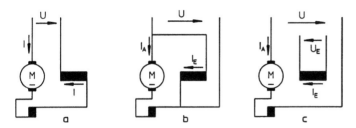

Figure 2.1: Circuit diagrams of DC motors: (a) series-wound motor, (b) shunt-wound motor, (c) separately excited motor.

with

n motor speed (rotational speed)
U terminal voltage
I_A rotor current
R_I internal resistance of the rotor winding
R_V possible ballast resistor
Φ magnetic flux
M engine torque

Speed adjustment of a DC motor is therefore possible in the following ways (Figure 2.2):

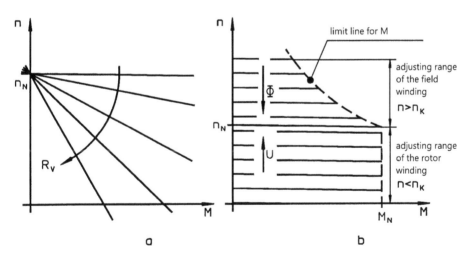

Figure 2.2: Characteristics of a separately excited and a self-excited shunt-wound DC motor: (a) lossy speed adjustment by series resistors R_V, (b) lossless speed adjustment by lossless change of terminal voltage or magnetic flux Φ.

–   Starting from the nominal speed, the speed can be reduced by connecting *resistors* R_V in the rotor circuit. In these resistors, electrical energy is converted into heat, so that this operating state is only useful and possible for very short periods; on the other hand, this also makes the speed more load-dependent (see Figure 2.2(a)). This type of speed adjustment is therefore ruled out; series resistors may only have been used for the starting process of larger machines.

– Almost without loss, a reduction of the speed can be effected by *reducing the terminal voltage U* at the rotor. In the past, this reduction of the terminal voltage was carried out using rotating converters, the so-called Ward-Leonard converter, by field weakening of a DC generator (Figure 2.3(a)), or a variable transformer with rectifier

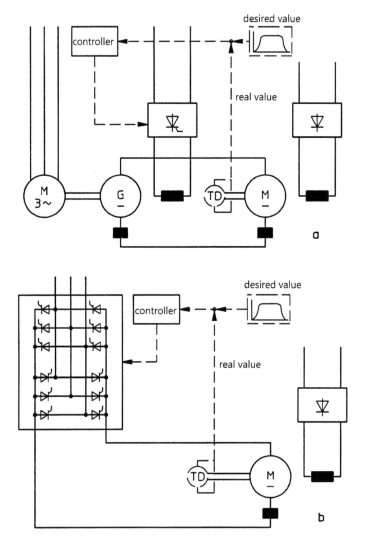

Figure 2.3: DC drive open-loop controlled or feedback controlled (dashed line): (a) with Ward-Leonard converter, (b) with converter in three-phase bridge circuit.

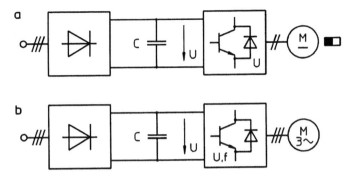

Figure 2.4: Possibilities of speed control – schematic sketches: (a) lossless control of a DC motor with a DC chopper, (b) lossless control of a three-phase motor with a frequency converter (with an intermediate direct current link).

was used. Today, static converters are used with modern power electronics and the voltage is reduced by phase angle control. By switching off the supply voltage over short time intervals, the effective value of the voltage is changed as an energetically effective voltage value (Figure 2.3(b)). The resulting characteristic curves are hardly load-dependent and are shown in Figure 2.2(b). Figure 2.4(a) shows the principle circuit of a DC controller for variable-speed operation of a DC motor. Externally excited DC motors with 4-quadrant thyristor controllers were the most frequently used electric drives in stage technology, but – in line with the general trend – are increasingly being replaced by three-phase AC drives.

- An adjustment into higher speed ranges could be achieved by *field weakening*, i. e., by reducing the magnetic flux in the stator of the DC motor. From Eq. (2.2) it can be seen that with field weakening, however, the motor torque is also reduced, i. e., in the field weakening range the motor can only deliver a lower torque. Therefore, field weakening is only applicable for partial loads and is de facto out of the question for stage drives.
- Under these considerations, Eqs. (2.1) and (2.2) can be simplified by writing the following proportionalities:

$$n \approx U, \tag{2.3}$$

$$M \approx I_A. \tag{2.4}$$

Three-phase AC motor

In the three-phase AC motor, at least three electromagnet coils are arranged in the stator, offset by 120°. If these three coils are each fed with a conductor voltage phase of the three-phase system, then a magnetic field is generated in each coil, the time sequence of which is offset by one third period. A rotating magnetic field results from the individual coil magnetic fields. If a rotating magnet (the rotor) is placed in the center of this magnetic field, the rotating field causes the magnet to rotate. These three electromagnet coils, each consisting of a north and south pole, together form a pole pair.

If the stator windings of a three-phase motor are connected to the three-phase mains, the magnetic field (rotating field) rotates at a speed n_s, where

$$n_s = \frac{f}{p}, \quad \text{resp. } n_s^* = 60 \cdot \frac{f}{p}, \tag{2.5}$$

with
n_s synchronous speed [R/s], [rps]; n_s^* [R/min], [rpm]
f frequency [1/s = Hz]
p number of pole pairs [–]

With regard to the rotor behavior, there are basically different modes of operation.

In case of a *synchronous motor*, the rotor (also called the pole wheel in this motor) rotates at the synchronous speed n_s. Its stator winding generates a rotating field and the rotor, which is fed with direct current via slip rings, must follow as a result of the electrodynamic force effect. (The rotor can also be designed as a permanent magnet.) Under load, the rotor and stator poles shift relative to each other by an angle proportional to the load.

In motor operation, the rotor lags behind by this load angle; in generator operation, the pole wheel runs ahead of the stator field. However, load shocks can cause the pole wheel to oscillate, and under certain circumstances this can disrupt the synchronism and cause the pole wheel to "fall out of step." Such synchronous motors have only been used in a very few systems in stage technology, but did not prevail.

In case of an *asynchronous motor*, the magnetic field of the rotor must first be generated by induction. Therefore, the term *induction motor* is also used. Electrodynamic force can then only be generated by intersecting the field lines in a relative movement between the stator rotating field and the rotor, i. e., the speed of the rotor during motor operation must be smaller by a *slip σ* than the synchronous speed n_s, where

$$n = n_s \cdot (1 - \sigma),$$
$$\sigma = \frac{n_s - n}{n_s} \, [-] = \frac{n_s - n}{n_s} \cdot 100 \, [\%]. \tag{2.6}$$

The frequency of the rotor current is

$$f_R = f_S \cdot \sigma, \tag{2.7}$$

with
n_s synchronous speed (speed of the rotating field) [rps], [rpm]
n speed of the asynchronous motor [rps], [rpm]
σ slip [–]
f_R frequency in the rotor winding [1/s]
f_S frequency in the stator winding [1/s]

If the motor is driven by the load, i. e., being a generator, an oversynchronous speed with negative slip will occur.

Figure 2.5 shows the speed–torque diagram of a three-phase motor.

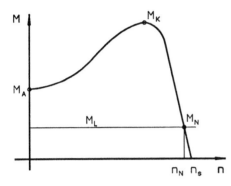

Figure 2.5: Characteristic curve of an asynchronous motor in the M/n diagram: M_A: starting torque, M_N: rated torque, M_K: break down torque, M_L: load torque (in the diagram $M_L = M_N$), n_N: nominal speed.

For the torque delivered by a three-phase motor, the following proportionality applies:

$$M \approx I_R \cdot \Phi. \tag{2.8}$$

For Φ = const., it follows that

$$M \approx I_R. \tag{2.9}$$

In terms of design and function, two types of three-phase asynchronous motors are distinguished:

In the *slip ring motor*, the winding ends in the rotor are guided to the outside via slip rings, so that the rotor can be influenced, for example, by connecting resistors in series.

In *squirrel-cage motors*, the ends of the windings in the rotor are short-circuited and not led to the outside and access to the rotor winding from the outside is not possible. This type of motor is particularly robust and, in principle, maintenance-free.

When a three-phase motor is switched on directly, an inrush current occurs which is about 6 to 8 times the rated current. This can lead to an excessive demand of the main. In the case of slip-ring motors, the motor characteristics can be changed by briefly connecting resistors in the rotor in such a way that the inrush current is kept much lower. In addition, the torque behavior of the motor can be adapted to the requirements. Switching on a motor with series resistors and gradually switching off these resistors is shown in Figure 2.6.

Therefore, only those possibilities of speed adjustment are possible for squirrel-cage motors where only the stator is influenced (frequency, voltage, number of pole pairs). Therefore, it is not possible to start a squirrel-cage motor with resistors. However, one

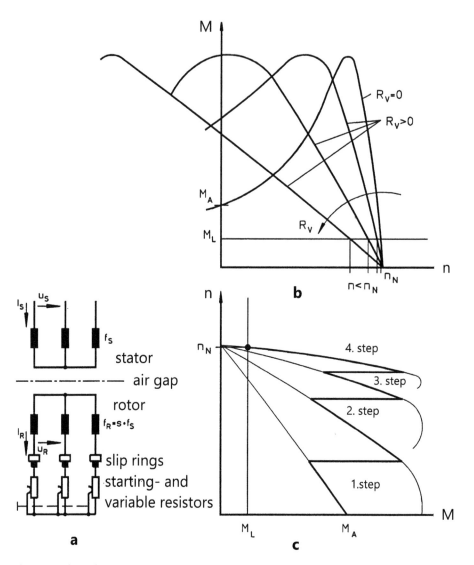

Figure 2.6: Three-phase slip-ring motor with resistors R_V in the rotor circuit: (a) circuit of the windings, (b) speed behavior shown in the speed–torque diagram M versus n, (c) start-up of a three-phase motor with starting resistors shown in the speed–torque diagram n versus M.

can help oneself by first connecting the stator winding of the motor to the mains in *star connection* and then switching over to *delta connection*. By this different chaining of the three three-phase windings according to Figure 2.7(a, b) and their connection to the three-phase mains, the inrush current is reduced to one third. However, when starting in delta connection, the motor then also delivers only one-third of the torque possible with delta connection. *Star-delta starting* can therefore only be implemented when starting with reduced load or in no-load operation. This is possible, for example, for starting an

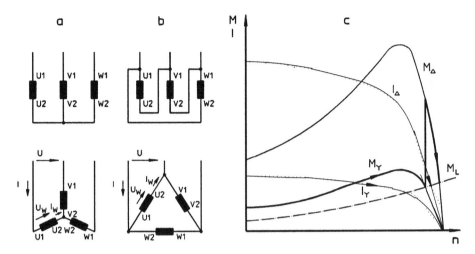

Figure 2.7: Squirrel-cage motor – star and delta connection: (a) star connection of the windings, (b) delta connection of the windings, (c) speed–torque diagram with starting characteristic.

electric motor for operating a hydraulic pump, if the start can take place with pressure-less circulation of the hydraulic medium. The characteristic curves of a squirrel-cage motor with star-delta starting circuit can be seen in Figure 2.7(c).

From Eqs. (2.5) and (2.6) again, the possibilities of speed adjustment of asynchrony motors can be read:

– Adjusting the speed n_s and thus also the asynchronous speed n by *changing the number of pole pairs p*. Different rated speeds can then be achieved with *pole-changing motors*. According to Eq. (2.5), for example, switching from 2 to 4 pole pairs (4 to 8 poles) and feeding from the 50 Hz three-phase supply results in the speeds n_s = $(60 \cdot 50)/2 = 1500$ rpm and $(60 \cdot 50)/4 = 750$ rpm.

For elevators in houses, the pole-changing method is very often used to achieve sufficient stopping accuracy when stopping at the selected floor by a short distance at creep speed.

– The speed of an asynchronous motor can also be reduced by increasing the slip. *Slip control* is possible by changing the rotor current or changing the stator voltage. For example, the field in the stator is weakened by applying a smaller voltage through a phase controlled modulator so that the motor can only deliver the required torque at a reduced speed (Figure 2.8). In the past, *phase controlled modulator systems* were used in stage technology, especially for drives in the lower stage, but only a relatively small control ratio of about 1:50 could be achieved.

A similar effect can be achieved by reducing the voltage in the rotor, also by phase angle control or with series resistors. However, it is only possible to influence the current flow in the rotor if the winding ends of the rotor are led to the outside via slip rings. This is the case with the slip ring motor. This type of motor therefore

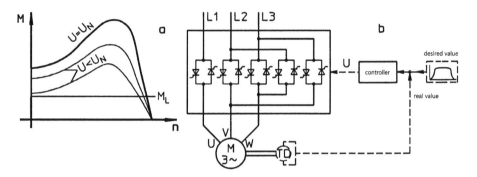

Figure 2.8: Speed adjustment via slip variation when adjusting the stator voltage by phase angle control: (a) speed–torque diagram, (b) schematic diagram with three-phase AC converter.

offers this very simple possibility of influencing slip. Preswitching resistors bring losses and load dependence, as already mentioned for rotor preswitching resistors in the DC motor, and are therefore mainly used only for starting (cranking) larger motors.

– Adjusting the speed by *changing the supply frequency f*. In the past, rotating frequency converters (three-phase motor with three-phase generator) were used. A slow speed could thus be achieved almost without loss by providing, for example, a 6 Hz supply in addition to the usual 50 Hz supply. This allowed operation at a "creep speed" of about 1/10 of the rated speed. In modern *frequency converter control systems*, the three-phase current taken from the mains is first rectified and converted back into three-phase current from a DC link using an inverter. The frequency generated in the inverter can be adjusted. In this way, a stepless and almost lossless speed adjustment of the asynchronous motor is possible.

Nowadays, such frequency inverter controls are state of the art and are used in stage technology. Figure 2.4(b) shows the principle circuit diagram of a *frequency converter* (with an intermediate direct current link). It should be noted that when the supply frequency of the motor is changed, the supply voltage must also be changed so that the motor can deliver the same torque at a lower speed.

Synchronous operation of several electric drives

Stage drives often require two or more drives to be synchronized as precisely as possible. For example, several podiums or several bar or point hoists must be moved at the same speed:

– Three-phase synchronous motors rotate exactly at the synchronous speed. Therefore, when using three-phase synchronous motors, synchronization of several drives is possible without any special control effort. However, a motor can "fall out of step" if overloaded (see above). In fact, synchronous motors were only used in few stage equipment in the past.

- Asynchronous motors can also be linked together in such a way that they run synchronously, e. g. by specifying the rotation via a so-called "wave guide machine". This is like switching an electric shaft, instead of mechanical coupling of several motors with a mechanical shafts. In the past, this method was often used in general materials handling technology, but rarely in stage technology.
- In synchronization control systems based on *the master-slave principle*, the adjustable actual speed of a DC or AC drive is specified as the setpoint for one or more other drives. In this way, synchronous control can be achieved with the integration of corresponding actual data acquisition with displacement or speed sensors.
- However, the *same setpoints* can also be specified for all drives to be synchronized – if they are all to be moved at the same speed – so that each drive must follow this setpoint specification. This is the synchronous connection with feedback control of several stage drives generally used today.

Apart from special cases, only the latter type of synchronization is used in stage technology today!

2.2.2 Servo motor technology

Servo motors are special electric motors that allow control of the angular position of their motor shaft, as well as the speed of rotation and acceleration. They consist of an electric motor, which is additionally equipped with a sensor for determination of specific features.

In Latin, "servus" means servant. The term servo motor is therefore intended to express that it is a particularly "good servant" motor that does exactly what is required of it.

Modern servo motor technology works with different types of motors. The most important types are mentioned below:

- *Brushless DC motor.*
  This is a synchronous motor with a permanently excited rotor field; the supply to the stator winding phases is dependent on the rotor position, which requires exact position detection of the rotor. The designation "direct current motor" is misleading, since a revolving rotating field is generated as in the three-phase motor, but the commutation for generating a rotating stator field is carried out electronically ("electronic commutation" – EC motors). Such motors were used, for example, in the upper machinery of the Munich State Opera.
- *Three-phase synchronous servo motor.*
  This is a synchronous motor with a permanently excited rotor field; the stator is supplied with sinusoidal voltage in so-called sinusoidal operation. The position detection of the rotor is somewhat more complex than with block operation.

– *Three-phase asynchronous servomotor with field-oriented control as inverter-fed drives.*
The special feature of this principle is that the stator is excited in a field-oriented manner. The magnetizing current and the torque-forming current are controlled separately. This results in conditions similar to those of the externally excited DC motor, where the rotor current, which is decisive for torque formation, and the excitation current can be adjusted independently of each other. In the asynchronous servomotor, the magnetizing current, on the one hand, and the rotor current required for torque generation, on the other hand, are changed in a similar way. However, this rotor current can only be generated by induction, i. e., by a relative movement of the field vector with respect to the rotating rotor. In this case, too, the position of the rotor must be precisely measured. With this control method, reactive and active current can be adjusted and limited completely independently of each other at any time. This provides a system equivalent to the DC motor, but which is characterized by freedom from maintenance and robustness and allows a much larger control ratio than with the DC motor.

This *field-oriented control* – also called *vector control* – allows the motor to be operated at full torque down to zero speed (standstill), thus giving the desired servo characteristics. The control itself is very complex and is implemented by a microprocessor built into the inverter. In addition to the speed, the torque can be controlled and a position control of the rotor can be performed. The novel control method is also characterized in particular by high control dynamics with settling times in the millisecond range, stepless control with high control quality and stiffness, and a control range of 1:1000 and more. The motor could even be controlled down to zero speed (standstill), but this is not required in stage applications. However, with this kind of control higher control ratios are possible than can be achieved with DC or hydro drives. In stage technology today, control ratios of 1:1000 are common.

The speed adjustment of a servomotor then takes place as follows: The three-phase current provided by the mains is converted into direct current of constant voltage via a converter circuit (DC link). From this DC voltage, an inverter generates a three-phase current of variable frequency to supply the motor. In contrast to the classic frequency converter control, the field-oriented control not only controls the frequency and current and voltage, but also their phase shift to each other depending on the rotational position of the rotor.

2.2.3 Linear motor technology

It can be assumed that as development in electric linear motor technology progresses, applications in the field of stage technology could also take place.

In principle, a distinction can be made between the following two solutions:

- With the brushless, permanently excited linear motor, the entire travel path must be equipped with expensive permanent magnets. Stage applications involve limited travel paths, so this criterion does not have the same importance as in traffic applications.
- In asynchronously operating linear motors, the secondary consists only of soft iron or short-circuit bars made of copper.

Electric linear direct drives can be used to create a backlash-free, low-inertia and mechanically stiff power transmission with high control dynamics and high load stiffness. The high travel speeds that can be achieved with these drives are of little relevance in stage technology.

2.3 Hydraulic drives

Hydraulic drives can work according to the following two principles:
- In *hydrodynamic drives*, the kinetic energy of a flowing fluid is used to perform mechanical work. In centrifugal pumps, for example, electric energy is converted into flow energy of the hydraulic fluid by driving the pump with an electric motor and converted back into mechanical work in turbines. A compact unit consisting of pump and turbine is the fluid turbo coupling. If a guide wheel is also integrated, it is referred to as a hydraulic converter. Such drives are not used in stage technology.
- In *hydrostatic drives*, the hydrostatic pressure of a fluid is used to perform work by applying pressurized fluid to hydraulic cylinders as linear drives or to hydraulic motors as rotary drives. The hydraulic fluid is brought to working pressure by positive displacement pumps driven by electric motors and supplies hydraulic cylinders or hydraulic motors directly or from a hydraulic accumulator station. The maximum operating pressure for stage equipment is rarely much higher than 160 bar. In industrial application, higher pressures of 300 bar and more are also used.

2.3.1 Components and their circuit symbols

Hydraulic power unit – pumps – motors – accumulators

In a hydraulic power pack, electrical energy (or energy from a combustion engine) is converted into pressure energy, i. e., the working capacity of a fluid.

A hydraulic power pack consists of *drive motor*(s) and *pump*(s), possibly also *hydraulic accumulator*(s), as well as the *switching elements* (valves) required for control and a *tank* for the hydraulic fluid. Figure 2.9 shows a hydraulic power pack in compact design for driving a scissor lift table.

There are several types of hydraulic pumps, their working principle shown in Figures 2.10 and 2.11. Depending on the of operation, these are gear, vane and screw pumps,

Figure 2.9: Hydraulic power pack in compact design. Photo: Bosch Rexroth.

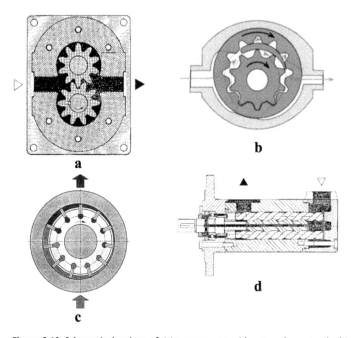

Figure 2.10: Schematic drawings of: (a) a gear pump with external gear teeth, (b) a gear pump with internal gear teeth, (c) a vane pump, (d) a screw pump. Picture credits: Bosch Rexroth.

axial piston and radial piston pumps. Cellular, axial and radial piston pumps can be designed not only as *fixed displacement pumps* (constant displacement at constant drive speed), but also as *variable displacement pumps* (variable displacement despite constant drive speed). The adjustment consists of a mechanical change of position of components

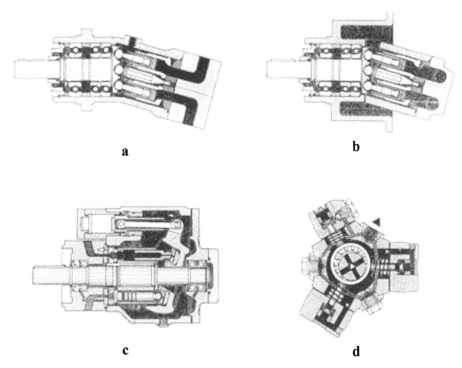

a

b

c

d

Figure 2.11: Schematic drawings of piston pumps: (a) axial piston fixed displacement pump with inclined axis, (b) axial piston variable displacement pump with inclined axis, (c) axial piston variable displacement pump with inclined disk (swash plate), (d) radial piston pump (fixed displacement). Picture credits: Bosch Rexroth.

in the pump (see Section 2.3.2) and can be carried out manually (e. g., via a handwheel), electrically or hydraulically by means of an adjustment cylinder. The adjustment mechanism can also be integrated into a control loop of a capacity, pressure or flow controller.

Pressurized liquid can also be taken from hydraulic *accumulators* (Figure 2.12).

A weight accumulator (weight load acts on the piston) is shown as *variant a*. Its advantage would be that the pressure in the accumulator is determined only by the constant weight, irrespective of the filling quantity in the accumulator. However, the required mass of the weight would be much too large to be able to realize required pressures and fluid quantities with this design.

In *variant b*, the pressure is determined by the pressure force of the spring. However, the spring force depends on the filling quantity, so that large pressures would be present when the accumulator is full and small pressures when it is almost empty. The pressure differences would be so great that this design is also not applicable.

In *variant c*, the mechanical spring is replaced by a gas spring. In this case, too, the pressure in the liquid depends on the filling volume. With this concept, however, it is possible to accommodate a very large gas volume in connected external gas cylinders of about 4 to 7 times the volume of the storage capacity for the pressurized liquid, so

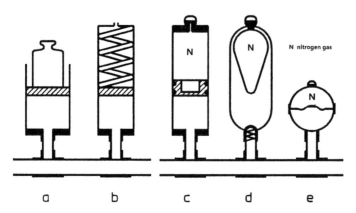

a   b   c   d   e

Figure 2.12: Types of hydraulic accumulators and circuit symbols: (a) weight accumulator, (b) spring accumulator, (c) gas accumulator as piston accumulator, (d) gas accumulator as bladder accumulator, (e) gas accumulator as diaphragm accumulator. Picture credits: Bosch Rexroth.

that the pressure differences between full and empty accumulators can be kept very low. For this reason, only such piston accumulators are used in the central hydraulic pressure stations of stage equipment (see Section 3.8.3).

The bladder and diaphragm accumulators shown as *variants d and e* are often used as small accumulators, but their size next to the piston accumulator in the illustration does not correspond to reality at all. While piston accumulators are often several meters high, bubble and membrane accumulators are less than one meter in size.

Typical for stage operations are operating sequences in which a large quantity of pressurized fluid is required during short intervals, e. g., to simultaneously lift all hydrostatically driven podiums or to drive several hoists. Afterwards, there is no or only little demand for pressure fluid for a longer period of time. If one wanted to provide this short-term large volume demand from direct pump operation, it would be necessary to install many large pumps to achieve a sufficiently large delivery volume. It is therefore much more economical to cover this peak demand from hydraulic accumulators and to design the pumps in terms of their flow rate so that the accumulators can be filled again in an operationally reasonable time.

This *combined pump/accumulator operation* has proved very successful in stage technology. Figure 2.13 shows a *central hydraulic accumulator pressure station* with piston accumulators and nitrogen gas containers (flasks). In order to keep the pressure drop as low as possible when liquid is removed from the storage tanks, nitrogen gas flasks are connected to a piston accumulator as gas accumulators with about 5 to 7 times the volume of the piston accumulator (see Section 3.8.3).

Working equipment
These are used to convert hydraulic energy into mechanical work.

Figure 2.13: Hydraulic accumulator facility at the Stuttgart State Theatre: (top) pump unit and tank, (bottom) piston accumulators and nitrogen cylinders. Photo: Bosch Rexroth.

Linear motion can be generated very easily with a *hydraulic cylinder*. Therefore, the designation *hydraulic linear motor* is also common. Various designs of hydraulic cylinders are shown in Figure 2.14. Figure 2.14(a) shows a *plunger cylinder* with end-position cushioning and hinged connecting plate, as used for lifting platforms, for example. Large cylinders that travel the entire stroke path are designed with end-position cushioning to

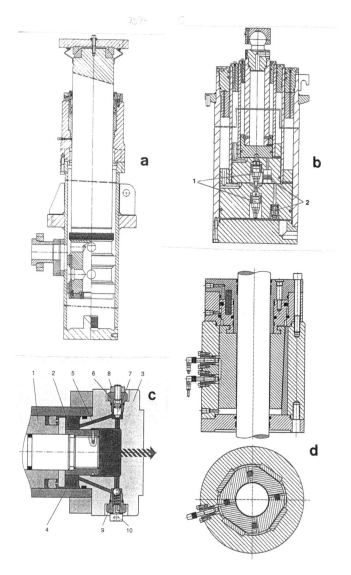

Figure 2.14: Hydraulic cylinder: (a) plunger cylinder for a lifting platform, (b) telescopic cylinder, (c) cylinder with end-position cushioning at the cylinder base, (d) cylinder with clamping device. Picture credits: Bosch Rexroth.

prevent the piston from hitting the end of the cylinder hard. There are several designs to achieve a throttled outflow of hydraulic fluid at the stroke end. In the variant shown in Figure 2.14(c), the fluid must escape from the piston chamber via adjustable throttle valves by moving a damping piston into the bore of the cylinder base.

Figure 2.14(b) shows a cylinder with three extendable elements working synchronous telescopic cylinder with hydraulic retraction, in which valves are used to adjust the effective piston forces so that the individual telescopic tubes each retract or extend simultaneously, so that the operating speed remains constant at a given volume flow throughout the entire stroke.

Figure 2.14(d) shows a cylinder with a *clamping device*, which can be used on lifting platforms to fix the position of the platform.

Rotating motion can be generated with a *hydraulic motor*. Just as an electric motor can change from motor to generator operation and motor and generator are, in principle, the same machines, so hydraulic pump and hydraulic motor are analogous units too; Figures 2.10 and 2.11 represent both pumps and motors in terms of their basic construction (with the exception of the screw pump). For stage drives, slow-speed radial piston motors are used particularly frequently. Often it is important that even at minimum speed they show a smooth running without jerking and stuttering movements.

Standardized symbols for hydraulic pumps and motors, accumulators and hydraulic cylinders are shown in Figure 2.15.

Control elements (valves)

The circuit symbols of the valves described below are summarized in Figures 2.15, 2.17 and 2.18.

- Valves can have a switching or regulating effect, i. e., defined switching positions are specified (*black/white technology*) or continuous changes of a control variable are possible (*continuous valve technology* or *proportional valve technology*).
- Valves can also be designed as *poppet valves* = *spool valves* = slide valves. Poppet valves close absolutely tightly (leakage-free) and have short response times due to small travel ranges. Spool valves are equipped with fine control edges with higher control quality. Figure 2.16 shows a schematic drawing of a pressure relief valve in both designs.
- For small nominal diameters or small flow rates, *directly controlled valves* are used; for large nominal diameters for large flow rates, *pilot-controlled valves* with an intermediate hydraulic amplifier are used. Figure 2.18 shows, for example, a pilot operated directional control valve: An electromagnetically moved directional controlled valve of small nominal size switches a hydraulic control circuit, the control liquid hydraulically actuates a directional control valve of large nominal size.

The most important types of valves are listed below with brief explanations, classified according to their function:

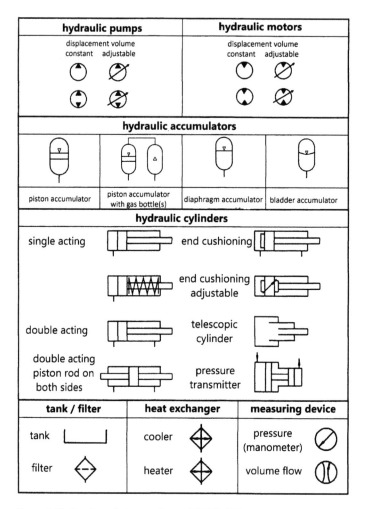

hydraulic pumps		hydraulic motors	
displacement volume		displacement volume	
constant	adjustable	constant	adjustable

hydraulic accumulators

piston accumulator	piston accumulator with gas bottle(s)	diaphragm accumulator	bladder accumulator

hydraulic cylinders

single acting		end cushioning	
double acting		end cushioning adjustable	
		telescopic cylinder	
double acting piston rod on both sides		pressure transmitter	

tank / filter		heat exchanger		measuring device	
tank		cooler		pressure (manometer)	
filter		heater		volume flow	

Figure 2.15: Circuit symbols according to DIN ISO 1219.

Locking valves (nonreturn valves)

The function of *locking valves* is to block a flow in one direction and allow free flow in the opposite direction. In the case of *hydraulically pilot-operated check valves*, the shut-off position can be reversed by hydraulically opening the poppet, thus releasing the flow in the originally shut-off direction. This is a very simple way of preventing a load from sinking (falling) in the event of a pipe rupture or burst hydraulic hose.

For this purpose, a *hydraulically pilot-operated nonreturn valve* is mounted on the discharge opening of the cylinder. Unblocking (release of the liquid outflow) takes place with the aid of the pressure in the discharge pipe, which is generated, for example, by a pressure relief valve set to a low pressure. In the event of a pipe rupture, this pressure drops and the check valve closes, so that further downward movement is no longer

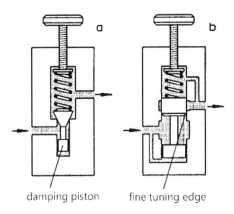

damping piston   fine tuning edge

Figure 2.16: Schematic drawing of a pressure relief valve designed as: (a) a poppet valve (seat valve), (b) a gate valve (slide valve).

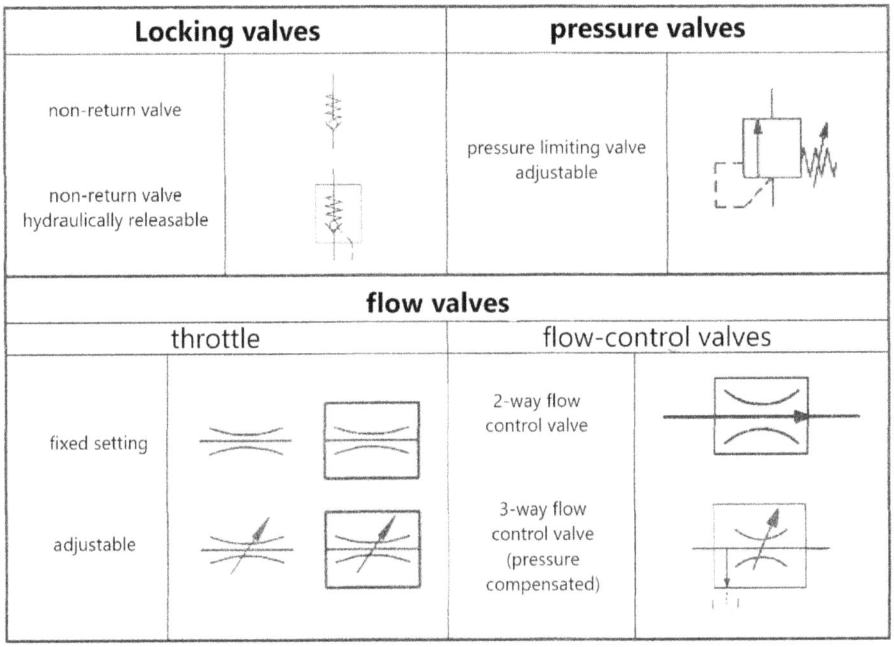

Locking valves		pressure valves	
non-return valve			
non-return valve hydraulically releasable		pressure limiting valve adjustable	
flow valves			
throttle		flow-control valves	
fixed setting		2-way flow control valve	
adjustable		3-way flow control valve (pressure compensated)	

Figure 2.17: Circuit symbols according to DIN ISO 1219.

possible. In stage technology, analogous to the requirement for the installation of two redundant brakes on hoists, two redundant pilot-operated nonreturn valves are also required in this case (see relevant standards or regulations).

Pressure valves

A distinction is made between switching pressure valves, which trigger an opening or closing at a certain control pressure, and regulating pressure valves, which influence

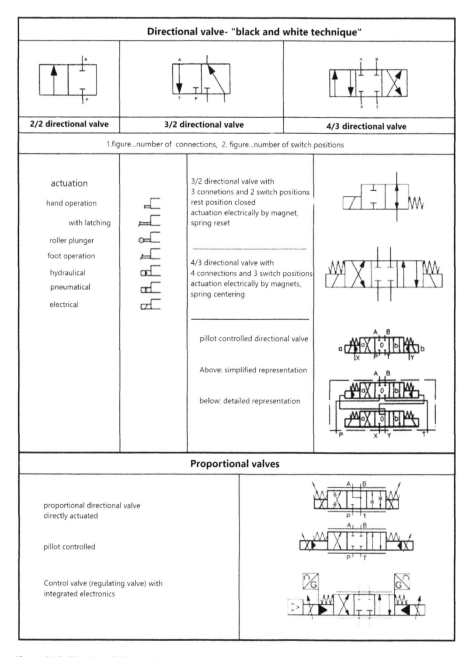

Figure 2.18: Circuit symbols according to DIN ISO 1219.

the system pressure of a plant by continuously changing a throttle cross-section. A *pressure relief valve* limits the pressure in a system; a *pressure reducing valve* reduces the pressure for certain consumers in a part of the piping system.

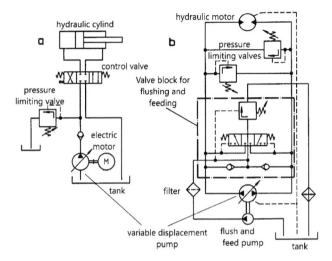

Figure 2.19: Schematic diagram for speed adjustment: (a) of a double-acting cylinder with a variable displacement pump in an open circuit, (b) of a hydraulic motor with a variable displacement pump in a closed circuit, valve block for flushing and supply.

In Figure 2.16, two types of pressure relief valves – poppet and spool – were shown. Figures 2.19 and 2.20 show the use of pressure relief valves to limit the working pressure in simple circuits.

Electrical systems are always protected by overcurrent protection devices that limit the maximum current (fuses, circuit breakers). Similarly, hydraulic systems are protected by overpressure protection devices that limit the maximum pressure (pressure relief valves).

Flow valves

They reduce the volume flow by means of a cross-sectional restriction (throttle, orifice). By changing the throttle cross-section, the operating speed of hydraulic cylinders or motors can be adjusted.

A distinction is made between throttle valves and flow control valves. The actual volume flow through throttle valves depends on the pressure difference at the throttle and the density, respectively viscosity (see Section 3.8.1). Thus, a throttle in a hydraulic drive acts analogously to an electric resistor in an electric actuator and in both cases this results in a dependence of the lifting speed (flow rate) on the size of the load, with the effect that a large load is lifted more slowly than a small load.

In *flow control valves*, the flow rate is made independent of the load pressure by means of a control mechanism inside the valve (installation of a pressure compensator) by keeping the pressure difference at the throttle determining the flow rate constant, thus there is no load dependence.

Depending on the mode of operation, a distinction is made between 2- and 3-way flow control valves. With the 2-way flow control valve, the excess quantity delivered by

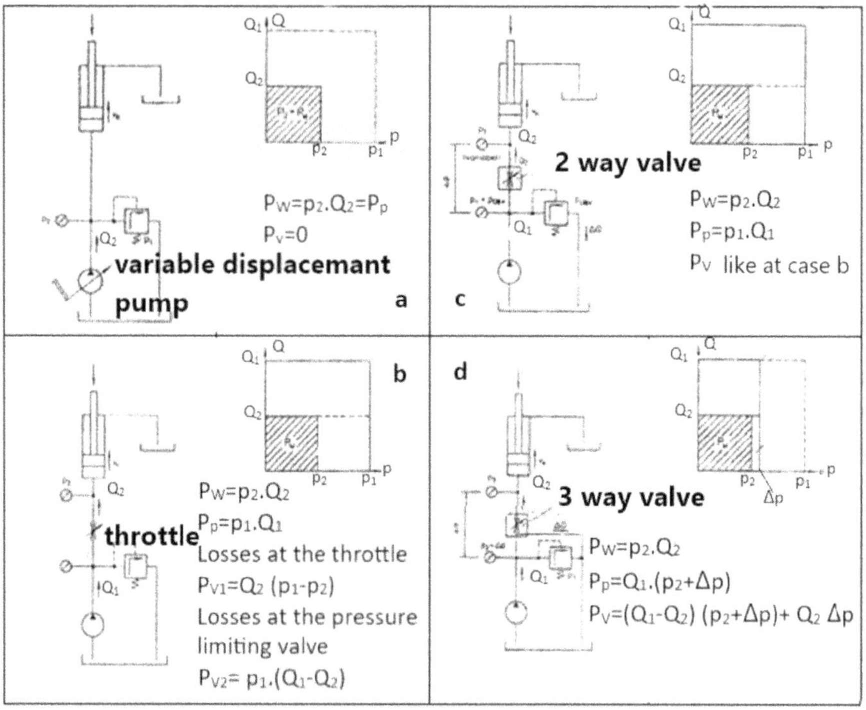
Pw=p2.Q2=Pp
Pv=0
variable displacemant pump
a
2 way valve
Pw=p2.Q2
Pp=p1.Q1
Pv like at case b
c
b
d
throttle
Pw=p2.Q2
Pp=p1.Q1
Losses at the throttle
Pv1=Q2 (p1-p2)
Losses at the pressure limiting valve
Pv2= p1.(Q1-Q2)
3 way valve
Pw=p2.Q2
Pp=Q1.(p2+Δp)
Pv=(Q1-Q2) (p2+Δp)+ Q2 Δp

Figure 2.20: Adjustment of the lifting speed with: (a) variable displacement pump, (b) throttle, (c) 2-way control valve, (d) 3-way control valve. Circuit diagram and power diagram Q over p, where P_W is active power (useful power), P_P is pump power and P_V is power loss.

the fixed displacement pump must be discharged into the tank via the pressure relief valve and this can lead to high power losses during partial load operation. In the case of the 3-way flow control valve, the excess flow is routed to the tank at a pressure that is only slightly higher than the actual working pressure. Therefore, the losses at the throttle and pressure relief valve are the same for the 2-way flow control valve, but smaller for the 3-way flow control valve. The differences in the power losses can be seen in the *Q*/*p diagrams* in Figure 2.20 (see Eq. (3.77)).

Directional control valves

They have the task of blocking or releasing various hydraulic lines against each other. Switching symbols can be found in Figure 2.18. Figure 2.19(a) shows the functional diagram of a 4/3-way valve as a switching element in a simple circuit for actuating a double-acting hydraulic cylinder. In the case of directional control valves of smaller nominal size, an electromagnet actuates the valve directly; in the case of larger nominal sizes, pilot-operated directional control valves are used in which, as already explained, the solenoid only actuates a small directional control valve in an upstream control circuit and the actual large valve is switched hydraulically in the operating circuit.

Proportional, control and servo valves

Unlike valves, which in *black-and-white technology* only assume two or three switching positions, proportional, control and servo valves are *continuously operating valves* that change a hydraulic parameter proportionally to an electrical input signal. Continuously operating valves therefore allow controlled switching transitions, avoid pressure peaks and therefore permit fast motion sequences.

Proportional valves have a controllable DC solenoid operating in oil (solenoids of normal directional control valves operate in the dry), either stroke-controlled with analog stroke-current behavior and a stroke of approx. 3–5 mm, or force-controlled with defined force–current behavior and a stroke of approx. 1.5 mm. However, proportional valves have a relatively low cut-off frequency (10–70 Hz), large switching overlap and large hysteresis (approx. 1 %). They are therefore mainly used in open control circuits.

Servo valves are very complex controllable converters. The cut-off frequency is approx. 100–200 Hz, there is zero overlap and the hysteresis is approx. 0.1 %. They are therefore ideal for use in feedback control systems.

Since servo valves are very sensitive and expensive, improved proportional valves, also often referred to as *control valves*, have been developed from the original proportional valves. They come close to the characteristics of servo valves. The cut-off frequency is approx. 50–150 Hz, there is zero overlap, and the hysteresis is approx. 0.1–0.2 %. They can therefore also be used in feedback control loops.

In continuous valve technology, also a distinction is made between directional, flow and pressure valves.

The terms used here for proportional valves are not always used with the same meaning in the literature and by manufacturers. The term proportional valve technology is often used as a synonym for continuous valve technology without referring to a specific valve type.

Other components of a hydraulic system

Figure 2.15 also shows symbols for some other important components of a hydraulic system. For example, every hydraulic system must be protected against contamination by *filters*; it must be possible to return the hydraulic fluid to a *tank*. If the hydraulic fluid is heated up too much as a result of losses, *coolers* must be used.

2.3.2 Possibilities to change the working speed

The stroke speed of a hydraulic piston is

$$v_K = \frac{Q}{A_K} \quad \text{(see Eq. (3.81))}$$

and the rotational speed of a hydraulic motor

$$n_M = \frac{Q}{V_M} \quad \text{(see Eq. (3.84))},$$

with

Q flow rate; volumetric flow rate to the implement (cylinder or hydromotor) [m³/s]

A_K piston area of the hydrocylinder [m²]

V_M hydromotor displacement, i. e., the volume of liquid taken up or discharged by the hydromotor during one revolution [m³/U] = [m³]

v_K speed of movement of the piston [m/s]

n_M rotation speed of the hydraulic motor [rps]

The speed of the piston of a hydrocylinder v_K can be varied by changing the volume flow Q supplied to the cylinder. The piston area A_K is given by the geometry of the cylinder.

The magnitude of the speed n_M of a hydraulic motor can also be adjusted by changing the volumetric flow Q supplied to the hydraulic motor, but also by changing the absorption volume V_M of the hydraulic motor if a *variable displacement motor* is used. However, when V_M is reduced, the torque of the hydromotor is also reduced (see Eq. (3.85)).

Changing the working speed by means of variable displacement pumps and variable displacement motors

The change in speed can be achieved by changing the flow rate supplied to the device (cylinder, hydraulic motor) per unit of time. In the case of hydrostatic drives, this is possible almost without loss by changing the flow rate of the pump by changing its delivery volume per revolution (this corresponds to the displacement in the case of the motor). Analogous to the above equation for the motor, the following applies to the pump:

$$Q = V_P \cdot n_P \quad \text{(see Eq. (3.83))},$$

with

V_P absorption volume (swallowing capacity) of the pump per revolution [m³/U]

n_P pump speed [rpm]

Varying the speed n_P of the pump is generally uneconomical because all the expenses for adjusting the speed of an electric motor would then be required. Variable displacement pumps offer the possibility of varying the flow rate Q by changing the displacement V_P. The mechanical adjustment consists, for example, in swiveling the swash axis (Figure 2.11(b)) or the swash plate (Figure 2.11(c)) of an axial piston pump, which varies the piston stroke of the hydraulic cylinders installed in the pump.

Therefore, if $Q_{\text{nenn}} = Q_{\text{max}} = V_{P_{\text{max}}} \cdot n_P$, the motor speed n_M can be reduced by reducing the displacement V_P at the pump (*primary adjustment*); for $V_P = 0$, $n_M = 0$. When the swash plate or swash axis is swung through, the direction of the flow is reversed. The hydraulic motor then rotates in the opposite direction.

If the variable displacement motor $V_{M_{nenn}} = V_{M_{max}}$, the motor speed can be increased by reducing the displacement V_M of the hydraulic motor (*secondary adjustment*). However, it should be noted that this reduces the motor torque M_M, as can be seen from Eq. (3.85). In a system already operating at maximum operating pressure, power can also be increased in the primary adjustment range as speed increases, while in the secondary adjustment range power can no longer be increased as speed increases and therefore torque is reduced. With the exception of special cases, variable-speed motors are not used in stage technology.

According to Eq. (3.77), the power to be delivered by the pump is $P = p \cdot Q_P$ and is thus adapted to the actual demand, because the pressure p is determined by the required force effect and the flow rate Q by the desired operating speed.

Figure 2.19(a) shows as an example the circuit diagram of the control of a hydro-cylinder in open circuit and Figure 2.19(b) that of the control of a hydro motor in closed circuit.

There is an analogy between the possibility of an almost lossless speed adjustment of a hydraulic motor and that of a DC motor, if one considers, for example, the speed adjustment of a DC motor described as "Ward-Leonard control" according to Section 2.2.1 (Figure 2.3(a)).

The hydraulic pump corresponds to the DC generator, the hydraulic motor to the DC motor. A reduction of the displacement of the hydraulic pump is equivalent to the field weakening at the DC generator to reduce the supply voltage of the electric motor. A reduction in the displacement of the hydromotor is equivalent to the field weakening at the DC motor. Just as a hydraulic motor can only deliver a reduced torque when its absorption volume is reduced, an electric motor can only deliver a reduced torque in the field weakening range (see Eq. (2.2)). Thus, analogy exists between the pressure p and the rotor current I, the displacement V and the magnetic flux Φ as well as between the volume current Q and the rotor voltage U.

Speed regulation with variable displacement pumps or variable displacement motors is generally not applicable for stage drives, since the many drives of a stage system are fed by a pump–storage system. This principle of speed adjustment is frequently used for individual drives in general materials handling technology, e. g., also for industrial trucks, but only in special cases in stage operation. This type of speed adjustment was therefore only mentioned for the sake of completeness.

Changing the speed by throttling the amount of liquid supplied to the consumer
The flow rate Q_P supplied by a fixed displacement pump can also be reduced to Q_D by a *throttle valve*. The differential flow $Q_{PRV} = Q_P - Q_D = Q_1 - Q_2$ is returned to the tank via the *pressure relief valve* in the hydraulic circuit shown in Figure 2.20(b). If the pressure relief valve is set to p_{DBV} (determined by the required pressure for the largest load),

the pump will bring the total flow rate Q_P to the pressure p_{DBV}, thereby expending the power

$$P_P = p_1 \cdot Q_1 = p_{DBV} \cdot Q_1,$$

but only P_W will be utilized as mechanical power,

$$P_W = p_2 \cdot Q_2,$$

and P_V is lost as dissipation,

$$P_V = P_{V1} + P_{V2} = Q_2 \cdot (p_1 - p_2) + p_1 \cdot (Q_1 - Q_2),$$

where
$Q_P = Q_1$ pump flow rate [m^3/s]
$Q_D = Q_2$ flow rate through the throttle [m^3/s]
P_P power to be provided by the pump [kW]
P_W active power = useful power for the consumer [kW]
P_V power loss (P_{V1} at the throttle, P_{V2} at the pressure relief valve) [kW]
$P_{PRV} = p_1$ set pressure at the pressure relief valve [N/m^2]
p_2 working pressure at the consumer [N/m^2]

The pressure p downstream of the throttle is determined by the actual force or torque demand, so it will vary as required, while p_{DBV} is specified as a constant value to protect the system from overload and must be selected slightly higher than the maximum required working pressure. However, the flow rate of a throttle depends on the pressure difference according to Eqs. (3.75) and (3.76). Therefore, a load-dependent working pressure p causes different flow rates and thus load-dependent operating speeds of the drive.

This load dependence can be avoided by using a flow control valve, by keeping the pressure difference at the throttle constant with the help of a pressure balance. When a 2-way flow control valve is used, the power loss P_V is maintained (Figure 2.20(c)); when a 3-way flow control valve is used, it can be greatly reduced, as already mentioned in Section 2.3.1 (Figure 2.20(d)).

There is also an analogy between the effect of a throttle in a hydraulic system and the effect of an ohmic resistor in an electrical system.

Changing the operating speed of hydraulic pressurized storage systems
As described in Section 2.3.1, hydraulic storage systems are used in large-scale hydraulic stage equipment. Hydraulic accumulators are repeatedly filled by automatically controlled switching on and off of fixed displacement pumps and/or the operation of pressure-controlled variable displacement pumps. Therefore, the hydraulic cylinder or motor is not directly supplied with a flow rate determined by the pump, but a quantity

of liquid corresponding to the demand is withdrawn from the hydraulic storage system. To actuate a drive, the quantity of liquid supplied to this consumer is precisely metered using modern continuous valve technology. Depending on whether open- or closed-loop control is required, suitable proportional, control or servo valves must be used for this purpose (see Section 2.3.1).

The pressure of the accumulator system prevails upstream of the control valve (in the case of piston accumulators with gas spring, depending on the filling state of the accumulator), and the pressure determined by the load prevails downstream of the control valve. The control causes the control valve position (the throttle cross-section or an equivalent measurement) to be selected on the basis of the comparison of the setpoint and actual velocities so that the necessary quantity of liquid is supplied to the drive in accordance with the velocity requirement.

When lowering a load, the energy originally expended as lifting work, reduced by efficiency components (cf. Section 3.5), must be expended again as braking work by throttling with the control valve. This means that, for example, relatively large amounts of heat are generated by throttling in large podium systems, which heat up the hydraulic fluid to an excessively high temperature when it is returned to the tank. In this case, a cooler, e.g., as a heat exchanger with water cooling, is required. Figure 2.21 shows a schematic diagram of the pressurized hydraulic storage system of the Vienna State Opera (the cooling circuit is also visible at the tank) and Figure 2.22 shows photos of this central hydraulic accumulator pressure center.

2.4 Hydrostatic drives compared to electric drives

After dealing with both drive variants, the question arises as to which of the two drive variants should be given preference. This question cannot be answered in general terms, since the development of electric and hydraulic drive technology is progressing rapidly and many arguments for and against a particular solution can lose their validity after only a short time. The range of technical products and their prices are constantly changing, so that for the same technical requirements, sometimes one solution and sometimes the other can be the more economical.

In historical terms, the following general statement can be made:

In the years of beginning conversions from manual to mechanical drives in the upper stage, hydrostatic drives were far superior to electric drives of classic design. They have proven themselves from a technical/economic point of view and can therefore be found in many stages.

In recent years, however, such revolutionary developments have taken place in the field of electric drive technology that it may be expedient to include both an electric and a hydraulic variant in the planning considerations in order to make the right technical and economic decision depending on the general conditions.

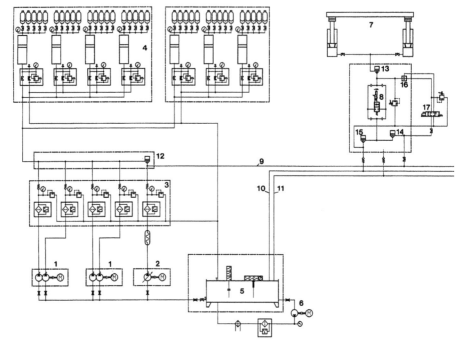

1   fix displacement pump, controlled depending on the filling level of the piston accumulators
2   pressure controlled variable displacement pump
3   filters and pressure protection
4   Piston accumulators and nitrogen cylinders, pressure protection
5   tank with liquid level interrupter
6   cooling circuit with water cooler
7   lifting platform with plunger cylinders
8   control valve to regulate the speed when hoisting or lowering

9   pressure pipeline
10 return pipe
11 leakage pipe
12 2-way directional control valve (seat valve)
13 switching valve to block the flow to and from the lifting cylinders
14 switching valve for hoisting operation
15 switching valve for lowering operation
16 unidirectional restrictor valve hydraulically unlockable
17 4/3 multiway valve as emergency control valve

Figure 2.21: Schematic diagram of the hydraulic accumulator facility in the Vienna State Opera House: Picture credits: Bosch Rexroth.

The following is an attempt, without claiming to be exhaustive, to present some criteria and arguments for decision-making:

–   While the electric motor converts electrical energy directly into mechanical work, a hydraulic circuit is interposed in hydrostatic drives. The electric motor drives a hydraulic pump, transfers the energy to a pressure fluid and performs mechanical work on hydraulic cylinders or hydraulic motors. This "detour" via a hydraulic system generally means a loss of efficiency and additional maintenance costs in addition to the additional technical effort required for the use of the equipment (including the cost of a hydraulic medium).

–   The energy requirement for lifting a load is in principle given by the size of the load and by the lifting distance (cf. Eq. (3.23)), the power requirement by the size

Figure 2.22: Central hydraulic pressure station of the Vienna State Opera: (right) pumping station with 2 internal gearing pumps (fixed displacement pumps) and one pressure-regulated axial piston pump (variable displacement pump); tank for hydraulic fluid with cooling system (pump, water cooler), (left) accumulator station consisting of 7 piston accumulators of 1000 l (each 7000 l) and $7 \times 5 = 35$ nitrogen tanks (each 900 l), a total of 31500 l. Photos: Waagner-Biró.

of the load and the desired lifting speed (cf. Eq. (3.26)). Due to unavoidable losses in the drive chain, the actual energy or power requirement is always greater than the power ultimately to be delivered for moving the lifting load and is taken into account in the calculation considering the efficiency (see Section 3.5). The interposition of a hydraulic system generally means a certain reduction of the efficiency, unless one compares, for example, a cylinder drive with a spindle or worm drive, in which particularly high frictional losses occur, especially in the case of self-locking drives (see Sections 4.3.2 and 4.4.1). The situation just described for a hoist is, of course, also present in an analogous form for a trolley or slewing gear.

– Very large forces can be generated with hydraulic cylinders and very large torques with slow-running hydraulic motors with a large absorption volume (swallowing capacity). With electric motors, large linear forces can only be generated via mechanical gear elements (spindle/nut, pinion/rack), and for larger torques and smaller speeds, transmission gears must be interposed, since electric motors are usually operated at high nominal speed. However, there are also low-speed high-torque electric motors, the use of which will become more common.

– Pressure energy can be stored in hydraulic accumulators. This means that the peak power requirement for drives that are used simultaneously for only a short time can be taken from such accumulators. The peak demand for electrical drive power can therefore be kept very low in hydraulic storage systems, since the drive power of the hydraulic pumps for filling the hydraulic accumulators can be far lower than the peak power required for a short time on individual drives of stage equipment.

- Whether this argument is of decisive importance will also depend, among other things, on how high the demand for electrical power for lighting, etc., is in comparison and under what financial framework conditions the billing of electrical energy consumption takes place, taking into account short-term peak demand.

- Electrical energy can also be stored in accumulators. Up to now, however, this has only been used in stage technology if a cable supply is to be avoided in certain operating states, e. g., for stage wagons.

In principle, however, it would also be perfectly conceivable to install electrically rechargeable battery control centers analogous to hydraulic pressurized storage centers. Just as hydraulic pumps with low power charge hydraulic accumulators, this can also be done with electrical energy accumulators. A positive aspect of battery storage systems also arises when a load is lowered (e. g., lowering the podiums). The braking energy in a hydraulic storage system is not used because the hydraulic fluid is directed into the tank through throttling and is lost as thermal energy, whereas in an electric storage system braking power could be used as battery charging power (or as drive power for another drive).

Such a technique could be particularly interesting when using frequency inverter controls for three-phase motors, since the DC link (see Section 2.2) can then be fed from the mains via a rectifier circuit on the one hand, but also from the battery system on the other.

Only time will tell whether such electrical storage systems, as are common in UPS systems ("uninterruptible power supply"), can also be used in a technically/economically sensible way in stage operations.

- Hydraulic equipment such as hydraulic cylinders and low-speed hydraulic motors operate almost silently, while electric motors in the usual nominal speed range generally emit more noise. The extent to which this criterion is relevant depends on local conditions and the spatial arrangement of the drives. In many cases, the winches of bar and point hoists are housed behind a wall in acoustically shielded areas and the ropes are guided through small openings in the wall.

Slow-running hydromotors operate without disturbing noise emission, but working with very low rotation velocity they show no smooth running but jerking and stuttering movements. Therefore, they do not allow a large control ratio between minimum and maximum speed.

If very high control ratios are required, hydromotors with a higher nominal speed in combination with a gearbox for speed reduction must therefore also be used for hydraulic rotary drives. However, this results in higher noise emissions, and noise protection measures may then also have to be taken for these drives, because high-speed hydromotors have much higher noise emissions.

- There are also slow-running multipole electric motors with very low running noise, which nevertheless allow high control ratios using the latest control technology.
- Hydraulic pumps driven by high-speed electric motors operate with relatively high noise emission. However, hydraulic power units or the central hydraulic pressure station can usually be housed in a secluded room without any problems. Small hydraulic power units can also be designed to be very low-noise through special equipment selection and design measures.
- In certain applications, which are also relevant for stage technology, the compressibility of the hydraulic fluid can lead to adverse effects. Compressibility of a fluid column means elasticity and thus the ability to vibrate. However, the compressibility of the hydraulic fluid also causes changes in the position of components supported on a fluid column when the pressure conditions change due to a change of the load.
- Thermal expansion of the hydraulic fluid can also lead to adverse effects. In the case of hydraulic linear drives of hoists in the upper stage with block and tackle systems, temperature changes can result in relatively large changes in the position of suspended decorative elements. However, this problem can be eliminated by metered refilling of pressurized fluid when the medium cools down (see Section 1.8.3 – bar hoists with hydraulic cylinders – hydraulic linear hoists).
- The laying of electrical lines in particular is far less complex than the laying of pipelines and flexible hose lines. The transmission of electrical current via slip rings to rotating components is also simpler and less problematic than rotary connections for hydraulic lines.

As a *summary*, the development of the last few years can be summarized as follows:

Hydraulic drives are used, for example, in the lower stage as linear drives with hydraulic cylinders for large elevator systems, with the pressure fluid being provided from hydraulic accumulators housed in a separated pressure storage center.

In the upper stage, only electric drives with servomotors are generally used for bar and point hoists. One of the main reasons for this is that the control ratios between maximum and minimum operating speed that are common today cannot be achieved with hydraulic motors, as low-speed rotors would have to be used for this purpose for reasons of noise emission, but these are not suitable for these control ratios.

2.5 Operating the stage drives

First, the topic is treated in general terms, but reference is also made to the conception of specific companies in Sections 2.5.6, 2.5.7 and 2.5.8.

2.5.1 Basic types of control

Whenever the control of the drives is mentioned in this context, the term "control" is used in the general sense of influencing the motion behavior of the drive.

In the strict and narrower sense, one must distinguish between an "open-loop control" and "closed-loop control = feedback control."

In *open-loop control*, I specify a certain frequency for a drive, e. g., a three-phase motor, and the motor will rotate at a speed that depends, among other things, on the load, and this will result in a certain operating speed. In *closed-loop control*, a specific setpoint is specified for the speed or operating speed, the actual value is measured and the frequency is changed so that the setpoint and actual value ultimately match.

For a DC motor a higher voltage, for a three-phase motor a higher frequency, and for a hydromotor or hydrocylinder, a larger flow rate results in a higher speed.

Regardless of whether the drives are electric or hydrostatic, the switching functions, open- and closed-loop control commands, mutual interlocks and linkages are performed electrically.

In the classic *contactor control system*, the electric drives or electric switching elements of the hydraulic devices were switched via relays with a control voltage of approximately 24 to 220 V, and the switching circuits are implemented via wiring in the control cabinets in a plant-specific manner depending on the functional requirements. An auxiliary contactor is mainly used for switching control voltages, while a power contactor (main contactor) can switch the load circuit of a three-phase motor, for example.

Apart from the control of subordinate individual, these contactor controls have been superseded by modern *programmable logic controllers*, abbreviated to *"PLCs."* In this system, the logical linking of the circuits is not done by wired auxiliary contactors, but by electronic switches in the millivolt voltage range by programming microprocessors. Thus, universally designed microprocessor units can be programmed for specific needs. A change of the circuit-technical connections is possible by reprogramming, i. e., in the software without changes in the hardware. Such PLC modules can be networked via bus communication to very complex systems.

In stage technology, the high requirements with regard to their possible applications can only be realized with such computer-based controls. For this reason, larger stages are now equipped with *computer controls* as standard, as described in the following sections.

2.5.2 Requirements for the conception of operation

In a stage equipped with modern technology, it must be possible at the operating level to control all drives in the upper and lower stages, taking into account the following aspects, among others:
– It must be possible to operate individual drives independently of each other, but also several drives, e. g., several podiums, load bar or point hoists, together in variably selectable groups. With regard to *group operation*, there are various operating modes, which will be explained in more detail a little later.

- When designing the *control panel*, it must be taken into account that a large number of drives must be controllable simultaneously, but an operator has only two hands available, and two control levers to be operated must also be arranged in a position that can be reached simultaneously. Therefore, large control panels, as they were sometimes built years ago, are a thing of the past. On the contrary, today control panels can be built very small because they often work with touchscreens and the menu-driven screen displays only the areas that are currently required.
- *Complex motion sequences* often have to be processed in very short time intervals and this places high demands on the operator. It must therefore be as easy as possible to assign the control levers to individual drives or drive groups and to enter data, such as target positions, etc. The data must be stored in the computer and retrieved during the performance. Modern systems with computer control offer the possibility of storing this data and calling it up during play.
- For safety reasons, it must be possible to follow the motion sequence on sight, possibly via video camera on a screen. In addition, the movement can usually also be observed graphically on a monitor in *analog display* by moving symbolic picture elements. Furthermore, the displacement is also displayed in numerical values – *digital display*.
- Drives of stage equipment move podiums and can cause dangerous differences in the levels of the stage floor. They move stage wagons and make turntables rotate, they lift or lower loads in the upper stage above the heads of the performers. This generally poses a safety risk. Therefore, the operator must have sufficient overview of the movement sequence. This is not always possible from a central control point. In most cases, therefore, *peripheral* or *variable-position control panels* must also be available.
- The operator must be able to interrupt a movement sequence once it has been initiated at any time. Furthermore, in the sense of a dead man's control, the control lever must immediately move automatically to the zero position when released. A movement sequence must therefore not be started by pressing a button and the stop command given by pressing another button; instead, the movement must take place, for example, via a control lever whose deflection from the zero position results in a movement. Any special provisions for protective curtains or exceptions for very long continuous movements (see later) are not referred to here.
- It must be ensured that drives do not overrun geometrically or functionally determined limit positions. Therefore, an automatic shutdown must take place in these limit positions if a shutdown has not been performed in time via the control lever or by programming.
  This refers, for example, to the following limit positions: a rope drum may only be unwound to such an extent that at least two reserve windings remain on the drum; a load bar may only be lifted to such an extent that it does not collide with the grid structure.

- Irrespective of the shutdown in limit positions just described, modern operating concepts are designed in such a way that the drive is automatically brought to a standstill at the entered target point via a set deceleration ramp, even if the control lever remains deflected.
- In modern systems, the acceleration of a drive also takes place according to a specified acceleration ramp, i. e., even a very fast deflection of the control lever does not result in any impermissible acceleration. However, a very slow deflection of the control lever can cause a smaller acceleration. The same applies to deceleration.
- If the permissible tolerance is exceeded during synchronous operation of several drives, an automatic shutdown must also be performed to prevent overloading of individual drives, unintentional position changes, etc.
- In the case of computer-controlled systems, redundancies can be built into the hardware and software. However, for every motorized drive, whether electric or hydraulic, linear or rotary, there should be available a *multilevel emergency operating system*. Its design depends on the construction and use of the drive. As a first step, for example, a switch is made from a two-channel mutually monitoring control system to a single-channel one. As a further step, e. g., a switch-off of the computer level is enabled, then the transition from controlled group operation to directly controlled individual operation. Depending on the situation, operation with auxiliary motors or a manual emergency drive will also have to be provided.

For control units of mechanical stage equipment, there are partly similar problems as for control systems of the lighting technology. Therefore, similar concepts can also be applied in some respects. However, there are very decisive differences in the safety-related requirement profile of both control technologies. In lighting technology, a large number of lighting devices usually have to be controlled individually and in variable group assignments and often very complex time sequences with regard to light intensity, color tone, spatial alignment, etc. In terms of safety, this is a matter of control engineering. From a safety point of view, this only involves changes to spotlight settings that do not endanger anyone. In the case of the operation of mechanical equipment such as hoists, podiums, etc., however, there is a relatively large potential hazard.

For this reason, standards also demand a high level of functional safety for the control system so that the risk of a malfunction can be minimized. The *safety integrity level* (*SIL*) according to EN 61509 serves this purpose. Safety integrity level 4 represents the highest level of safety integrity and safety integrity level 1 the lowest. For stage control systems, usually SIL 3 is required. (For more details, refer to the relevant standards.)

Manual control level

Since only two control levers can be operated by one person at a time, there are usually only two control levers on the control panel, or possibly four for sequential operation or for operation by two people.

The control panel has keys arranged in a structured manner on the panel, input options via touchscreen (or via a computer keyboard), with which a drive can be assigned to one of the travel levers. For *group travel*, several drives can be assigned to a single travel lever. A specific drive or a specific group of drives can therefore be operated with one travel lever.

Pressing the dead man's button and deflecting the control lever then means the initiation of a movement (lifting or lowering, driving or turning in one direction or the other). Depending on the size of the deflection of the control lever, the speed of movement of the drive or of the drives included in a group can be changed continuously, if the drives are equipped with stepless speed adjustment – which has generally become self-evident in stage technology. It is also possible to enter target values for the travel distance for automatic control.

In the case of very slow, long-lasting movements, some control systems enable the initiation of the movement by pressing the button and periodically pressing the deadman's pushbutton; in the case of actuators for completely nonhazardous movements, there is no need to press a deadman's pushbutton.

Computer control level

If very complex motion sequences of several drives are to take place, different movements of several groups are to be carried out simultaneously or at short intervals one after the other, this can no longer be managed from the manual control level. In this case, it makes sense to pre-program travel lever assignments, group formations, speed and travel specifications, etc., store them as stored data and then call them up in the corresponding step sequence.

In this way, complete shows can be programmed in the sequence of the drive control and stored on a memory medium (*scene libraries*). When required, its contents are read back into the computer.

Programming, i. e., data input, may be menu-driven via the normal computer keyboard, but usually via a special keypad in the control panel. In "*teach-in mode*," however, positions actually approached with the drive can also be adopted as setpoints.

It is important to emphasize that "computer control" in this application does not mean that the movements of the stage machinery run automatically at predetermined time intervals and speeds. This would be firstly far too dangerous and secondly unsuitable from the point of view of performance technology, because not every performance runs at the same tempo.

With computer control – as already explained – only data for motion sequences are provided on demand in time. As with manual control, however, the movement itself must be controlled via the travel lever(s). As in the manual control level, full deflection of the control lever now also means movement at the preprogrammed maximum speed. If the lever is not fully deflected, the speeds are reduced in proportion to the lever deflection. If the control lever is moved to zero position, the drives are stopped. Preset programmed target positions are automatically approached, even if the travel lever remains in the deflected position, and the drive is automatically shut down and stopped in good time. Furthermore, preprogrammed acceleration values (deceleration values) are not exceeded even if the travel lever remains deflected.

The absolute maximum value for the speed at maximum deflection of the travel lever is limited by the dimensioning of the drive and by safety considerations or regulations (standards, accident prevention regulations, etc.). In addition to this maximum value, which is basically anchored in the control system, lower maximum values for the full deflection of the control lever can also be set for certain movements.

Furthermore, maximum values for the acceleration (deceleration) of the drives are specified for the drives, which can also be reduced for specific movements according to the performance requirements.

However, there are also motion sequences that are so complex that the setpoints for the individual drives must first be determined in a calculation process. This is the case, for example, when several drives have to cover different paths in the same time interval and the required setpoint speeds or the ratio values of the required speed increments have to be calculated.

As an example of such a complex motion sequence, which can certainly only be carried out via computer control, a transformation process in the understage can be cited. A horizontal flat stage surface is to be transformed into an inclined flat stage surface, whereby the flatness of the stage is to be maintained during the transformation. This means that the lifting movement of the podiums must be performed at different speeds, but also a change in the inclination of the podium ceilings must be superimposed. In the case of a stage with rectangular podiums arranged one behind the other, only an inclination about the transverse axis of the stage is possible, but in the case of a checkerboard stage and podiums with decks that can be inclined in any direction, a flat surface of any spatial inclination could also be created.

As an example of a complex motion sequence in the upper stage machinery, the tilting of a ceiling suspended from point hoists into any spatial position or a wavelike lifting and lowering of load bar hoists could be mentioned.

Some systems also offer the possibility of providing shorter motion sequences in such a way that not only various movements are performed simultaneously, but movements are also staggered according to specific time or path demands as motion sequences; this is also referred to as *"profile travel."*

When operating a system at the computer control level, the following basic operating settings are generally possible:

In the *basic operating mode*, all travel motions are preprogrammed by the manufacturer with system-specific limit values. This means that when the travel lever is fully deflected, the operationally defined maximum speeds are reached. When the travel lever is swiveled rapidly, acceleration or deceleration is performed with likewise preprogrammed maximum values. In addition, travel paths are limited mainly due to geometric operational conditions and are also protected by limit switches so that the drives are switched off automatically when the limit positions are reached.

In the *programming mode* – also called *setup mode* – imagination-specific target speeds, reduced maximum speeds at full travel lever deflection, different acceleration ramps, etc., if necessary, can be specified and synchronous groups can be formed.

In the *performance mode*, preprogrammed motion sequences are played back step by step via the travel levers as an *automatic operating mode*, with speed adjustments possible via the travel lever.

In a simulation operation – "*simulation mode*," motion sequences can be played virtually, i. e., the movements can be followed on the screen, but the drives themselves are not moved. This is particularly valuable if motion sequences are to be played back or tested only virtually for the time being. This can also be the case if the stage is used for other purposes or cannot be used due to maintenance or repair work. This simulation mode is also very useful for training purposes.

With regard to drive technology, it should also be mentioned that the control concepts described can be applied at the operating level for both electromechanical and hydraulic drives.

Almost only *absolute encoders* are now used as position measuring devices in a control loop; relative encoders (*incremental encoders*) that measure relative to an initial position are sometimes used as monitoring devices.

When approaching a target point, high repetition accuracy is required, and for synchronous group travel, high dynamic synchronization accuracy is also required.

2.5.3 Organization of control stations

It is often common to arrange the drives of the upper stage and the lower stage in two differently positioned *main control panels*; one for the lower stage, e. g., in a portal tower, one for the upper stage on a lateral working gallery. Sometimes, manually or motorized movable consoles are used as main control consoles, which can be placed in positions favorable to visibility for monitoring hazardous areas.

In general, control should be possible from several locations, and therefore *auxiliary control* panels are often provided in addition to these main control panels. However, depending on the product and type of such secondary control panels, it may not be possible to perform all the same operations as from the main control panel.

Auxiliary control panels can be installed at other locations in a fixed position, but can also be movable in a variable position, e. g., on wheels. They can then be connected to permanently wired sockets with flexible cables.

Portable auxiliary control panels, possibly with somewhat limited operating functions, can also be used. They can either also be connected to sockets provided for this purpose or the control commands are transmitted wirelessly. For safety reasons, the radio transmission takes place in redundant form via two independent radio systems.

For exceptional operating situations, especially for emergency operation and for repair and maintenance purposes, it is advisable to provide a *"local individual control facility"* for each drive in its immediate vicinity or at locations with particularly good visibility conditions.

In addition to the above-mentioned control consoles, which are used for the drives of the upper and lower stage during the performance, there are of course also independent individual control consoles and control boxes for other equipment, such as the portal bridge, the protective curtain and the smoke flaps.

In this context it should be pointed out once again that a well-designed drive concept should offer as many provisions as possible for maintaining operation during a performance. Redundancy elements can serve this purpose. Otherwise, emergency operation should be made possible. For example, it must be possible to bypass malfunctions in the computer control system during show and rehearsal operation by switching to a control level independent of the computer, accepting unavoidable operational restrictions. As a further step, it should be possible to bypass closed-loop control while taking safety considerations into account and to run open-loop controlled drives only.

If operating faults occur, the software of modern control systems provides detailed information on how to locate the fault, but the fault messages are also stored for later retrieval so that targeted repair work can be carried out after the performance.

2.5.4 Operating modes for group travel

In *single travel*, the travel of a drive assigned to the control lever is performed. The travel is initiated manually with the control lever and controlled and terminated manually over the entire travel, or it is followed by automatic control when the target point is reached.

As already explained, however, it must also be possible from the control station not only to control each drive individually, but also to move several drives together as a group, e. g., several lifting platforms or several load bar or point hoists.

When traveling in the group, several operating modes of *group travel* can be distinguished, as explained below. (Unfortunately, the designations mentioned below are not uniformly defined.)

In terms of monitoring group travel, a further distinction can be made between the formation of asynchronous groups with and without group shutdown and synchronous groups.

There are the following variants of group trips:

Asynchronous group travel
Travel in which all drives of a group travel after the common start without mutual influence and dependence. (The travel of the devices in the group can be controlled with or without feedback control.) Reaching the travel range limit or triggering of a safety

device must lead to the standstill of the machine concerned (group travel without group shutdown) or to the shutdown of the entire group (group travel with group shutdown). It must be indicated which device has led to the shutdown of the group.

Synchronous group travel

Travel in which all the mechanical equipment of a group travel in a controlled manner (distance or time synchronous) after the common start with mutual influence and dependence. Synchronization must be monitored and the specified synchronization tolerance must not be exceeded in any operating state. If the travel range limit is reached or a safety device of a device is activated, the entire group must come to a standstill; the same applies if the synchronization tolerance limits are exceeded. In addition, it must be possible to identify or display which device has caused the group to stop.

The following operating situations are possible with synchronous group travel:

Synchronous travel – time synchronous

Initially, this actually only means that all drives start and end their travel at the same time. In general, however, it also means that the individual drives must travel different specified distances in this time interval.

These different paths can be freely selected depending on the application or they are geometrically dependent. If, for example, a ceiling suspended from point hoists is to be tilted from a horizontal position when it is raised, each hoist must travel certain distances depending on its coordinate position. The computer can calculate the necessary speed grading of the individual drives if the coordinate positions of the point hoists are known.

Synchronous travel – distance synchronous

All drives, e. g., flies or podiums of a group, travel through a specified distance simultaneously and in the same amount of time, regardless of the respective starting position of the individual drives, i. e., the travel distances of all drives are the same. If the travel lever is deflected, all drives move up with the same acceleration, continue to move at constant speed if the travel lever position remains unchanged, and are simultaneously decelerated and brought to a standstill when the travel lever is returned.

Thus, all drives have to traverse the same paths at the same speeds, so their motion is also speed-synchronous and, since all drives are moved at the same time intervals, they are also time-synchronous (but this is not reflected in the designation).

Synchronous travel – speed synchronous

All drives move at the same speed. If all drives move at the same speed in the same time periods and therefore cover the same distances, this corresponds to the term synchronous travel.

Sequence run (profile travel)

A cumulative travel of groups (individual travels and/or group travels) can be performed with a combined control device. The respective group characteristics (e. g., synchronous travel, shutdown in the event of a fault) of the individual devices are retained.

Some controllers also offer the possibility to run a motion sequence by hand in teach-in mode, and the motion sequence recorded in this way is then precisely reproduced in presentation mode.

2.5.5 Emergency control options

Of course, as explained above, such complex plant systems must also offer several emergency control options.

Here, too, there are different concepts depending on the manufacturer, especially in the hardware arrangement – integrated into the control panel or as a separate port console.

– An accident mode has already been mentioned. It consists of dispensing with the monitoring function driving computer – monitoring computer and driving single-channel instead of dual-channel.
– If a fault occurs on an axis computer, it is possible to switch over to an axis computer that is available as a reserve.
– The next step can be to bypass the computer control and perform operation including group formations at the travel lever at PLC (programmable logic controller) level.
– Often, systems also offer the option of controlling a single drive via an emergency control cassette, which is connected to the terminal box of the drive or can be connected as a mobile unit.
– As a further step, as described in Sections 2.1 or 2.5.1, the possibility of attaching a motorized auxiliary drive or switching over to manual drive is often also offered.

2.5.6 C·A·T control system by Waagner-Biró Stage Systems

The C·A·T Computer Aided Theatre control system from Waagner-Biró Stage Systems was first installed in the Oberhausen Stadttheater in 1989. Since 2003, the C·A·T V4 generation has been installed as an SIL 3 control system and has been described in previous editions of this book. In 2017, the fifth and technically completely revised version, called C·A·T V5, was launched and is described below.

With the new hardware and software platform developed for C·A·T V5, all the safety functions specified in EN 17206 can be implemented completely independently of each

other and, for the first time, without drive or system-wide looped electrical safety circuits at the required safety integrity level of SIL3 or PLe. As there is no longer any need
to bridge safety loops in the installation, the handling of safety problems and faults in
the sensor system can be fully integrated into the user interface of the control panels with graphical user guidance. Accordingly, the tools required for brake tests, load
tests and other maintenance tasks for expert and periodic inspections are also included
as guided user dialogues in the C·A·T STUDIO on the main control panels. This means
that C·A·T V5 does not require a separate panel for repairing or testing during operation.

To achieve this powerful and flexible troubleshooting concept, C·A·T V5 makes use
of the ability to selectively disable individual sensors or safety functions for a limited
time in the event of a fault or during test scenarios. For example, overload monitoring
is temporarily disabled to lift the test load during the brake test, while all other safety
functions, such as maximum speed or travel range monitoring, remain active at full
safety level during the test.

Even in the event of sensor failure, e. g., load cell failure, the operator can use these
muting functions and perform the necessary movements from the main control panel
as part of the show. The operator is responsible for ensuring that alternative (organizational) measures are in place to maintain the required level of safety. For example,
the operator must ensure that the payload to be moved is known and within the permitted range, and that no collisions with adjacent decoration or system parts can occur.
In case of doubt, the movement can be monitored by another person on site and authorized by an additional external approval button. Accordingly, only specially trained and
experienced operators are authorized to access the muting system.

The structure of the drive control system

Operating system
The C·A·T control system uses Linux as the operating system and modern HTML5 technology for the user interface. The operating system is real-time, allowing it to scale to
the biggest sites with defined cycle time. Linux is used throughout the C·A·T system at all
levels: In consoles, servers and the network, it guarantees a stable basis for application-
specific software.

High availability was an important objective in the development of the C·A·T system:
even if individual components fail, the overall system should remain as functional as
possible. So-called "single points of failure" are critical, i. e., components whose failure
would have a negative impact on several other components: servers, network switches,
central racks, etc. Thanks to its modular design and multiple levels of redundancy, the
C·A·T control system avoids such bottlenecks: Identical parts can be replaced quickly
and easily. Critical central components are redundant or designed separately for each
drive. The result is high availability without compromising safety.

In the C·A·T control system, each network node has two separate network connections. Each axis controller and console is permanently connected to two different networks and can continue to communicate via the other network seamlessly and without interruption to a move if one network component fails.

The basic system structure of the C·A·T V5 control unit is shown in Figure 2.23.

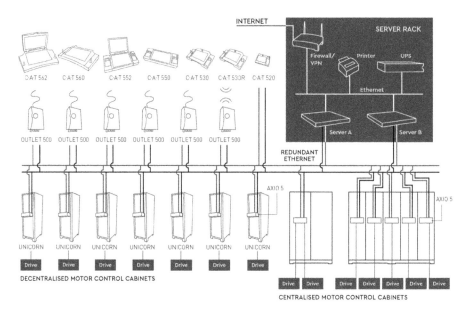

Figure 2.23: C·A·T V5 system structure. Picture credits: Waagner-Biró Stage Systems.

System design – modular structure

Each drive is considered to be an independent unit. For example, if an axis controller fails, only a single drive is affected and the others can continue to operate. The modular design of the system also makes future extensions extremely easy. Sites can be migrated step by step, adding a few more drives to smaller sites year after year.

The C·A·T system consists of the following components (see Figure 2.23):
– Central Server and Network System
– Control panels
– Outlets for control panels
– UNICORN cabinets or central control cabinets
– I/O controller – AXIO 5
– Additional emergency stop button if required

Central server

The central server system is based on powerful servers that can be scaled according to the size of the system and the required cycle time. The server calculates and coordinates the motion sequences and performs central tasks such as managing the show database, telemetry data, log files and remote maintenance access.

The central server is redundant: A second server runs in parallel to the main server and mirrors the data changes on its hard drive. If the main server fails, the system switches to this backup server. This is done completely automatically, but can also be done manually by the operator. The control system continues to run with full functionality on the other server, so there is no interruption to the stage performance. The entire server hardware is duplicated. This offers greater reliability than, for example, simply mirroring the hard drive on the same server.

The server cabinet includes a UPS system to protect the power supply to the control system, as well as a dedicated additional Ethernet network for connecting peripherals, but also for interfacing with other control systems or other systems such as lighting controllers or show controllers.

The end-to-end Ethernet network concept makes it possible to access the C·A·T Remote Service for support or maintenance tasks via a VPN connection not only to the servers but also to all C·A·T control system devices. In this way it is also possible to support the commissioning of new drives or entire sites via remote maintenance without the need for a C·A·T specialist to be present on site.

Control panels

The control panels are used to control as many of the site's drives as possible. It is possible to specify which drives can be controlled from which outlet and by which user. In this way, dangerous blind runs can be avoided right from the start. A high user experience is achieved with the help of graphics, integrated videos, etc.

Waagner-Biró Stage Systems classifies its control desks into 3 different categories with their corresponding control panels:

Main control panels: C·A·T 562 (see Figure 2.24), C·A·T 560, C·A·T 552 (see Figure 2.25), C·A·T 550

Wall-mounted control panels: C·A·T 520 with a 7-inch touchscreen

Mobile desks: C·A·T 530 and C·A·T 530(R) (see Figure 2.26)

The wired consoles are connected to the OUTLET500 sockets distributed around the stage area (see Figure 2.27), with the C·A·T 530R control panel communicating via a Wi-Fi network connection to the OUTLET500 sockets.

A multitouch screen with gesture support is integrated into the console surface. Depending on the type of panel, a second identical screen is also integrated into the panel cover. This allows the two interfaces of the C·A·T software to be displayed simultaneously, e. g., C·A·T CONTROL, optimized for programming and running the show, and C·A·T VIEW, optimized for machine overview. Both views can be used to select and visualize runs.

Figure 2.24: C·A·T 562 Main console on console stand C·A·T ROVER COMFORT. Picture credits: Waagner-Biró Stage Systems.

Figure 2.25: C·A·T 552 on C·A·T GALLERY rail system in Dramaten, Stockholm. Picture credits: Waagner-Biró Stage Systems.

Access to all control panels is granted by an RFID token or card. The access authorization is based on different user roles and can be configured individually for each user account. This allows the operators to use their preferred settings at each of the different consoles. In addition, user roles can be used to predefine typical configurations. Different people have access to different functions according to their allocated user rights.

Figure 2.26: C·A·T 530 portable mobile console. Picture credits: Waagner-Biró Stage Systems.

Figure 2.27: OUTLET 500 in the Copenhagen Opera. Picture credits: Waagner-Biró Stage Systems.

Authorization for dangerous movements and access to sensitive data are clearly regulated, increasing the security of the system (see Figure 2.28).

Each joystick module has two programmable push buttons with an integrated OLED display showing the selected function and current motion parameters in real time. The buttons can be used to select, deselect and start cues (in conjunction with the higher

Figure 2.28: C·A·T V5 User login with personalized profiles. Picture credits: Waagner-Biró Stage Systems.

level approval button). When working with the joystick, however, the approval button integrated into the joystick is used. The control levers are backlit in color and can be freely assigned. This means that several moves can be controlled independently. If text or numbers need to be entered, an appropriate keypad appears on the touchscreen. It is also possible to connect a standard USB keyboard or mouse to the console.

The C·A·T 562 and C·A·T 560 have two/one 24" multitouch displays and the standard version has four sensitively scaled C·A·T GEARS JOYSTICK INTERACTIVE control lever modules. However, as with all C·A·T V5 control panels, the C·A·T GEARS motion control and input modules can be arranged in many different combinations to suit the user's requirements. The C·A·T 560 series consoles can be expanded to integrate up to 12 different control elements in one console, which is particularly useful for complex productions with many overlapping cues. This allows the operator to individually adjust or stop each cue as it runs, and to schedule emergency or evacuation moves in the show as required, ready to be activated at the touch of a button. When using the C·A·T GEAR PLAYBACK WHEEL, it is possible to individually control six running cues without having to reach around with one hand.

All main consoles can be mounted on the specially designed C·A·T ROVER BASIC, C·A·T ROVER COMFORT or C·A·T ROVER RADIO mobile console stands. While the basic model C·A·T ROVER BASIC offers maximum stability and good mobility with manual height adjustment, the C·A·T ROVER COMFORT also offers electric height and tilt adjustment for all sitting and standing positions. An integrated storage cabinet and various mounting options for tools and equipment aim to further optimize and ergonomically design the operator's working environment, even beyond the graphical user interface. The top of the range C·A·T ROVER RADIO, with its wireless, battery-powered operation and long autonomy, allows you to quickly take up or leave the right position on a busy stage without the need for cables.

The C·A·T 552 and C·A·T 550 main control panels are identical to their larger siblings, the C·A·T 562 and C·A·T 560, in terms of performance and features, but have been specially designed to create an ergonomic workstation even in a narrow gallery. To optimize installation depth, they are equipped with two respectively one 20-inch multitouch display and control buttons arranged entirely on the side. However, the most important space-saving feature is the C·A·T GALLERY arm, which can be moved along the gallery and along multiple axes. In contrast to fixed gallery brackets, the position, height, orientation and, most importantly, the orientation of the control panel on the narrow gallery can be adjusted literally on the fly to suit the operating and space conditions.

C·A·T 530 gives users mobile control. It has a 12" HD touch screen and two control modules. The console functions are accessed using the same RFID cards as the C·A·T 562. The software interface of this console is the same as that of the C·A·T 562, which has been adapted to the different sizes of the display. The console has been designed primarily for use on the stage, in the rigging area and for maintenance in the machine rooms, but it can also be used to run preprogrammed shows. However, due to the smaller screen size, the ergonomics are slightly inferior to the main control panels for complex programming operations.

The C·A·T 530R is even more flexible. The console supports secure data transmission via radio and battery operation, and connects to the sockets either via cable or Wi-Fi. As a result, it offers almost unlimited mobility.

The C·A·T ROVER ONE mobile console stand is available for both mobile consoles. The consoles are screwlessly supported in the robust WALLMOUNT and can be removed if required. The carrying strap makes both models easy to carry.

The C·A·T 520 is a local control panel with a 7" touch screen and a single control lever module. It is used when access to the show database and its cues is required at a fixed control point, but also as a common control point with graphical position display for a number of basic up and down moves of hoists. Typical applications are the orchestra pit control point, the curtain zone control point, or the control of side and rear stage rigging hoists. However, the C·A·T 520 can also be used as a control point in the artist rigging zone for (3D) aerial applications, where the operator is required to rig the performers or actors into the flying harness (flying corset) and move them to the start position. In all these situations, the C·A·T 520 offers an easy to understand user interface, clear visual feedback on local system status and, if required and authorized, quick access to the show's ride parameters and preprogrammed rides.

With the C·A·T 500 desktop console, Waagner-Biró offers a tabletop console for offline simulation with a laptop or large control stations with numerous external screens. It provides the safe and convenient operating hardware of the C·A·T main control consoles, where programming and visualization are carried out via external computers and screens.

C·A·T GEARS Control elements for all control panels

Waagner-Biró also offers the possibility of configuring the control panels and their control elements using C·A·T GEARS modules (see Figure 2.29). These are different, modular control elements, of which up to 12 can be used on one console, depending on the type of control panel. In addition to various types of motion control modules (C·A·T GEAR JOYSTICK, C·A·T GEAR PLAYBACK WHEEL), the customer can also select haptic input modules, such as a programmable macro button, a 3D mouse or jog wheels, to complement the fully touchscreen user interface. A total of nine different modules are available to the customer.

Figure 2.29: C·A·T GEARS PLAYBACK WHEEL at Dramaten, Stockholm. Picture credits: Waagner-Biró Stage Systems.

Universal sockets, radio connection and emergency stop zones

Each mobile control panel is connected to an OUTLET 500 socket via cable or Wi-Fi (see Figure 2.27). Wired consoles are connected to the outlets using the proven and flexible CABLE 500 system cable. Typically, several OUTLET 500 are distributed across the stage area.

If required, to streamline configuration, outlets can be assigned to zones with specific functional permissions or drive groups. The configuration for each zone is automatically applied to each connected control panel. This allows administrators to grant or deny access to specific machines or special functions, such as relocating a point hoist, muting safety functions or sensors for troubleshooting, and much more, depending on the location.

The OUTLET 500 can also be assigned to one or more emergency stop zones of a larger stage installation via software configuration. In this case, the emergency stop on the connected control panel or directly on the OUTLET 500 will only affect the drives assigned to that emergency stop zone. For example, in order not to interfere with rehearsals or performances on the main stage, the emergency stop of an outlet in the rear stage can, depending on the risk analysis, only affect the stage wagon system in this area and the rigging hoists. Especially in the en-suite operation of a musical theatre, it can be beneficial to define and configure these emergency stop zones individually for each production and set.

The operating range of the wireless connection depends on the conditions on stage. Indoors, a range of up to 100 meters is possible with the radio technologies used. If the radio coverage is still inadequate, roaming is also available: as there are usually several outlets distributed around the stage, the system automatically switches to an OUTLET 500 with a better signal if the signal to the C·A·T 530R becomes too weak. Today, however, there are so many devices sharing the limited radio bandwidth that it is difficult to ensure reliable radio operation in a fully occupied auditorium with hundreds of individual mobile devices. Just like a road with cars, radio frequencies and channels cannot be assigned arbitrary bandwidths without causing traffic overloads and interruptions that make it difficult to operate a machine reliably in real time. For these reasons, even in the age of smartphones and laptops, wired operation remains the safest alternative for high availability of performance. In practice, however, this is hardly a limitation, thanks to the thin and flexible C·A·T system cables and cable management solutions integrated into the console stands.

Each of the outlets can be fitted with an emergency stop button, which can also be covered with a protective ring, depending on the installation situation. Deactivating and unplugging the control panel does not result in an emergency stop in the system, but any drives still selected from the console will be safely stopped and deselected.

Switch cabinets

Waagner-Biró Stage Systems offers an advanced control cabinet system with its specially developed UNICORN decentralized control cabinet. Each machine has a dedicated control cabinet installed close to the machines themselves. The cabinet is fully equipped and tested at the production site. All connections to the outside are made using plug-in connectors. This greatly reduces commissioning time. In the event of a failure, the cabinet can be replaced quickly and easily.

Integrated in the *Unicorn control cabinet*:
- Frequency inverter.
- Mains filter.
- Braking resistor (mounted on the rear).
- Main switch.
- Main contactor.
- Brake contactor.
- Power supply.
- AXIO 5-axis controller.

The separate control cabinets (see Figure 2.30) are independent of each other and do not influence any other drive or control cabinet in the system. They can be used for all

Figure 2.30: Typical system setup with UNICORN and AXIO 5 in winch room. Picture credits: Waagner-Biró Stage Systems.

variable speed drives in the range of 1.5 up to 45 kW of the upper stage. The cabinets are so small and light that they can be installed next to the winch to be controlled.

This eliminates the need for a separate cabinet room. All cables between the winch and the control cabinet are plug-in connections – this means that all cables can be pre-assembled and tested at the factory. On-site installation is reduced to connecting pretested cables. This reduces errors on site and makes maintenance easier as there are no screw terminals to tighten.

Each cabinet is supplied with 400 V three-phase power and Ethernet only. This significantly reduces the amount of cabling required compared to conventional centralized cabinet technology. Special busbars can be used to further reduce the amount of cabling required. These busbars are mounted on top of the UNICORN.

Wherever it is not possible or desirable to install decentralized UNICORN control cabinets next to the machines, centralized control cabinets are used. These can be placed in a specially designed electrical room. The required electrical and electronic components are then grouped together in standard cabinets for multiple drives.

Axis controller

The AXIO 5 axis controller connects the C·A·T V5 control system with the machine equipment and the outside environment. It is responsible for all safety-related functions as well as drive control and motion monitoring. It combines an SIL 3/PLe certified safety motion controller and freely assignable safety inputs and outputs in a plug & play interchangeable black box. AXIO 5 translates console commands into the electrical signals required to control all types of machinery, including high-speed winches, chain hoists, turntables, stage wagons, stage lifts, hydraulics and much more.

The axis controller has several position encoder channels which are used to calculate and monitor the safe position of the drive. Derived values such as the speed and acceleration of the drive are also determined and monitored based on the particular encoder combination. If an inadmissible deviation is detected, the drive is stopped.

If an axis controller fails, only the corresponding drive is affected. The rest of the system remains operational. The simplest and most cost-effective redundancy is to have a spare AXIO 5 axis controller available. The AXIO 5 is installed in a very maintenance-friendly manner by simply plugging it onto two contact strips. The AXIO 5 can therefore be installed and removed quickly and without tools. If one AXIO 5 stops working, it can easily be replaced with a new AXIO 5. The new AXIO 5 is automatically detected, coupled to the detected device by the operator at the touch of a button on the control panel, and automatically configured for the appropriate drive. No configuration by maintenance personnel is required. Each drive is identified by an ID chip which is read by the AXIO 5 upon start-up. This means that any AXIO 5 can be used as a replacement for any other.

Ethernet network

All components of the C·A·T system are connected via an Ethernet network based on the IP protocol. This flat structure provides excellent and cost-effective transmission speed and scalability. The clear and consistent communication structure does not require additional data compression or protocol conversions between different subsystems.

The C·A·T V5 network architecture is designed for 2,000 network nodes without segmentation and is certified for use in a safety control system. Waagner-Biró Stage Systems had to develop its own data protocol with special error detection and correction options in order to demonstrate the low error probability required for SIL3 systems with such a large number of network nodes. Parallel sensors with complex cable routing or hard-wired interlocks between machines are no longer required.

Because of its importance to the availability of the site, the network is designed with parallel redundancy. In this variant, the server, axis controllers and consoles have two network connections and all network components are duplicated. The control system communicates in parallel over both networks, so that if one line fails, the network remains fully operational without interruption.

Operating concept

During the development of C·A·T V5, a modern user experience was and is at the center of all considerations. In addition to general solutions to ensure ergonomics and safety in the workplace, this includes above all the realization that (most) users of the C·A·T V5 control system are not programmers or engineers who think in terms of numbers and rules. Accordingly, the new C·A·T V5 user interface largely dispenses with the tabular overviews and numbering systems found in older generations of stage control systems, and is oriented towards the simplest possible implementation of more natural concepts such as

- Linguistic instructions: "Drive a little faster" instead of "Please drive at 1.37 m/s".
- Repetitions: "The same position as in the second scene".
- Practical sorting: Dragging and dropping an item directly to another position instead of adjusting sort parameters.
- Everyday knowledge: Use of everyday symbols and gestures.
- Free working method: The control system does not impose fixed sequences for processing tasks and tolerates temporary inconsistencies. caused by individual working methods. However, any problems that arise are controlled and clearly visualized so that they can be resolved later.

C·A·T V5 uses the following techniques as key elements of this operating philosophy:
- Display only information that is currently important.
- Use familiar gestures and operating logic that are commonplace and therefore easy to learn.
- Use a graphically appealing design that is not cluttered, is fun to use, and allows you to do what you need to do in a focused way.
- Explain screens and expected input values immediately as they are used.

This contemporary approach also means that the console's operation is explained not only in the form of a user manual, but also in simple, short instructional videos on the control panels themselves.

Main element of the user interface
The C·A·T user interface is designed throughout as a multiuser platform, allowing multiple users to work simultaneously at different control panels in the same or different shows. For this reason, all operations are performed directly on the server and mirrored to all control panels. Advanced locking strategies ensure that all operators see changes to simultaneously loaded objects immediately, but only one user is allowed to make changes at a time. All this is done in the background, without the user having to perform any special operations.

User administration
To make it easy to identify which operator is working where on the user interface, the individual user accounts are personalized with names and photos, which the operator can create or upload directly from the control panel using the built-in webcam. In addition to the personal settings and default behavior that the user can set, centrally assigned user and group rights are managed by the system administrator.

Show data management
Shows created in the system can be accessed via the repertoire management. In addition to favorites markers, an incremental search function helps to quickly find the desired show. Shows can be archived, duplicated or printed.

C·A·T CONTROL – page view
The page view is the counterpart to the C·A·T CONTROL screen of previous C·A·T generations. This is where cues are programmed, displayed and driven in page views. The view of the machines and cues automatically adapts to the available screen space: the more machines and moves are loaded, the less detail is displayed.

The page view distinguishes between temporary ad hoc moves, which correspond to a spontaneous move without saving in the show, and saved cues. This concept makes it very easy to combine equipment and performance control on one screen, without the operator having to switch views or sacrifice functionality.

C·A·T VIEW – machinery view
The machinery view is a 2D representation of the site (see Figure 2.31) that is customized for each installation and venue, sometimes even for a specific production. It corresponds to the C·A·T VIEW of an earlier generation, but unlike that, it is not just a status display. In the machine view, machines (or objects suspended in them) can be selected directly in free run or with a target specification and assigned to a control lever.

For a better overview of the stage machinery, the machinery view not only shows the machines, their height levels and movement parameters, but also the associated decoration, including height, name and selected icon. In this way, the machinery view provides a comprehensive rig plan for the stage tower at all times.

3D C·A·T VIEW/3D visualization
Depending on the site configuration, a simple 3D visualization of the machines or a high quality, realistic 3D visualization with lighting simulation can be displayed on the control panel (see "Simulation and 3D visualization" section).

C·A·T STUDIO
The C·A·T Studio is the configuration and maintenance environment for the C·A·T control system. Depending on user rights, it allows access to the muting functions for safety functions and sensors, to the built-in wizards for recurring tests (semiautomatic brake test, overload tests, limit switch tests) or to the administration pages. C·A·T STUDIO also includes comprehensive configuration and commissioning tools for commissioning engineers from Waagner-Biró Stage Systems and authorized partner companies.

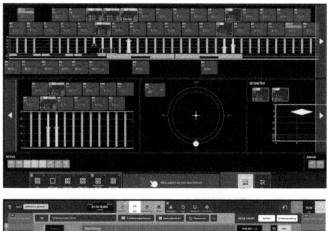

Figure 2.31: (top) C·A·T VIEW machinery view with various objects; (bottom) C·A·T CONTROL side view with various rides and cues. Picture credits: Waagner-Biró Stage Systems.

Alerts

C·A·T V5 continuously checks the system status and the consistency of the active shows on the control panel in the background. Errors in the system or inadmissible program parameters are immediately and clearly indicated on the screen with an explanatory text. The alert view lists all active alerts with navigation markers, allowing the operator to jump to the relevant point and rectify the problem at the touch of a button. In addition, messages between operators and notes on specific issues (e. g., why a drive is locked) are recorded and displayed. The system's error messages and warnings are displayed in the logbook over time.

Telemetry

The system's telemetry is integrated into the C·A·T V5 control system, which logs all relevant input and output data, status changes, operator actions and screen views of the installation over a period of several months. Several hundred gigabytes of data are stored so that faults or historical developments can be analyzed retrospectively in the context of the respective operating and system situation.

Help pages
The online help included in the C·A·T V5 user interface contains not only the user manual and troubleshooting tips, but also video tutorials for first-time or infrequent operations, software update change notices and, of course, contact information for your local service representative.

Synchronized moves and cues

Synchronization options
C·A·T offers various options for specifying the motion behavior of the drives within a group – *group travel*:

If *time asynchronous movement* is selected, all the drives in the group move at the speed set by the operator. They reach their target positions earlier or later, depending on the distance to be travelled.

With *time synchronous movement*, all the drives in a group reach their target positions at the same time, regardless of the distance they have to travel. To do this, C·A·T calculates the speed in relation to the distance travelled for each individual drive.

In a *synchronized distance (path) movement*, all the drives in a group move a rigid object at the same time. To avoid damaging the suspended object, the drives in a synchronized positioning group always keep exactly the same distance from each other.

Multiple movements and sequences of movements, which are therefore on a common control lever, can be aligned so that they start or end together – regardless of whether the drives within a movement are asynchronous, time synchronous or position synchronous. This means that the operator only needs to set the appropriate time or speed for each movement. Movements that end together are automatically delayed, regardless of their actual start position, so that they arrive at their destination at the individually preselected speed or time.

In a *sequence run*, several runs can be played automatically one after the other, even with different drives. The user can decide whether the machines should stop at the transition from one sequence step to the next, or move seamlessly to the new speed or direction.

Moves and cues
C·A·T V5 combines all movements that are started and controlled together by a movement command device (control lever, button or dial) as one cue. Several cues started at the same time or within a short period of time are grouped together on one page. These pages show the chronological sequence of a show.

A move is displayed as one line in the page view. Fields for motion parameters such as acceleration behavior, repeats, delays, conditions, etc., are usually hidden as long as they contain default values. This means that only the start position, target position, speed and travel time, as well as the object name and its icon – if configured – are displayed next to the moves.

If only a part of the programmed show is to be performed as part of a rehearsal, the operator must create the correct start position for this starting point of the rehearsal. For this purpose, C·A·T V5 offers an automatically generated page containing all the necessary moves to the last position before the starting point.

Objects and decoration

Opera houses, in particular, use large scenery sets, some of which have to be set up and dismantled on a daily basis. C·A·T V5 offers extensive support for objects (see Figure 2.32), which makes it easier and safer to handle the scenery parts, and makes the cued positions more reproducible. When working with objects, targets are no longer entered and managed as the position of the drives, but as the position of the object relative to the stage floor. This allows the exact position of the scenery to be reproduced even if the connection to the load bar changes: When rigging the object into the drives, the operator references the height once at the touch of a button and can then safely move to the set targets without any further modifications to the programming.

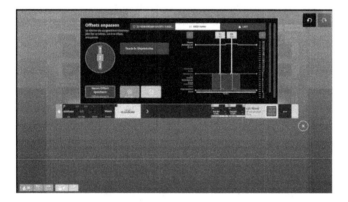

Figure 2.32: User guidance when referencing the new backdrop heights. Picture credits: Waagner-Biró Stage Systems.

There are also many other features associated with objects to ensure safety and clarity:
–   The height of the decoration is not only used to represent the decoration in the 2D/3D display, but also automatically limits the travel distance in free movement towards the stage floor and the grid floor. As the object can be attached to different drives at different distances, C·A·T V5 offers a clear and interactive configuration dialogue for object height and attachment, as well as for referencing.
–   The actual load of the object on the machines can be measured and stored. C·A·T V5 can therefore not only implement specific underload and overload detection for the object per machine and in total, but also ensure automatic load compensation to the preset load values per machine.

- For large, heavy or fragile objects, it is often useful to define maximum values for the speed and acceleration of the object. These defined maximum values supersede the permissible limit values for the drives in C·A·T V5 as soon as the object is suspended.
- A name and an icon identify the object in the page view and in the machinery view.
- Depending on the 3D visualization version, a 3D model of the object can be saved by the user or selected from a model palette.

Simulation and 3D visualization

Virtual stage(s)

C·A·T V5 offers one or more virtual stages, integrated in the system servers, to which the user can connect the control panel by means of a menu item. On the virtual stage, the user can simulate programmed movements without actually moving the machines, either alone or in combination with other control panels. The current status of the real stage can be copied at the touch of a button.

3D visualization

There are two different use cases for 3D visualization in stage automation control systems. C·A·T V5 offers different solutions for each:
- The operator or stage manager wants a more realistic picture of the space used by the decoration and equipment during a performance, including objects left over from standard installations or other productions. The 3D visualization is used for the spatial overview and is provided entirely internally in the C·A·T V5 system. The 3D visualization on the control consoles shows only the machinery and the decorations hanging in it, in a simplified image of the stage space.
- The creative team, including set designers, lighting designers and production designers, want to get a realistic impression of the visuals and cues on stage. The 3D visualization is used in preproduction and is available both on the control consoles and on a standalone production workstation used by the set and production designers. Separate 3D servers provide high-quality 3D rendering of the building model, flown object models, manually placed object models, lighting and media content projected onto objects or played on LED walls, in addition to the C·A·T V5 system servers. Images are individually calculated for the different C·A·T V5 consoles, providing not only their own camera positions, but also allowing a different context to be displayed on each console (real stage with existing structures or virtual stage with a selected image of a show). In addition to this system-integrated rendering, a separate 3D production workstation is connected to the external C·A·T V5 network.

While the user interface of the C·A·T V5 consoles is optimized for the workflows of an automation operator (e. g., loading a decoration from the prepared repertoire or a standard decoration, changing a view angle, switching from show light to work light) and automatically follows the workflow on the console, the user interface of the 3D production workstation is made for production designers and allows them to load new decoration objects, place them on the stage, create subscreens for manual handling of objects, save changed lighting presets and much more. To make all this preproduction work directly available to the automation team, the show databases of C·A·T V5 and the 3D production system are automatically synchronized. Lighting and media playback can also be received in real time from the lighting console and played back in the simulation.

3D flying

The C·A·T V5 control system supports 3D flying rigs on multipoint hoists in the form of 3D groups, which are popular in acrobatic theatres in China, but are also available for various theatres in Europe. To create a 3D group, the user must first configure the precisely measured 3D coordinates of the pulleys used and the rope guide to them on the control panel, so that the system can automatically calculate the permissible 3D movement space with the available rope lengths. The user must also specify how the ropes are attached to the object being flown. After this preparation, the user can enter the 3D coordinates for the object directly on the control panel and fly towards it. Depending on the configuration of the flight system with 2 to 8 point hoists, movement in and/or rotation around the axes are possible.

For more complex movements, a separate 3D control stick can be used to fly and record the movements freehand, or they can be sketched and edited in a specially developed 3D editor. The editor can display a 3D model of the stage, including rigging, to improve orientation when editing flight movements. If multiple 3D rigs are being used at the same time, the editor can be used to detect and correct collisions between people or objects being flown and the rigging systems. The flight curves from the 3D Editor are available in the C·A·T V5 control system as recorded profile runs, which can be started and controlled from the control panel in the show just like any other run.

Interfaces to other trades

Real-time position data

To facilitate the tracking of objects moved by the stage control system in the lighting control system and the correct alignment of LED and projection content on moving objects, C·A·T V5 implements various standard protocols for real-time transmission of machine and object positions, including PosiStage.Net, s/ACN and ArtNet. However, Waagner-Biró Stage System can also implement customized protocols in consultation with the customer.

Show automation
As the C·A·T V5 stage control system is also used in applications where a predefined show runs automatically, C·A·T V5 supports various options for starting cues via external signals (e. g., SMPT/E timecode) and automatic switching within a show, but in most cases the operator must also release the movements via the approval button in the desk.

2.5.7 COSTACOwin – control system of the company SBS

Structure of the stage control system
The COSTACOwin® control system includes the entire hardware and software for controlling all the drives on the stage. It controls all movements of the overstage and understage machinery and the stage technology including on-off drives, regardless of whether these are electrically or hydraulically powered, variable or fixed speed. The modular design ensures the highest possible safety. The COSTACOwin® control system meets Safety Integrity Level 3 (SIL3) according to EN 61508 and is certified accordingly. Its flexibility also makes extensions easily possible. All components – such as the main computer, axis control computers or network – can be designed with redundancy, in the overall system or also individually. The architecture with main server level allows external control systems (e. g., chain hoist control systems, PLCs, safety equipment etc.) to be incorporated and can support an unlimited number of connected operating panels or controlled axes.

The COSTACOwin® control system is divided into three levels:
– user level
– server level, and
– drive level

User level

The user level forms the interface between man (user) and the COSTACOwin® control system (see Figure 2.33). Four different *control panels* are available for users:
– the main panel SCOUT Eagle
– the secondary control panel SCOUT Milan
– the portable SCOUT Hawk
– with a wireless version – SCOUT Hawk radio
– and the SCOUT Merlin maintenance panel

The operating panels run the operating system Windows® embedded, which has proven its stability and reliability in many industrial applications.

The user interface is designed for operation on multiple screens and is consistently tuned for touch screen operation. Due to the clear structure and up to four configurable

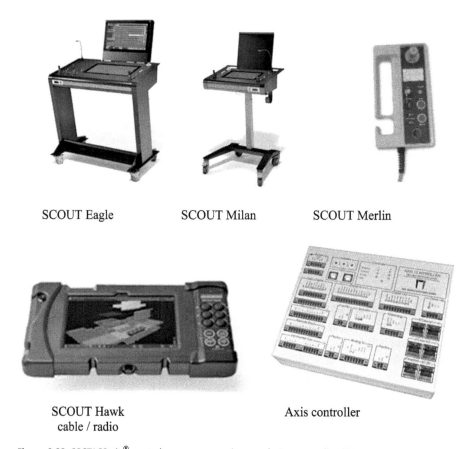

SCOUT Eagle   SCOUT Milan   SCOUT Merlin

SCOUT Hawk
cable / radio

Axis controller

Figure 2.33: COSTACOwin® control system – operating panels. Image credits: SBS.

widgets per screen, the operator always maintains the overview. The system guides the operator through the various views and facilitates simple individual movements as well as complex sequences to be programmed with just a few command inputs. Individual elements like machines or programmed movements can be easily exchanged via drag and drop between the different widgets. Functions and parameters currently not required are hidden.

Up to four widgets per screen configurable in content and size allow users to navigate the control system without losing orientation. The user interface offers various views such as movement table, side and topography view and makes it easy to change between them. This gives the best view of the current process while allowing a fast change to the overall view.

SCOUT Eagle is a modular-design main operating panel. It can be modified to meet customer requirements. External controls such as an intercom and building services can be easily integrated. It is equipped with at least four joysticks and two independent screens to enable multiuser operation.

SCOUT Milan complements the main operating panel as a mobile console, trolley-mounted as an option. It can also be used as the main operating panel on small and medium-sized stages.

SCOUT Hawk is a light portable auxiliary control panel. Long battery runtimes enable continuous work. The wireless SCOUT Hawk radio operates as a fully-functioning console via wireless signal exchange.

SCOUT Merlin is a local operating terminal developed for testing and maintenance. It is connected directly to the drive and enables simple movements controlled on sight, completely bypassing the computer control system. Individual parameters can be entered via pushbuttons and potentiometers.

The **SCOUT 100 universal outlet** connects the different operating panels to the control system. Connection and disconnection of consoles is possible during operation. Every universal outlet is fitted with contacts for the emergency-off signal of the panel's e-stop button. COSTACOwin® allows using an unlimited number of universal outlets to be used within one installation, so that the SCOUT operating panels can be deployed very flexibly.

Server level

The server level components include:
- the main computer
- database
- remote maintenance
- data logger, and
- network

Main computer

The main computer processes all the cross-system computing and safety processes. A real-time operating system especially designed for safety-critical industrial applications is used. Parallel to the main computer, a safety computer calculates the movement target values using different calculation methods. Subsequently the results of the two calculations are compared and the system is shut-down in case of an error.

Data is managed in a *database system* which stores all system, standard and plant configuration data as well as user data. Additionally, software updates can be distributed and service data is provided for download.

Remote maintenance

Remote maintenance enables efficient access by SBS specialists to quickly analyses and remedy problems. This applies to all computer-based components such as servers, database systems, operating panels and axis control computers.

Data logger

The data logger is a memory module for recording all movement, usage and error data of the machine installation over a longer period of time (configurable; several weeks, months depending on the scenario of use). This enables targeted usage and error analyses of individual machines or the entire installation to be carried out in order to gain a better understanding of how the machinery is used. The data is accessible by means of a separate operating program and thus is independent of the performance and rehearsal operation.

Network

The network provides a secure connection between all control system components. It consists of two sectors:
–   a standard Ethernet connecting the server and user interfaces, as well as
–   an industrial real-time network connecting the server and drive level

Drive level

The core component of the drive level is the *axis controller*. It controls the machine movements based on the operator's target values and the individual machine states. This safety controller was specially designed for the use in entertainment technology. Furthermore, the axis controller offers a redundant bus communication interface and can be replaced within a few minutes. Due to the ring structure of the real-time network only one axis is affected in the event of an error; there is no repercussion on the system as a whole. External systems (electrical, hydraulic, safety equipment, etc.) can also be integrated.

The dual-channel axis controller features two independent controllers that perform the calculation of all motion and safety functions in parallel. It acquires all analogue and digital input and output signals of the respective drive and triggers the corresponding motion and safety functions. The axis controller can be used universally and is designed to be easy to maintain. All electrical connections are pluggable, which makes it easy to replace the device. Together with the main computer/safety computer of the server level, the axis controller(s) form a deterministic system with multiple backups, that eliminates systematic errors.

Control cabinets

The electric plant can be concentrated, with cabinets at a few central locations, or distributed throughout the venue. The **SCUBE** drive cabinet is the basis for the distributed system structure. Its compact design allows it to be mounted directly at the drive. Each drive can be coupled to its respective cabinet via plug-in connector. The main advantages of SCUBE are: variable applicability, low installation costs, short commissioning times and minimized interference.

Some details on the user interface:
-   The typical system and performance management functions are to be found in the main menu, a. o. system maintenance, managing user data, as well as functions for import and export, loading and saving performance data.
-   The status bar shows which mode of the control system (live, performance or simulation) is currently active and which user is logged in, as well as status information about the main computer and the database.
-   The navigation bar allows the user to switch between the different views with a single click. In addition to the true-to-scale side view with a configurable line of sight 3D views are available as an option, as well as views specific to a particular customer and installation. The most important views for the theatre operations are topography view and programming view. The latter is used to program or modify the intended sequence of machine movements of a performance.
-   Functions and information which are necessary irrespective of the current display setting are permanently visible at the bottom of the screen in the global function bar. This includes the selection and deselection of joysticks as well as system messages.

Performance structure

The programming of a performance follows a clear logical system. This comprises four levels:
-   Level 1 is the lowest level. It contains every drive with all its parameters.
-   In level 2, drives are aggregated into groups and assigned with movement parameters and dependencies. A group can contain any number of drives.
-   In level 3 (cues, transformations), groups are combined into cues and specified further. A cue can contain any number of groups.
-   Level 4 (= performance)
    A performance contains a sequence of cues, which are executed one after another. The number of cues to be integrated is unlimited. If desired by the operator or stage manager, cues can also be executed in a free sequence. Furthermore, it is possible to skip individual cues, repeat them or execute them at a different time. Thus, the system offers a wide range of flexibility to respond to the requirements of the individual performance.

Movement functions

The basic functions available in each view include:
-   Movement between operating limits
-   Movement between any programmed targets
-   Movement to a target position (numeric input or marker)
-   Differential movement (specified distance from the current position)

Any movement functions that go beyond the basic functions can be implemented in the movement table view. This applies to, for example, synchronized and effect movements.

Synchronized movements – group travel

Asynchronous movement means that all the drives of the group are run at the preset speed. Each drive reaches its final target in accordance with its movement parameters, regardless of the other drives in the group. An asynchronous group movement can also be equipped with a monitoring function at group level.

Time-synchronous movement means that all associated drives arrive at their targets simultaneously, regardless of the individual distances. The respective speeds are set by the operating system.

Path-synchronous movement means that all the drives in the group travel the same distance. The distances between them are precisely maintained according to the programming, for example, in order to move a scenery element that is connected to several drives.

Effects

These are group movements, the dependencies of which are defined to achieve certain movement effects. The standard program includes the following effects:

A *beam movement* combines several drives of a group. The user only needs to define the movement parameters of the first and last drive. The start and target values, as well as the movement speeds of the remaining drives between them, will be adjusted automatically.

A *base point movement* aggregates individual drives or a group of drives and controls their movement according to the programmed time/speed/position profile. Graphical representations of the resulting movement curves and speed profiles over time for each drive indicate progress and result of the base point movement. Parameterization is done via dialogue and/or by graphical editing of the individual base points.

An *oscillation movement* directs individual drives or a group of drives between the start and the target position within a specified interval (oscillation factor). Parameters to be entered by the operator are the number of repetitions or the duration of the oscillation.

Advanced features

COSTACOwin® offers a large number of additional functions:

The *decoration management* administrates stage decorations together with their specific parameters in a database. The decorations may be allocated to one or more performances, allowing for them to be quickly found and reused.

The *installation plan* represents the desired state of the stage with regard to the decorations used. It links drives to the different decorations. Whether the drive or decoration is used in the current performance is irrelevant.

The *user administration* holds the data of all users, their respective access authorizations and code allocations for the contactless authentication system at the operating panels by means of magnetic key, transponders and other methods. COSTACOwin® also features convenient switching of languages at runtime of the software.

Print function, individual panel messages, calculator, context-sensitive help system, on-board documentation and a flexible USB interface for import/export functions complete the scope of this stage control software.

VISTOR – 3D visualization

The VISITOR visualization environment offers the possibility to virtually display and test all movements of the stage setting throughout the entire performance, including stage lighting. Thus, for example, collisions can be detected and avoided by changing the programmed movements. VISTOR therefore can be used both, during development of a performance sequence as well as for checking the programming.

The 3D visualization includes all elements: the stage machinery, the stage setting and decorations, the scenic lighting, as well as the stage building with the auditorium. Thus, it enables a realistic impression of a future performance at an early stage. VISTOR simulates how real objects behave on stage based on physical laws. The 3D visualization also includes stage lighting. The visualization emulates the behavior of the stage machinery equipment and can be coupled with the COSTACOwin® stage control software.

The 3D visualization is characterized by:
- Real-time visualization.
- Realistic physical simulation.
- Extremely accurate details.
- Freely selectable viewing position.
- Simple decoration management (decoration import from CAD data).
- Collision detection between all elements.
- Easy performance and scene management (scene import).
- Various, freely selectable views.
- Possibility of overlaying a visualization with real images.
- Optimal integration of the scenery of visiting productions and road show equipment.

Automatic tracking
COSTACOwin® provides a solution that enables lighting, audio and video equipment to elegantly interact with the stage control systems' position information and to respond to it.

COSTACOwin® provides this position information in appropriate formats, such as Art-Net for lighting and video, and OSC for audio and camera protocols. The interlocking with stage movements offers the possibility of automatic adjustment of lighting, audio and video to changes on stage. The automatic tracking, for example, allows a video wall to move while the projection is running. The automatic tracking changes the light and color of the moon as required although it moves across the stage as it rises and sets.

Conclusion

The COSTACOwin® stage control system offers all functionality required for safe and productive stage operations in theatres, opera houses, concert halls, and musical theaters. Performance sequences can be adapted to the requirements of the particular performance by means of a large number of different order levels and movement parameters. In addition, special effects offer programming macros that can be used to parameterize complex movement sequences quickly and easily. Due to its decentralized structure, COSTACOwin® can be operated and programmed simultaneously by several operators. Various console series are available. This means that all user requirements, from municipal theaters to musical theaters, can be met.

2.5.8 SYB 3.0 control technology from Bosch Rexroth

For more than four decades, Rexroth has stood for customized stage technology solutions worldwide – from pioneering new cultural buildings to functional and substance-preserving renovations and modernizations during short breaks in performance.

Figure 2.34 illustrates the development of Bosch Rexroth stage control technology, and the SYB 3.0 control system is explained in more detail below.

The illustration shows the evolution from mechanical switch elements to the modern touch screen and multitouch system.

The SYB 3.0 system combines electric, hydraulic and hybrid drive technology with modern electronic control technology specially designed for stage technology. The system is decentralized with redundant components. The drive and control technologies have been developed in-house and guarantee high safety and availability in Bosch quality.

Thanks to its strictly modular design, the control system covers the requirements of every stage and also allows for future expansions with minimal effort. Up to 700 drives and 16 control panels can be integrated.

SYBNet is a specially developed deterministic industrial Ethernet network. With a transmission rate of up to 1 Gbit/s, it offers real-time communication with high bandwidth for the exchange of information and data. This creates room for extended functionality. The SYB 3.0 also supports standard fieldbuses for seamless integration of ad-

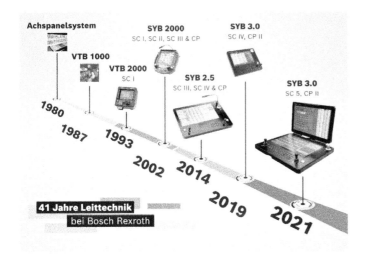

Figure 2.34: Forty one years of Bosch Rexroth control technology. Picture credits: Bosch Rexroth AG.

ditional components. The result is high synchronization accuracy for optimally coordinated moves.

The stage control technology also offers all the options for easy integration of lighting, sound and other components outside the actual stage automation.

The SYB 3.0 control system is divided into the operating, control and drive levels (see Figure 2.35).

The operator level or HMI is the interface between the operator and the SYB control system and consists of the following components:
– SC5 and CPII operator panels.
– RC room computer and SM status monitor.
– VPN remote access.
– Central storage (NAS – RAID 1).

The control level includes:
– Console computer (Redundant Master Controller MC).
– Axis Controller BLE (Redundant Axis Controller AC).
– Bus Master (BM).
– Secure storage (Data Concentrator DC).

The drive level is shown below, which can be:
– Frequency controlled electric motors (induction and servo).
– Frequency controlled electromechanical linear actuators (EMA).
– Frequency controlled hydraulic linear axes (EHZ, Cytro Force, stand-alone axes).
– Valve-controlled hydraulic motors.
– Valve-controlled hydraulic cylinders.
– Frequency controlled hybrid drives.

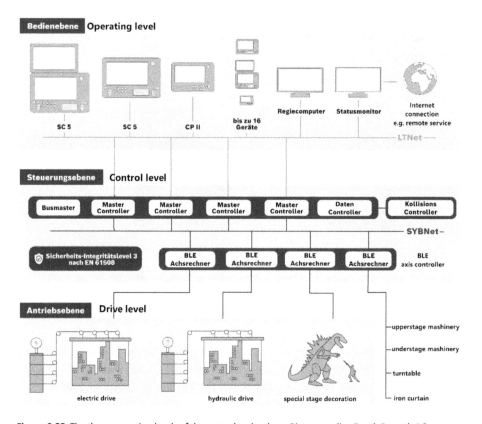

Figure 2.35: The three operating levels of the control technology. Picture credits: Bosch Rexroth AG.

The system allows the synchronization of more than 128 axes per group. A particular advantage of this structure is that it reduces complexity and does not require any on-site adjustments when control modules are changed. All drive and control technologies have been developed by Bosch Rexroth.

The entire control technology is operated via the very compact 22" SC5 *control panel* and the smaller CPII version. The compact CPII Desk offers the same user interface and functionality (with the exception of one control lever and no SYB keys) as the SC5 control panel and is also available as a Wi-Fi version (Figure 2.36).

SYB SC5 at a glance:
– 22" 10-point capacitive multitouch with gesture support.
– Full HD resolution.
– Compact dimensions W × H × D in mm: 600 × 460 × 60.
– Extremely powerful with Intel Core i7 processor.
– Four start and start-inverse buttons to start programmed moves. Additional slow down/acceleration of individual moves (override function).

SYB SC5   SYB CPII

Figure 2.36: SYB SC5 and CP II control panels. Picture credits: Bosch Rexroth AG.

- Completely silent and maintenance-free without a fan thanks to an innovative cooling concept, optimized for the acoustically demanding stage area.
- Expandable with a second 22" multitouch monitor, gooseneck lamp and configurable SYB frames.
- SYB Keys → context controlled, tactile LCD keyboard with 32 keys.
- Two safe joysticks with Hall sensors for starting movements and for slowing/accelerating programmed moves (sum override function).
- Standard VESA mount for mounting on brackets (mobile trolleys, rail systems, swivel brackets, etc.).
- Weight: 13.3 kg.

SYB CPII at a glance:
- 13.3" capacitive 10-point multitouch with gesture support.
- Full HD resolution.
- Extremely compact dimensions W × H × D in mm: 402 × 320 × 62.
- Four start and start-inverse buttons to start programmed movements. Additional slow down/acceleration of individual moves (override function).
- One safe control lever with Hall sensors for starting moves and for braking/accelerating programmed moves (total override function).
- Wired or Wi-Fi.
- Integrated handle for mobile use.
- Standard VESA mount for attachment to mounting brackets (mobile trolleys, rail systems, swivel mounts, etc.).
- Weight: 4 kg.

The SYB SC5 base unit can be expanded at any time with a second 22" multitouch display and various SYB frames (see Figure 2.37).

With the SYB keys, the SC5 offers an integrated, context-controlled LCD keypad to support the previous touch operation.

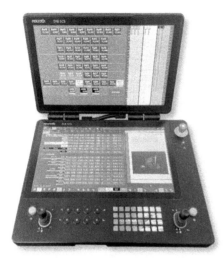

Figure 2.37: SYB SC5 with second 22" multitouch display. Picture credits: Bosch Rexroth AG.

The multipurpose assignment of the LCD keyboard (Figure 2.38) combines the advantages of touch operation and the haptic keyboard. On the one hand, context-sensitive, flexible operation and, on the other hand, haptic feedback and short distances.

Figure 2.38: Context-controlled LCD keyboard.

In addition to this development of the operator panels, Bosch Rexroth uses modular expansion modules, so-called SYB frames (Figure 2.39). The control panels can also be upgraded with SYB frames. In addition to standardized variants with additional control levers, rotary potentiometers, signal lamps, gooseneck lamps, etc., customized SYB frames are also possible.

In particular, SYB 3.0 control technology offers the ability to program, simulate and archive a variety of scenic moves up to complete shows. It allows several people to work

Figure 2.39: SC5 control panel including second screen and frames.

simultaneously on one or more shows. This means that one or more operators can program movement sequences, while movements are executed in parallel. Multiuser capability on all control panels adds flexibility, saves time and increases efficiency. Intelligent functions support the safe and fast realization of complex cues and shows. For example, the autocorrection function ensures that the decoration is always positioned exactly as required, even if the height of the drives has changed due to other lifting equipment. This eliminates the need for time-consuming manual adjustment of movement sequences within preprogrammed shows. The setup function helps the user to jump between existing moves and programmed cues during rehearsals by displaying all the drives and their targets that are not yet in their correct but required position, allowing the user to load these moves at the touch of a button and correct them so that the rehearsal can be continued in a new position.

Numerous other functions, such as the scenery database, grid floor management, flexible cue structures and scenery-oriented programming and operation, help to simplify the implementation of complex processes and thus the handling of the technology. The possibility of connecting to external systems is made available by means of appropriate additional inputs. This enables synchronized motion sequences of the machinery with lighting and sound technology via common triggers (e. g., time code), as well as networking with other systems via standardized interfaces such as Streaming ACN (sACN) or PosiStageNet (PSN).

Rexroth's SYB 3.0 platform control technology offers maximum safety in any configuration. Sensors that are adapted to these requirements and a large number of software functions of the Rexroth safety system prevent dangerous situations from arising. The collision assistant, for example, helps the operator to carry out movements safely. The software alerts the operator to potential conflicts before and during the move, giving him enough time to respond.

The stage control system displays the operating status of the entire site on all the control panels as well as at a central location. It collects and manages operator input, operating data, system information and fault messages. The integrated expert system uses rule-based analysis to simplify the identification of fault causes, thereby increasing availability. In the event of a fault, a secure Internet connection can be used for quick remote diagnosis and on-site operator support.

A contactless transponder system protects against unauthorized access, grants different levels of authorization and individualizes the associated control panel. The operator can interrupt his work, log out and, if necessary, continue at another console after logging in as required. Authorizations are defined and set by the customer using an administration tool. It is possible to restrict both functions and access to drives.

Three modes of operation cover all tasks:

- Manual mode provides the basic functions and parameters required for simple stage work and setup.
- Automatic mode provides the functions and parameters for programming, execution and supervision of complex movements in performance mode as well as in test and setup mode. Movement data, cue data and scenery data can be created and stored in a structured manner in a performance database.
- Offline mode offers most of the same functions and parameters as automatic mode. However, no movements can be executed. Complex movements as well as cue and scenery data can be created and stored in a structured way in performance databases without the need to occupy the system.

Bosch Rexroth pays particular attention to functional safety based on the current state-of-the-art and thus fulfils the applicable normative regulations, in the first place EN 17206. The primary objective is to ensure the safety of all participants on and around the stage as well as the highest possible availability of the system for smooth operation.

The Information Center contains the PDF Viewer. Here, the Functional Description, Brief Description, System Description and Help functions are included as standard. These documents and user-specific PDF documents can be accessed easily and intuitively. Different display modes allow documents to be viewed while working or in full screen mode.

The message service is also to be found in the information center. Messages can be sent to individuals or groups of people (e. g., shift change, etc.). As soon as a user logs in to a control panel, their new messages or other information are displayed.

With the selective emergency mode, it is possible to bypass individual safety functions. This is used when external sensors are faulty, safety devices detect a fault and/or collision monitoring of individual drives is to be deliberately disabled so that the affected drives can still be moved. In this mode, individual checks (safety functions) can be bypassed or deactivated for individual damaged drives. Controlled movement is possible even if monitoring functions are selectively bypassed. It is therefore still possible to carry out controlled movement with the same operation and functionality as in normal operation. Targeted and grouped travel is also possible without restriction, provided the safety instructions are observed.

Bosch Rexroth pursues a holistic approach to safety. Risk analyses are the basis for identifying a wide range of potential hazards. Technological responses to these potential hazards are formulated and implemented. These are the so-called safety functions, which are primarily implemented in the control system in order to eliminate the potential hazards "in a fail-safe manner." In SYB3.0, Safety Integrity Level 3 (SIL3) is implemented for all safety functions. The dual-channel axis controller and bus structure ensures that the overall system will continue to operate safely even if individual components fail.

In addition, all components used in the control system (encoders, sensors, load measuring devices, etc.) and their functional integrity must be set to at least Performance Level PLD, corresponding to the safety functions of the control system. EN 13849, parts 1 and 2, is the standard to be considered for this.

3 Fundamentals of mechanics (mechanics of solid bodies and fluid mechanics)

This chapter is not intended to replace a textbook on mechanics, nor does it claim to be scientifically exact. Rather, it is intended to remind readers about a corresponding background of some aspects of the field of mechanics, which serve for a better understanding of statements made in other chapters of the book. In addition, it is intended to provide a collection of formulas for calculations tailored to this specific need, without, however, giving exact indications for the range of validity of the formulas. Therefore, their application presupposes a basic knowledge, or one must revert to appropriate technical literature (see also references for supplementary literature to the parts 2, 3 and 4).

3.1 The International System of Units

Before some basic principles and formulas of mechanics are explained, the International System of Units shall be introduced in the aspects relevant here. This system uses so-called *coherent units*. This means that formulas can be worked with without using conversion factors if the parameters occurring in them are used in these units.

Without going into the physical principles in more detail, it should be noted that first so-called "base units" (see Table 3.1) were defined, from which the other units (see Table 3.2) can then be derived according to physical laws.

Table 3.1: Coherent basic units of mechanics.

Mass	[kg]	kilogram	m
Length	[m]	meter	s
Time	[s]	second	t
Current	[A]	ampere	I
Temperature	[°K]	degree Kelvin	θ
	[°C]	degree Celsius	$\theta\,°K = \theta\,°C + 273.15$

(For the difference between two scale values, 1 °K = 1 °C = 1 grd.)

In the following lists, the unit designations are given in square brackets, and the letter designation most often used for the term in formulas is also given. For example, a mass is often abbreviated with the letter "m". The unit of a mass is kilogram, abbreviated "kg".

Multiples and parts are formed according to Table 3.3.

https://doi.org/10.1515/9783111366968-003

Table 3.2: Derived units.

Speed	[m/s]		v (velocity)
Acceleration	[m/s^2]		a (acceleration)
Angular velocity	[1/s]		ω
Angular acceleration	[1/s^2]		ε
Area	[m^2]		A (area)
Volume	[m^3]		V (volume)
Specific mass (density)	[kg/m^3]		ρ
Moment (torque)	[Nm]		M
Rotation angle	[rad], [1]		φ
Force	[N]	Newton	F (force)
Work	[J]	Joule	W (work)
Power	[W]	Watt	P (power)
Pressure	[Pa], [bar]	Pascal	p
Electrical voltage	[V]	Volt	U

Table 3.3: Formation of multiples and parts (excerpt).

Factor	Prefix	Prefix sign
10^9	giga	G
10^6	mega	M
10^3	kilo	k
10^2	hecto	h
10	deka	da
10^{-1}	deci	d
10^{-2}	centi	c
10^{-3}	milli	m
10^{-6}	micro	μ
10^{-9}	nano	n
10^{-12}	pico	p

3.2 Fundamental terms of kinematics

Motion sequences, including the working movements of stage equipment, can be broken down into time intervals in which:

– The movement takes place with variable speed (phase of acceleration or deceleration).

– The movement proceeds at constant speed (in equal time intervals Δt, equal distances Δs are traversed).

For the acceleration and deceleration phase, the special case in which the velocity increases or decreases by the same value Δv in equal time intervals Δt is of particular interest. This is then called a *uniformly accelerated* or *uniformly decelerated motion*, since the value of acceleration or deceleration remains constant over time. Such motion situations always occur when forces of constant magnitude have an accelerating or decelerating effect. This is at least approximately the case in most applications.

For example, the motion sequence of a motor-driven load bar of a bar hoist can be decomposed during lifting:

- Into a motion section in which the velocity is uniformly increased from standstill, i. e., $v_0 = 0$, to the velocity v_1 (acceleration phase with constant acceleration a_c);
- Into a movement section in which the load bar is moved at constant speed $v_c = v_1$; and
- into a motion section in which the speed of the load bar is again decelerated uniformly from the speed v_1 to $v_0 = 0$, i. e., to a standstill (phase with constant deceleration a_c); a_c can be of different quantity during acceleration and deceleration, but is usually of the same magnitude.

The example just given represents a linear motion – a so-called *translation*.

An analogous situation is, of course, also given with rotating motion – so-called *rotation*.

Instead of a distance Δs [m], the angle of rotation $\Delta \varphi$ [rad] is then decisive, instead of the velocity v [m/s] the angular velocity ω [1/s] and instead of the acceleration a [m/s^2] the angular acceleration ε [1/s^2].

The following formula relationships exist for these kinematic units (translation and rotation):

3.2.1 Translation

(a) For a linear motion (translation) with constant velocity v_c [m/s], one has according to Figure 3.1(a):

$$v = \frac{\Delta s}{\Delta t} = v_c = \text{const.},$$
$$\Delta s = v_c \cdot \Delta t, \tag{3.1}$$
$$s = s_0 + \Delta s.$$

(b) For a motion with constant acceleration a_c [m/s^2], starting from an initial velocity v_0 at time $t = 0$, the following applies according to Figure 3.1(b, c):

$$a = \frac{\Delta v}{\Delta t} = a_c = \text{const.},$$
$$\Delta v = a_c \cdot \Delta t,$$
$$v = v_0 + \Delta v = v_0 + a_c \cdot \Delta t = \sqrt{v_0^2 + 2 \cdot a_c \cdot \Delta s}, \tag{3.2}$$
$$\Delta s = v_0 \cdot \Delta t + \frac{1}{2} a_c \cdot (\Delta t)^2 = \frac{v_0 + v}{2} \cdot \Delta t,$$
$$s = s_0 + \Delta s.$$

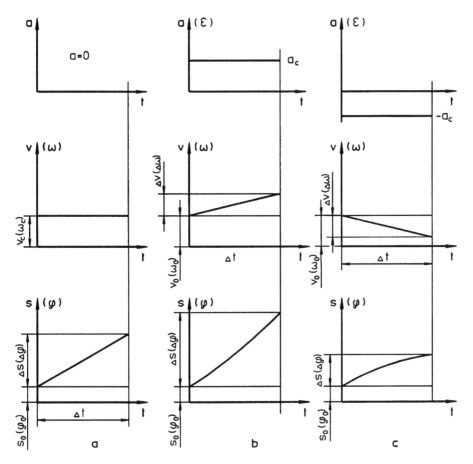

Figure 3.1: Time diagrams for motion with constant: (a) velocity, (b) acceleration, (c) deceleration. (Travelled distance s, velocity v and acceleration a for translation; angle of rotation φ, angular velocity ω and angular acceleration ε for rotation.)

For the special case of initial velocity $v_0 = 0$ (accelerating from standstill and $s_0 = 0$),

$$v = \Delta v = a_c \cdot \Delta t = \sqrt{2 \cdot a_c \cdot \Delta s},$$

$$\Delta s = \frac{1}{2} a_c \cdot (\Delta t)^2 = \frac{v^2}{2 \cdot a_c} = \frac{v}{2} \cdot \Delta t, \qquad (3.3)$$

$$\Delta t = \frac{v}{a_c},$$

where
Δt time interval [s]
s_0 position mark at time $t = 0$ [m]
s position mark at time t

Δs distance traveled in the time interval Δt [m]
v speed [m/s]
a_c acceleration [m/s^2]
  $a_c > 0$ positive acceleration means that the speed increases
  $a_c < 0$ negative acceleration (= deceleration) means that the speed decreases

For *free fall* with the acceleration due to gravity $a_c = g$ with the initial velocity $v_0 = 0$, the velocity after falling a height of $\Delta s = h$ is therefore

$$v = \sqrt{2 \cdot g \cdot h} \quad \text{after the time } \Delta t = \frac{v}{g} = \sqrt{\frac{2h}{g}}. \tag{3.4}$$

In the case of an *upward throw* with initial velocity $v_0 > 0$, g has a decelerating effect and the velocity after a time interval Δt *is*

$$v = v_0 - g \cdot \Delta t.$$

The velocity $v = 0$ is reached after the time interval Δt,

$$\Delta t = \frac{v_0}{g}.$$

From $\Delta s = v_0 \cdot \Delta t - \frac{1}{2}a_c \cdot (\Delta t)^2$ one obtains for $a_c = g$ and $\Delta t = \frac{v_0}{g}$ the throwing height h:

$$\Delta s_{\max} = h = \frac{v_0^2}{2 \cdot g}, \tag{3.5}$$

where h is the falling or throwing height [m].

3.2.2 Rotation

(a) For a rotation with *constant angular velocity* ω_c [1/s], one has according to Figure 3.1(a):

$$\omega = \frac{\Delta \varphi}{\Delta t} = \omega_c = \text{const.,}$$
$$\Delta \varphi = \omega_c \cdot \Delta t, \tag{3.6}$$
$$\varphi = \varphi_0 + \Delta \varphi.$$

(b) For a rotation with *constant angular acceleration* ε_c [1/s^2], from an angular velocity ω_0 at time t_0, the following equations are valid according to Figure 3.1(b, c):

$$\varepsilon = \frac{\Delta\omega}{\Delta t} = \varepsilon_c = \text{const.},$$

$$\Delta\omega = \varepsilon_c \cdot \Delta t,$$

$$\Delta\omega = \omega_0 + \Delta\omega = \omega_0 + \varepsilon_c \cdot \Delta t = \sqrt{\omega_0^2 + 2 \cdot \varepsilon_c \cdot \Delta\varphi}, \tag{3.7}$$

$$\Delta\varphi = \varphi_0 \cdot \Delta t + \frac{1}{2}\varepsilon_c \cdot (\Delta t)^2 = \frac{\omega_0 + \omega}{2} \cdot \Delta t,$$

$$\varphi = \varphi_0 + \Delta\varphi.$$

For the special case $\omega_0 = 0$ (acceleration from standstill) and $\varphi_0 = 0$, the above equations become:

$$\omega = \varepsilon_c \cdot \Delta t = \sqrt{2 \cdot \varepsilon_c \cdot \Delta\varphi},$$

$$\Delta\varphi = \frac{1}{2}\varepsilon_c \cdot (\Delta t)^2 = \frac{\omega^2}{2 \cdot \varepsilon_c} = \frac{\omega}{2} \cdot \Delta t, \tag{3.8}$$

where
Δt time interval [s]
φ_0 angle mark at time $t = 0$ [m]
φ angle mark at time t
$\Delta\varphi$ angle passing through in the time interval Δt [rad]
ω angular velocity [1/s]
ε_c angular acceleration [1/s^2]
    $\varepsilon_c > 0$ positive acceleration means that the speed increases
    $\varepsilon_c < 0$ negative acceleration (= deceleration) means that the speed decreases

An angle a, so also the angle of rotation φ, can be indicated either in *radians* or *degrees*. In the above formulas it is to be entered in radians [rad]!

An angle a of 360° is 2π in radians and thus corresponds to the circumference $u = 2r\pi$ of the unit circle with $r = 1$.

Therefore, in general, for an angle a [°] = $\frac{180}{\pi} \cdot a$ [rad].

The speed with which rotating movement takes place can be specified with the term *angular velocity* in [1/s]. However, it is also possible to describe the movement by specifying the number of revolutions per second or minute, the so-called *rotational speed (rotational frequency)* with the unit [rps = 1/s] or [rpm = 1/min]. Or the speed v in [m/s] at a defined distance r from the center of rotation is specified as the *circumferential speed*. Thus, the numerical value of angular velocity can be interpreted as circumferential velocity in [m/s] and the numerical value of angular acceleration as acceleration in [m/s^2] measured at radius 1 m. The following formulas result between these three quantities describing the rotation (Figure 3.2):

$$\omega = \frac{v}{r},$$

$$n = \frac{\omega}{2 \cdot \pi} = \frac{v}{2 \cdot r \cdot \pi} = \frac{v}{d \cdot \pi}, \quad \text{resp.} \ \omega = 2 \cdot \pi \cdot n, \tag{3.9}$$

$$v = d \cdot \pi \cdot n,$$

$$n^* = \frac{30 \cdot \omega}{\pi} = \frac{30 \cdot v}{r \cdot \pi} = \frac{60 \cdot v}{d \cdot \pi},$$

$$v = \omega \cdot r = d \cdot \pi \cdot n = \frac{d \cdot \pi \cdot n^*}{60}, \tag{3.10}$$

$$\varepsilon = \frac{a}{r},$$

where

ω angular velocity [1/s]

v peripheral speed [m/s] at radius r or diameter d

a acceleration [m/s^2] at radius r resp. diameter d

ε angular acceleration [1/s^2]

n speed [rps = 1/s]

n^* speed [rpm = 1/min]

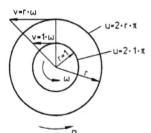

Figure 3.2: Terms for rotational speed: circumferential speed, angular speed, and rotational speed.

3.3 Fundamental terms of dynamics

In this section, the most important basic formulas of mechanics will be summarized again, which have already been used in Table 3.2 to explain the "derived units" of the International System of Units. They can be used to investigate the motion behavior of simple systems under the influence of forces and moments.

The plane motion of a rigid body can be described by the *principle of linear momentum* (*center of gravity theorem*) for translation

$$F = m \cdot a \tag{3.11}$$

and the *principle of angular momentum* for rotation

$$M = I \cdot \varepsilon \tag{3.12}$$

where

F resultant of all forces acting on the rigid body [N]
M moment of external forces around the axis of rotation [Nm]
m mass [kg]
I mass moment of inertia with respect to the axis of rotation (perpendicular to the plane of motion) [kg m^2]
a acceleration [m/s^2]
ε angular acceleration [1/s^2]

3.3.1 Kinetic energy – energy of motion

The *kinetic energy* represents the working capacity of a moving mass. Expressed in formulas, this means, applied to a mass moving translationally at the speed v,

$$E_{kin} = \frac{m \cdot v^2}{2}, \tag{3.13}$$

and applied to a mass rotating with angular velocity ω with mass moment of inertia I,

$$E_{kin} = \frac{I \cdot \omega^2}{2}. \tag{3.14}$$

The *energy theorem* (*work theorem*), applied to a rigid body, states that the increase in kinetic energy of a body in any time interval is equal to the work done by the external forces in that time interval,

$$\Delta E_{kin} = E_{kin2} - E_{kin1} = W_{1,2},$$

where
E_{kin} kinetic energy [Nm = J]
W work [Nm = J]
Index 1, 2   Condition 1, 2

3.3.2 Potential energy – energy of the position

If a mass m is lifted against the effect of gravity by the height h, the work to be done for this must be

$$W_{1,2} = m \cdot g \cdot \Delta h.$$

Thus, *potential energy* (work capacity) is supplied to the mass:

$$\Delta E_{pot} = E_{pot2} - E_{pot1} = W_{1,2},$$
$$\Delta E_{pot} = m \cdot g \cdot \Delta h, \tag{3.15}$$

where Δh is the height difference [m].

3.3.3 Braking work

If a mass m is decelerated from the speed v_1 at the braking distance s_{brems} to the speed v_2, the braking work W_{brems} must be performed, which is converted into heat. When braking to a standstill, set $v_2 = 0$.

With the effect of a braking force F_{brems} during the time t_{brems}, the braking work can be determined as (see Figure 3.3):

$$W_{brems} = \int_0^{t_{brems}} F_{brems} \cdot v \cdot dt = F_{brems} \cdot \int_0^{t_{brems}} \left(\frac{v_2 - v_1}{t_{brems}} \cdot t + v_1 \right) dt$$

$$= F_{brems} \cdot \frac{v_1 + v_2}{2} \cdot t_{brems},$$

$$W_{brems} = F_{brems} \cdot \frac{v_1}{2} \cdot t_{brems} \quad \text{if } v_2 = 0.$$

(3.16)

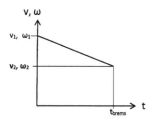

Figure 3.3: Deceleration with constant braking force or braking torque.

Similarly, if a braking torque M_{brems} acts during time t_{brems}:

$$W_{brems} = \int_0^{t_{brems}} M_{brems} \cdot \omega \cdot dt = M_{brems} \cdot \int_0^{t_{brems}} \left(\frac{\omega_2 - \omega_1}{t_b} \cdot t + \omega_1 \right) dt$$

$$= M_{brems} \cdot \frac{\omega_1 + \omega_2}{2} \cdot t_{brems},$$

$$W_{brems} = M_{brems} \cdot \frac{\omega_1}{2} \cdot t_{brems} \quad \text{if } \omega_2 = 0.$$

(3.17)

3.3.4 Application examples

Free fall of a mass m

The final velocity after a free fall over the height h according to Eq. (3.4) can therefore also be derived by equating the potential energy released with the increase in kinetic energy, i. e.,

$$m \cdot g \cdot h = \frac{m \cdot v^2}{2}.$$

From this follows, as given in Eq. (3.4),

$$v = \sqrt{2 \cdot g \cdot h}. \tag{3.18}$$

Lowering of a load with constant speed *v*

If, for example, a podium is to be lowered at constant speed, the amount of potential energy released must be applied as braking energy. Since the podium is lowered at constant speed, the kinetic energy is retained,

$$\Delta E_{pot} = m \cdot g \cdot \Delta h.$$

The braking work to be performed in the time interval t_{brems} at constant velocity v is given by

$$W_{brems} = F_{brems} \cdot v \cdot t_{brems}.$$

The braking force must correspond to the weight of the load, i. e., $F_{brems} = m \cdot g$, and with $t_{brems} = \frac{\Delta h}{v}$ becomes

$$W_{brems} = m \cdot g \cdot v \frac{\Delta h}{v} = m \cdot g \cdot \Delta h = \Delta E_{pot}.$$

If this is done, for example, by throttling a podium moved by a hydraulic cylinder, this energy is converted into heat and leads to heating of the hydraulic fluid at the throttle. The heating can be calculated with Eq. (3.94).

Deceleration of a load from a lowering speed to standstill

If the braking process takes place on a braking distance s_{brems} with deceleration a_{brems}, the braking work must correspond to the kinetic energy to be dissipated and the potential energy released by the lowering (see Eqs. (3.3), (3.13) and (3.16)),

$$E_{ges} = \frac{m \cdot v^2}{2} + m \cdot g \cdot s_{brems} = \frac{m \cdot v^2}{2} + m \cdot g \cdot \frac{v^2}{2 \cdot a_{brems}} = \frac{m \cdot v^2}{2}\left(1 + \frac{g}{a_{brems}}\right).$$

The same result is obtained by considering the braking work according to Eq. (3.17). The required braking force in this case is given by Eq. (3.3):

$$F_{brems} = m \cdot (g + a_{brems}),$$
$$t_{brems} = \frac{v}{a_{brems}}, \tag{3.19}$$
$$W_{brems} = F_{brems} \cdot \frac{v}{2} \cdot t_{brems} = m \cdot (g + a_{brems}) \cdot \frac{v}{2} \cdot \frac{v}{a_{brems}} = \frac{m \cdot v^2}{2}\left(1 + \frac{g}{a_{brems}}\right).$$

Braking a load during the lifting movement with a winch

If braking takes place with a deceleration greater than the gravitational acceleration g and the winch comes to a standstill in the time t_{brems}, the load hanging on the rope will be moved even further upwards due to its mass inertia (throw upwards) and subsequently fall into the slack rope.

The magnitude of the resulting drop height can be determined as follows, assuming that the brake does not apply with a delay:

According to Eq. (3.5), the throwing height with an initial velocity v until standstill under the effect of gravity is

$$s_{max} = \frac{v^2}{2 \cdot g}.$$

If the winch is braked with a_{brems} (calculated according to Eq. (3.35) or Eq. (3.36)), the resulting braking distance according to Eq. (3.3) is

$$s_{brems} = \frac{v^2}{2 \cdot a_{brems}}$$

(when determining a_{brems}, the inertia of the load must not be taken into account, since the rope cannot absorb compressive forces).

This results in a drop height h of

$$h = s_{max} - s_{brems} = \frac{v^2}{2} \cdot \left(\frac{1}{g} - \frac{1}{a_{brems}} \right). \qquad (3.20)$$

The same result is obtained with the following energy consideration:

During the braking process at the winch, the lifting speed of the load is decelerated from v to $\bar{v}$. With the energy equation $\bar{v}$ is calculated to

$$\frac{m \cdot v^2}{2} - \frac{m \cdot \bar{v}^2}{2} = m \cdot g \cdot s_{brems},$$

$$\bar{v} = \sqrt{v^2 - 2 \cdot g \cdot s_{brems}}.$$

With the remaining kinetic energy, the load is lifted further by h,

$$\frac{m \cdot \bar{v}^2}{2} = m \cdot g \cdot h,$$

$$h = \frac{\bar{v}^2}{2 \cdot g} = \frac{v^2}{2} \cdot \left(\frac{1}{g} - \frac{1}{a_{brems}} \right).$$

The decelerations and force effects occurring during the fall into the slack rope can be calculated according to the formulas in Section 3.7.

3.3.5 Summary of the most important formulas

Mass × acceleration = force

$$m\,[\text{kg}] \cdot a\,[\text{m/s}^2] = F\,[\text{kg m/s} = \text{N}], \quad 1\,\text{N} = 1\,\text{kg m/s}. \tag{3.21}$$

When substituting for a the acceleration due to gravity $g = 9.81 \approx 10\,\text{m/s}^2$, we obtain the (dead) weight of the mass m on earth,

$$m \cdot g = G.$$

Mass moment of inertia × angular acceleration = torque

$$I\,[\text{kg m}^2] \cdot \varepsilon \left[\frac{1}{\text{s}^2}\right] = M\,[\text{kg m/s}^2 \cdot \text{m} = \text{Nm}], \tag{3.22}$$

M Moment (torque) = force × lever arm.

Force × displacement = work (= energy = amount of heat)

$$F\,[\text{N}] \cdot \Delta s\,[\text{m}] = W\,[\text{Nm} = \text{J}], \quad 1\,\text{J} = 1\,\text{Nm} = 1\,\text{Ws} \tag{3.23}$$

(when measuring energy consumption, work is also expressed in Ws or kWh because of $W = P\,[\text{W}] \cdot t\,[\text{s}]$).

Moment × rotation angle = work

$$M\,[\text{Nm}] \cdot \Delta\varphi\,[\text{rad}] = W\,[\text{Nm} = \text{J}]. \tag{3.24}$$

Work/time = power

$$\frac{W\,[\text{J}]}{t\,[\text{s}]} = P\,[\text{J/s} = \text{Nm/s} = \text{W}], \quad 1\,\text{W} = 1\,\text{Nm/s}. \tag{3.25}$$

With $P = \frac{W}{\Delta t} = \frac{F \cdot \Delta s}{\Delta t} = F \cdot \frac{\Delta s}{\Delta t} = F \cdot v$ follows

Force × speed = power

$$F\,[\text{N}] \cdot v\,[\text{m/s}] = P\,[\text{W}]. \tag{3.26}$$

From $P = \frac{W}{\Delta t} = \frac{M \cdot \Delta\varphi}{\Delta t} = M \cdot \frac{\Delta\varphi}{\Delta t} = M \cdot \omega$ follows

Moment × angular velocity = power

$$M\,[\text{Nm}] \cdot \omega\,[1/\text{s}] = P\,[\text{W}]. \tag{3.27}$$

Force/pressurized area = pressure

$$\frac{F\,[\text{N}]}{A\,[\text{m}^2]} = p\left[\frac{\text{N}}{\text{m}^2} = \text{Pa}\right],$$

$$10^6\,\text{Pa} = 10^6\,\text{N/m}^2 = 1\,\text{N/mm}^2 = 1\,\text{MPa}, \tag{3.28}$$

$$10^5\,\text{Pa} = 10^5\,\text{N/m}^2 = 0.1\,\text{N/mm}^2 = 1\,\text{daN/cm}^2 = 1\,\text{bar}.$$

Before the introduction of the "International System of Units" just presented, other units were sometimes used in technology which did not represent a coherent system. These will also be explained briefly because they occur in older literature and the type of conversion should be known.

Units of the old engineering unit system and their conversion
- Old unit of force [kp] (kilopond)

$$1\,[\text{kg}] \cdot 9.81\,[\text{m/s}^2] = 1\,\text{kp},$$

i. e., the force effect of a mass of 1 kg in the gravitational field of the Earth (normal acceleration $g = 9.81\,\text{m/s}^2$) is 1 kilopond and represents its weight G.
- Old units for power [kpm/s], [hp]:

$$\frac{1\,\text{kp} \cdot 1\,\text{m}}{1\,\text{s}} = 1\,\frac{\text{kpm}}{\text{s}}, \quad \text{resp.} \quad \frac{75\,\text{kp} \cdot 1\,\text{m}}{1\,\text{s}} = 75\,\frac{\text{kpm}}{\text{s}} = 1\,\text{PS}.$$

- Old unit for a quantity of heat [cal] (calorie)
  1 [cal] = amount of heat (work) required to heat 1 g of pure water by 1 °C at atmospheric pressure (from 14.5 °C to 15.5 °C).

Conversions

$$1\,\text{W} = 1/9.81\,\text{kpm/s} = 0.102\,\text{kpm/s},$$

$$P\,[\text{W}] = P\,[\text{kpm/s}] \cdot 9.81 = P\,[\text{kpm/s}]/0.102,$$

$$P\,[\text{kW}] = P\,[\text{kpm/s}] \cdot (9.81/1000) = P\,[\text{kpm/s}]/102,$$

$$1\,\text{PS} = 75\,\text{kpm/s} = (75/102)\,\text{kW} = (1/1.36)\,\text{kW},$$

$$P\,[\text{PS}] = P\,[\text{kpm/s}]/75 = P\,[\text{kW}] \cdot 1.36,$$

$$1\,\text{kcal} = 427\,\text{kpm} = 427 \cdot 9.81\,\text{Nm} = 4187\,[\text{Nm} = \text{J}],$$

$$Q\,[\text{kcal}] = Q\,[\text{J}]/4187 = Q\,[\text{kJ}]/4.187.$$

3.3.6 Multimass systems

In most cases, a system will consist of translationally and rotationally moving masses.

If, for example, a stage wagon is to be set in motion, not only the entire mass m of the wagon including any payload masses must be accelerated in the sense of this translation, but also all rotating masses must be accelerated against the effect of their mass inertia.

The same applies to a lifting or lowering movement of a load. Not only the load mass m must be accelerated or decelerated accordingly, but also all rotating masses in the power transmission.

If we assume a rigid coupling of several individual masses for simplicity, the behavior of the system can be investigated as follows:

One chooses any translationally or rotationally moved system as a substitute system and converts all masses moved with other speed or angular velocity to this substitute system. This mass reduction is done in such a way that in an *energy balance* the masses to be reduced to the equivalent system are set in such a way that they have the same kinetic energy as in the actual system. Masses with very small kinetic energy, i. e., only slowly moving or very small masses, can be neglected. Since the velocity or angular velocity determines the magnitude of the energy with the square and the mass or moment of inertia influences the magnitude of the energy only linearly, thus especially slowly moving masses are negligible.

Similarly, forces or torques acting on the system must be reduced to the equivalent system by equating the powers in a *power balance*.

Thus, for example, if a force F (system 1) acts on a mass m moving with velocity v and a moment M (system 2) acts on a mass rotating with ω with moment of inertia I, the total system can be reduced either to system 1 or to system 2. In the following formulas, the energy and power balance ignores the efficiency influence. For systems with poor efficiency, the efficiency must be taken into account in any case.

Reduction to system 1

$$F + F_{red} = (m + m_{red}) \cdot a. \tag{3.29}$$

From the energy balance, according to Eqs. (3.14) and (3.15),

$$\frac{m \cdot v^2}{2} = \frac{I \cdot \omega^2}{2} \quad \text{results in } m_{red} = I \cdot \left(\frac{\omega}{v}\right)^2. \tag{3.30}$$

And from the current account, according to Eqs. (3.26) and (3.27),

$$F_{red} \cdot v = M \cdot \omega \quad \text{results in } F_{red} = M \cdot \frac{\omega}{v}. \tag{3.31}$$

Reduction to the system 2

$$M + M_{red} = (I + I_{red}) \cdot \varepsilon. \tag{3.32}$$

From the energy balance,

$$\frac{I_{red} \cdot \omega^2}{2} = \frac{m \cdot v^2}{2} \quad \text{results in } I_{red} = m \cdot \left(\frac{v}{\omega}\right)^2, \tag{3.33}$$

and from the current account,

$$M_{red} \cdot \omega = F \cdot v \quad \text{results in } M_{red} = F \cdot \frac{v}{\omega}. \tag{3.34}$$

Using Eqs. (3.21) and (3.22), respectively, the acceleration a and ε can thus be calculated, taking into account these reduced masses:

$$\begin{aligned}
a &= \frac{\sum F}{\sum m} = \frac{F_1 + F_{red}}{m_1 + m_{red}}, \\
\varepsilon &= \frac{\sum M}{\sum I} = \frac{M_1 + M_{red}}{I_1 + I_{red}}.
\end{aligned} \tag{3.35}$$

If the efficiency η is to be taken into account, the following applies:

$$\begin{aligned}
a &= \frac{F_1 + F_{red} \cdot \eta^w}{m_1 + m_{red} \cdot \eta^w}, \\
\varepsilon &= \frac{M_1 + M_{red} \cdot \eta^w}{I_1 + I_{red} \cdot \eta^w},
\end{aligned} \tag{3.36}$$

where

η efficiency [1]

w exponent of the efficiency [1]; w is to be set (+1) or (−1) in the numerator and denominator ($\eta^1 = \eta$ and $\eta^{-1} = 1/\eta$)

The sign of the exponent w is to be selected in such a way that the generally valid statement is met: With better efficiency, the acceleration becomes larger in magnitude (the acceleration time smaller) and the amount of deceleration smaller (the deceleration time longer). This can be easily checked by trial and error, e. g., by setting $\eta = 1$ and $\eta < 1$.

3.4 Friction

3.4.1 Types of friction

Sliding friction

If two bodies are in contact with the compressive force F_N and the *coefficient of friction* μ, which depends on the material of the two bodies and any lubrication condition, then for a relative displacement of the bodies according to Figure 3.4(a) a force

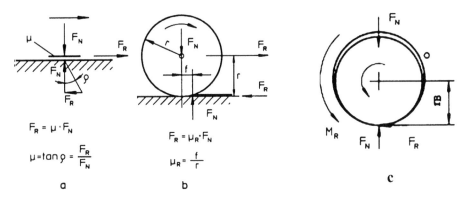

Figure 3.4: Friction: (a) sliding friction, (b) rolling friction, (c) bearing friction.

$$F_R = F_N \cdot \mu, \quad \mu = \tan \rho, \tag{3.37}$$

is required to overcome the friction.

With regard to the magnitude of μ, a distinction must generally be made between *sticking* (*static*) and *sliding* (*dynamic*) *friction*.

If there is no relative movement between the two bodies in contact, the frictional connection is determined by the *static coefficient of friction* μ_S, and a relative movement can only begin under a force effect $F > F_N \cdot \mu_S$.

In a state of relative motion, the required force F is determined by the *dynamic coefficient of friction* μ_D, which is generally lower than the static coefficient of friction μ_S. If a relative movement is already taking place, it remains as long as $F > F_N \cdot \mu_D$, where

F_N	compressive force [N]
F_R	frictional force [N]
μ	coefficient of friction [–]
μ_S	coefficient of friction for static friction [–]
μ_D	coefficient of friction for dynamic friction [–]
ρ, ρ_S, ρ_D	friction angle [°, rad]

In this context "static" means "without movement" and "dynamic" means "with movement between the two elements". If the static friction is noticeably greater than the dynamic friction, a so-called *slip–stick effect* can occur when a mass exposed to friction is moved by an elastic element. This can cause disturbing noise or structure-borne sound. See Section 3.10.3.

Rolling friction
This term refers to the resistance to movement of two bodies rolling against each other under a compressive force F_N resulting from plastic deformation of the pressure bodies. This situation is illustrated by the wheel rolling on a rail as shown in Figure 3.4(b).

As a result of the deformation, the action and reaction forces F_N form a pair of forces with distance f during rolling motion, resulting in a resistance to motion F_R at the wheel with radius r:

$$F_R = F_N \cdot \frac{f}{r} = F_N \cdot \mu_R, \tag{3.38}$$

where
f lever arm of rolling friction [m]
r radius of the wheel [m]
μ_R rolling friction coefficient [–]

In order to make sure that the wheel is rolling and not sliding, the condition $\eta_R < \eta_S$ must be met.

Bearing friction
A shaft guided in a bearing is also hindered in its rotary motion by frictional forces. In this case, too, there is a relationship between the bearing force and the frictional torque in the bearing (Figure 3.4(c)). The magnitude of the bearing friction coefficient μ_B depends mainly on whether the bearing is a sliding or rolling bearing:

$$F_R = F_N \cdot \mu_B,$$
$$M_R = F_R \cdot r_B, \tag{3.39}$$

where
r_B bearing radius [m]
M_R frictional torque [Nm]
μ_B coefficient of bearing friction [–]

Total resistance of a rolling wheel
This is composed of the rolling friction resistance at the wheel according to equation (3.38) and the bearing friction resistance according to equation (3.39) and amounts to

$$F_R = F_N \cdot \left(\frac{f}{r} + \mu_B \cdot \frac{r_B}{r} \right).$$

Often, a specific driving resistance w_R is calculated that includes both components and any additional resistances (such as the wheel flange friction),

$$F_R = F_N \cdot w_R, \tag{3.40}$$

where
w_R specific driving resistance [–], [N/N]
     (resistance in N per N wheel load, sometimes also given in [N/kN])

3.4.2 Adhesion condition

In the case of travel drives of stage wagons or, for example, in the case of rotary drives of revolving stages, the transmission of the driving force can be effected via frictional engagement between wheel and rail. The magnitude of the transmittable tangential force depends on the magnitude of the compressive force acting between wheel and rail and the coefficient of friction of the wheel–rail pairing, according to the relationship

$$F_{U_{\max}} = R_A \cdot \mu_A. \tag{3.41}$$

Therefore, the resistance to motion to be overcome by the drive wheel (see Section 3.6) must not be greater, and the adhesion condition follows

$$F_v^* + F_a^* \le R_A,$$

or with several drives

$$F_v^* + F_a^* \le \sum R_A \cdot \mu_A, \tag{3.42}$$

where
$F_{U_{\max}}$ peripheral force transmittable at the drive wheel
F_v^* resistance to movement in the inertia
F_a^* additional resistance during acceleration
R_A compressive force acting between drive wheel and rail
μ_A friction value at the drive wheel (adhesion), e. g.
     $\mu_A = 0.14$ steel wheel on steel rail
     $\mu_A = 0.2 \div 0.7$ Vulkollan wheel on steel rail

The additional symbol "*" for F_v and F_a is intended to indicate that, when determining these values, only those force components are to be taken into account which actually have to be transmitted to the drive wheels via frictional engagement. This means, for example, that F_v share from the wheel bearing friction or F_a share for the acceleration of rotating masses in the power train is relevant for the determination of the motor size, but not for the adhesion condition.

The force R_A can result from weight forces, as is the case, for example, with the travel drive of the stage wagon built as bridge carriage shown in Figure 1.128 or with the friction wheels of the stage wagon shown in Figures 1.138–1.140 or a turntable shown in Figure 1.150, or from a spring force, as is the case with the friction wheel drive of the turntable shown in Figure 1.144.

3.5 Efficiency

If a certain amount of energy is imported into a drive system in which friction occurs only a reduced amount of energy can be exported during the same time interval (*effective work*); the difference cannot be used as kinetic energy, but is lost primarily as heat energy due to friction effects.

If the time interval is considered to be the unit of time, this situation can also be described by the power instead of the energy. The *output power* P_{out} (*effective power*) is smaller than the *input power* P_{in} because of the *power loss* P_L,

$$P_{out} = P_{in} - P_L.$$

Only at a lossless drive (index 0) without any friction the following formula applies:

$$P_{0in} = P_{0out} = P_0.$$

The quotient "effective work divided by expended work" or "output power (effective power) to input power" is called *efficiency* η. The efficiency is therefore always less than 1 in a "real" lossy system (the numerator always has a smaller value) and only equal to 1 in an "ideal" lossless system:

$$\eta = \frac{P_{out}}{P_{in}} = \frac{P_{in} - P_L}{P_{in}} = \frac{P_{out}}{P_{out} + P_L} = \frac{P_{0in}}{P_{in}} = \frac{P_{out}}{P_{0out}}, \tag{3.43}$$

where

P_{in}	drive power in the lossy system
P_{out}	output power in the lossy system
P_L	power loss
$P_{0in} = P_{0out} = P_0$	input or output power in the lossless system
η	efficiency [–]

In order to determine the efficiency of a system, it is therefore necessary to compare, for example, the work or power that must be input in order to maintain a state of motion in the real lossy system as a result of friction and the work or power that would be input in an ideal lossless system without the effect of friction.

In the case of a lifting or lowering movement under the effect of gravity, a distinction must be made between two cases:

– When *lifting* a load with speed v, "*driving power*" must be provided. The force of gravity opposing the direction of motion must also be overcome in the lossless system. The frictional forces acting in the lossy system increase this resistance to motion. Therefore, in addition to the propulsive power P_0, a loss power P_{LH} must also be provided in the real system. The *lifting efficiency* η_H is therefore (index H for hoisting)

$$\eta_H = \frac{P_0}{P_H} = \frac{P_0}{P_0 + P_{LH}}, \tag{3.44}$$

$$P_H = P_0 \cdot \frac{1}{\eta_H}. \tag{3.45}$$

- When *lowering* a load with velocity v, "*braking power*" must be applied; otherwise, the load would fall under the effect of gravity, accelerating more and more. In a lossless system, therefore, the force of gravity driving in the direction of motion must be balanced. In a lossy system, this driving force is reduced by friction. Therefore, the braking power P_0 in the real system is reduced by the power loss P_{LL} and the *lowering efficiency* η_L is (index L for lowering)

$$\eta_L = \frac{P_L}{P_0} = \frac{P_0 - P_{LL}}{P_0}, \tag{3.46}$$

$$P_L = P_0 \cdot \eta_L = P_H \cdot \eta_H \cdot \eta_L. \tag{3.47}$$

Under the condition that the power loss P_L is the same in the lifting and lowering direction,

$$P_{LH} = P_{LL} = P_L$$

(see also Section 4.3.1), this results in the following relationship:

$$\eta_L = 2 - \frac{1}{\eta_H}. \tag{3.48}$$

Table 3.4 compares the values for η_H and η_S calculated according to Eq. (3.48).

Table 3.4: Efficiency by hoisting and lowering according to Eq. (3.48).

η_H	0.99	0.95	0.90	0.80	0.50
η_L	0.99	0.95	0.89	0.75	0

For very good efficiency values (such as for $\eta_H > 0.9$), $\eta_H = \eta_L = \eta$ can be set. For high efficiency values, therefore, one often does not distinguish between lifting and lowering efficiency at all, but sets for the lifting situation

$$P_H = P_0 \cdot \frac{1}{\eta_H},$$

and for the lowering situation, according to Eq. (3.47),

$$P_S = P_0 \cdot \eta_L \approx P_H \cdot \eta^2. \tag{3.49}$$

However, from the consideration of the braking situation as described for lowering a load, the following system behavior can also be derived: If the frictional forces effective in a drive unit are greater than the driving force effect due to gravity, then no braking power at all has to be applied by the drive unit, and in fact a driving power has to be applied for the lowering movement.

However, two cases must be distinguished:

– The need to drive even when lowering is only given for small, "non-pull-through" loads, while braking is very much required for larger loads.

– The requirement to drive even during lowering is inherent in the system and always present, regardless of the load size, i. e., even the largest load cannot cause a lowering process. This condition is called *self-locking*. A system is therefore self-locking if the lowering efficiency η_L is independent of the size of the load and is less than or equal to zero,

$$\eta_L \leq 0.$$

If Eq. (3.48) is valid, then it follows that this condition is equivalent to the condition

$$\eta_H \leq 0.5.$$

Section 4.3.1 shows the ratios for wedge and screw drives.

(Instead of gravity, another force can also act, e. g., the pretensioning force (clamping force) on a screw.)

3.6 Calculation of power requirement

To calculate the required power of a drive, it is first necessary to investigate which forces F oppose the desired movement as resistance to motion.

3.6.1 Movement resistance in the steady state (without acceleration)

These are resistances to motion that must be overcome in order to maintain a state of motion with a velocity v. In the case of translational motion, they can be described under the designation be summarized:

$$F_v = \sum F, \tag{3.50}$$

where F are forces effective as resistance to motion, for which some examples are given below:

– Effect of *gravity* with vertical lift of a mass m (see Eq. (3.1)),

$$F = G\,[\mathrm{N}] = m\,[\mathrm{kg}] \cdot g\,[\mathrm{m/s}^2]. \tag{3.51}$$

– *Driving resistance* acting on wheels during horizontal travel according to Eq. (3.40),

$$F = w_R \cdot F_N,$$ (3.52)

for w_R, see Eq. (3.40).
– In the case of moving a mass at a gradient angle α according to Figure 3.5, the resistance to motion results from the sine component of the weight and the driving resistance,

$$F = G \cdot \sin \alpha + w_R \cdot G \cdot \cos \alpha = G(\sin \alpha + w_R \cdot \cos \alpha).$$ (3.53)

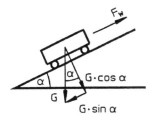

Figure 3.5: Driving resistance on inclined plane.

– Effect of a *static pressure p* on a surface A according to Figure 3.6 (see also Eq. (3.71)),

$$F = p \cdot A,$$ (3.54)

where
p hydrostatic pressure [N/m^2]
A pressurized area [m^2]

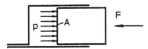

Figure 3.6: Force effect of a hydrostatic pressure on a surface.

– *The dynamic pressure of* a flowing medium, e. g., the wind pressure in an outdoor installation, can also act as a pressure force. The geometric shape of the flowed structure is taken into account by an *aerodynamic force coefficient* c_W,

$$F = c_W \cdot q \cdot A_W,$$ (3.55)

where
A_W area projected onto a surface normal to the direction of flow
q dynamic pressure [N/m^2], $q = \frac{\rho v^2}{2}$
v relative velocity between air and object [m/s]; this is the wind velocity when the object is stationary (or the velocity of a moved object)

c_W aerodynamic force coefficient (shape coefficient) [–] ($c \approx 0.8 \div 1.3 \div 1.6$)
ρ density of air [kg/m³] ($\rho \approx 1.3\,\text{kg/m}^3$)

For *rotating motions*, the resistances to motion can be represented as torques, and Eq. (3.50) can be expressed in the modified form,

$$M_v = \sum M. \tag{3.56}$$

3.6.2 Movement resistance during acceleration

To initiate a motion with velocity v, mass m must first be brought to velocity v by acceleration a [m/s²]. Due to its inertia, the mass m opposes the acceleration with an acceleration resistance denoted by F_a (Eqs. (3.11) and (3.29)),

$$F_a = m \cdot a, \tag{3.57}$$

respectively

$$F_a = (m + m_{\text{red}}) \cdot a.$$

Likewise, for rotating motion, torques M_a must be considered for acceleration of the system (Eqs. (3.22) and (3.32)),

$$M_a = I \cdot \varepsilon, \tag{3.58}$$

respectively

$$M_a = (I + I_{\text{red}}) \cdot \varepsilon.$$

3.6.3 Drive power

The required power of a drive, considering an efficiency η to account for additional losses (Eq. (3.43)), is thus composed of:
(a) The *steady-state power* P_v
  – To maintain a velocity v [m/s] against resistance forces F_v [N] during translation

$$P_v = F_v \cdot v \cdot \frac{1}{\eta} \ [\text{N·m/s} = \text{Nm/s} = \text{W}]. \tag{3.59}$$

  – To maintain an angular velocity ω [1/s] against a torque M_v [Nm] during rotation

$$P_v = M_v \cdot \omega \cdot \frac{1}{\eta} \ [\text{Nm} \cdot 1/\text{s} = \text{Nm/s} = \text{W}]. \tag{3.60}$$

(b) The maximum acceleration power P_a
  – For accelerating a translationally moving mass (Eqs. (3.11) and (3.26)) in the acceleration time t_a,

$$P_a = F_a \cdot v \cdot \frac{1}{\eta} = m \cdot a \cdot v \cdot \frac{1}{\eta} = m \cdot \left(\frac{v^2}{t_a}\right) \cdot \frac{1}{\eta} \; [\text{Nm/s} = \text{W}], \qquad (3.61)$$

for accelerating a rotating mass (Eqs. (3.12) and (3.27)),

$$P_a = M_a \cdot \omega \cdot \frac{1}{\eta} = I \cdot \varepsilon \cdot \omega \cdot \frac{1}{\eta} = I \cdot \left(\frac{\omega^2}{t_a}\right) \cdot \frac{1}{\eta} \left[\text{Nm} \cdot \frac{1}{\text{s}} = \text{Nm/s} = \text{W}\right]. \qquad (3.62)$$

In the acceleration phase, the maximum drive power (peak value at the end of the acceleration time) is therefore required,

$$P_{ges} = P_v + P_a; \qquad (3.63)$$

in the steady-state phase, a drive power P_v is sufficient.

The actual required rated power of an electric drive motor must be selected so that the starting torque and rated torque correspond to the demand. However, it must also be considered that the motor is designed thermally correctly, i. e., that it does not become too hot during operation. The thermal load depends above all on the duty cycle and the cooling conditions at the motor (see technical literature). At stage conditions this is not relevant in most cases.

Reduction of drive power with counterweights

If a load of mass m is to be lifted, the required lifting power can theoretically be reduced to zero if the mass is balanced with a counterweight, ignoring frictional resistance and inertia forces.

If a podium with its own mass m_E or its own weight E is to be lifted at the speed v, the power $P = E \cdot v$ is required. If the podium is also loaded with a mass m_Q or a payload Q, the total power $P = (E + Q) \cdot v$ is required.

This power requirement can be extremely reduced with the aid of a counterweight. The smallest lifting capacity is achieved if a counterweight of the size $G = E + Q/2$ is selected – as is generally used in traction sheave elevators in houses. In the case of the empty podium, the counterweight is then heavier by $Q/2$, and in the case of the fully loaded podium, the podium is heavier by $Q/2$. In both extreme cases, therefore, only the lifting load $Q/2$ has to be overcome and the power requirement is only

$$P = \frac{Q}{2} \cdot v.$$

3.7 Elastic components

The elasticity of a component – a bar, a spring, and a rope – can be described as follows (see also Section 4.1.1 – Ropes and rope drives):

If the elastic element stretches by Δl under the effect of force F, the *spring constant c* is defined (see Figure 3.7):

$$c = \frac{F}{\Delta l},$$
$$F = c \cdot \Delta l.$$
(3.64)

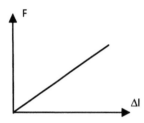

Figure 3.7: Linear spring characteristic.

For an elastic bar of cross-section A, Hook's theorem of strength theory states:

$$\frac{\Delta l}{l} = \frac{\sigma}{E} = \frac{F}{E \cdot A},$$
$$\Delta l = \frac{F \cdot l}{E \cdot A},$$
$$c = \frac{A \cdot E}{l},$$
(3.65)

where
F force [N]
A cross-section of the elastic component [m², mm²]
E modulus of elasticity of the component [N/m², N/mm²]
l length of the component before loading with F [m, mm]
Δl lengthening due to load F [m, mm]

For a wire rope with metallic cross-section A_m and *elastic modulus E_S*, one has

$$c = \frac{A_m \cdot E_S}{l}.$$
(3.66)

The product $A_m \cdot E_S$ is also called *rope module M*,

$$M = A_m \cdot E_S = c \cdot l.$$
(3.67)

Fall of a mass into a slack rope

If a mass m falls onto a spring or into a rope after a free fall height h, the kinetic energy generated in the free fall and the potential energy released during the stretching process of the spring or rope is converted into elongation work (see Eq. (3.4)),

$$\frac{m \cdot v^2}{2} + m \cdot g \cdot \Delta l = \frac{1}{2} \cdot F \cdot \Delta l, \quad v = \sqrt{2 \cdot g \cdot h},$$

$$m \cdot g \cdot (h + \Delta l) = \frac{1}{2} \cdot F \cdot \Delta l,$$

or, in other words, the potential energy released over the entire falling distance $h + \Delta l$ is converted into work for the extension of the elastic element. Here

m mass [kg]
h free fall height [m]
v velocity after a fall height h at an initial velocity $v_0 = 0$ [m/s]

With Eq. (3.64) for Δl, the quadratic equation follows:

$$F^2 - 2 \cdot m \cdot g \cdot F - 2 \cdot m \cdot g \cdot h \cdot c = 0,$$

$$F = m \cdot g \cdot \left[1 + \sqrt{1 + \frac{2 \cdot h \cdot c}{m \cdot g}} \right]. \tag{3.68}$$

With F as the force acting in the rope at rope elongation Δl,

$$\Delta l = \frac{F}{c} = \frac{m \cdot g}{c} \cdot \left[1 + \sqrt{1 + \frac{2 \cdot h \cdot c}{m \cdot g}} \right]. \tag{3.69}$$

The deceleration that occurs is

$$a_{ges} = \frac{F}{m} = g \cdot \left[1 + \sqrt{1 + \frac{2 \cdot h \cdot c}{m \cdot g}} \right] = g + a,$$

$$a = g \cdot \sqrt{1 + \frac{2 \cdot h \cdot c}{m \cdot g}}. \tag{3.70}$$

The force F will be smaller for a given fall height h the more elastic the rope is, i. e., for a given rope type, the longer the effective rope length l is.

In climbing technique, the *fall factor f* is defined as $f = h/l$.

The rope loaded with F can also be considered as a single mass oscillator; see Section 3.10.1.

In stage technology, this load situation can be relevant in the following cases, for example:

– Fall of an object (e. g., spotlight) into a safety rope.

– Fall of a load, abruptly braked during a hoisting operation in case of power failure, into the slack rope (see Eq. (3.20)). The "load" can be a decoration or a person hanging in a lifting device (performer flying).

3.8 Fundamental terms of hydraulics

3.8.1 Basic terms

Hydrostatic pressure
If the force F (Figure 3.8(a)) acts on an enclosed liquid via a piston surface A, this liquid is under the pressure

$$p\left[\frac{N}{m^2}\right] = \frac{F\,[N]}{A\,[m^2]}, \tag{3.71}$$

where
F force [N]
A area [m^2]
p pressure [N/m^2 = Pa]

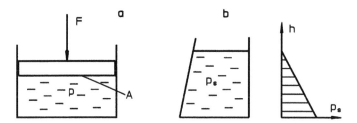

Figure 3.8: Hydrostatic pressure due to an external force and weight pressure: (a) pressure p due to an external force, (b) pressure distribution p_s due to fluid weight.

Weight pressure = gravity pressure
In addition to this hydrostatic pressure caused by external forces, the *weight pressure* or *gravity pressure* p_s also acts in a liquid column of height h, irrespective of the shape of the pressure column, according to Figure 3.8(b),

$$p_s = h \cdot \rho \cdot g, \tag{3.72}$$

where
h height of the liquid column [m]
ρ density of liquid [kg/m^3] (water $\rho \approx 1000$ kg/m^3, mineral oil $\rho \approx 900$ kg/m^3)
g gravitational acceleration, $g = 9.81 \approx 10$ [m/s^2]

Therefore, 10 m water column results in a gravity pressure of $p_s \approx 10\,[\text{m}] \cdot 1000\,[\text{kg/m}^3] \cdot 10\,[\text{m/s}^2] = 100{,}000\,[\text{kg m/s}^2 \cdot 1/\text{m}^2] = 100{,}000\,[\text{N/m}^2] = 1\,\text{bar}$.

Air and gravity pressures are negligible compared to the pressures used in hydro-static drive technology and are therefore usually not taken into account in hydrostatic drive technology.

Hydrodynamic pressure

A flowing fluid also generates force effects or pressure as a result of its inherent kinetic energy. This dynamic pressure, also called *dynamic pressure*, is the kinetic energy per unit volume and, if using $m = \rho \cdot V$ in Eq. (3.13) and dividing by V, results in

$$p_{\text{dyn}} = \frac{\rho \cdot v^2}{2}. \tag{3.73}$$

In hydrostatic drives, this hydrodynamic pressure effect is also negligible.

Continuity equation

If a liquid flows through the cross-section A with the velocity v, the volume flow per time unit (*flow rate*) is

$$Q\,[\text{m}^3/\text{s}] = A\,[\text{m}^2] \cdot v\,[\text{m/s}], \tag{3.74}$$

where

Q flow rate, volume flow of the liquid $[\text{m}^3/\text{s}]$
A cross-section $[\text{m}^2]$
v conveying speed, flow velocity $[\text{m/s}]$

Since hydraulic fluids can be assumed to be almost incompressible, the following applies to two cross-sectional areas 1 and 2, according to Figure 3.9,

$$Q = A_1 \cdot v_1 = A_2 \cdot v_2 = \text{const.}$$

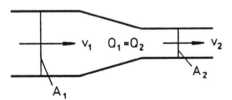

Figure 3.9: Flow velocity as a function of the flow cross-section.

Flow rate with throttling
Applying the theorem of conservation of energy for a flowing fluid (Bernoulli's equation; see technical literature), the relationship between flow rate and pressure drop at a constriction can be determined. Adding a correction value a_D to account for the geometric shape of the cross-sectional constriction and the viscosity of the fluid leads to the relationship

$$Q = a_D \cdot A \cdot \sqrt{\frac{2 \cdot \Delta p}{\rho}}, \quad A = \frac{d^2 \cdot \pi}{4}, \tag{3.75}$$

where
Q flow rate through the throttle [m³/s]
a_D flow coefficient [−], $a_D \approx 0.6 \div 0.9$
A cross-section [m²]
Δp pressure difference from the pressures upstream and downstream of the restriction [N/m²]
ρ density of the flowing medium [kg/m³]

Formula (3.75) is especially applicable for short cross-sectional constrictions like an orifice.

   If the throttle is in the form of a long thin pipe cross-section, the following formula is given in application of Hagen–Poiseuille's theorem (valid for laminar flow in thin pipes):

$$Q = \frac{\pi \cdot r^4}{8 \cdot \eta \cdot l} \cdot \Delta p, \tag{3.76}$$

where
η dynamic viscosity [kg/m·s]
l length of the throttle [m]
r radius of the throttle [m]
v kinematic viscosity [m²/s]
ρ density of the liquid [kg/m³]

Thus, it depends on the geometry of the constriction which equation provides values better corresponding to reality. Depending on the formula, a linear or nonlinear relationship between Q and Δp is given (see Figure 3.10).

3.8.2 Hydrostatic devices with linear and rotary work function

As already explained in Section 2.3, devices with *linear working motion* are the *hydraulic cylinder* as consumer and, in reversal of the working principle, the *piston accumulator*.

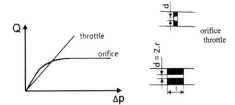

Figure 3.10: Flow rate as a function of pressure difference at orifice and throttle.

Devices with *rotating working motion* are the *hydraulic pump* as pressure generator and the *hydraulic motor* as load.

Power of a hydrostatic implement
Substituting Eq. (3.71) for the force F and Eq. (3.74) for the velocity v into Eq. (3.26), we obtain:

$$P = F \cdot v = p \cdot A \cdot v = p \cdot Q,$$
$$\boldsymbol{P}\,[\mathrm{W}] = \boldsymbol{p}\,[\mathrm{N/m^2}] \cdot \boldsymbol{Q}\,[\mathrm{m^3/s}],$$
$$P\,[\mathrm{kW}] = p\,[\mathrm{MPa}] \cdot Q\,[\mathrm{l/min}] \cdot 1/60 = p\,[\mathrm{bar}] \cdot Q\,[\mathrm{l/min}] \cdot 1/600, \qquad (3.77)$$
$$\boldsymbol{P}\,[\mathrm{W}] = \Delta\boldsymbol{p}\left[\frac{\mathrm{N}}{\mathrm{m^2}}\right] \cdot \boldsymbol{Q}\left[\frac{\mathrm{m^3}}{\mathrm{s}}\right].$$

If the hydrostatic implement is under p on the high-pressure side and under $p_0 > 0$ on the low-pressure side, only the pressure difference $\Delta p = p - p_0$ is to be taken into account for the power P, of course. (The back pressure $p_0 = 1\,\mathrm{bar}$ from the air pressure can generally be neglected in relation to the pressure p.)

Efficiency of a hydrostatic system
In a hydrostatic system, in addition to the *mechanical efficiency* due to frictional losses, the *volumetric efficiency* due to leakage oil must also be taken into account,

$$\eta_{ges} = \eta_m \cdot \eta_v, \qquad (3.78)$$

η_m [–] or [%] mechanical efficiency for consideration friction losses;

$$\eta_m \approx 0.8 \text{ gearpump (80 \%)}$$
$$\approx 0.9 \text{ piston pump;}$$

η_v [–] or [%] volumetric efficiency for consideration losses of pressure-effective volume due to leakage oil;

$$\eta_v \approx 0.85 \text{ gear pump}$$
$$\approx 0.95 \text{ piston pump.}$$

If the losses in a system are η_{ges}, then power $P' = P/\eta_{ges}$ must be put into the system for power P to be delivered; or if P is put into a lossy system, then only power $P'' = P \cdot \eta_{ges}$ is delivered.

Other relationship formulas

Applying these basic equations to hydrostatic devices results in the following formulas for linear and rotary devices:

Linear devices

With a piston area A [m^2] and a stroke h [m]:

–   *Weight-loaded piston accumulator* (Figure 2.12)

$$F = G = m \cdot g,$$
$$p = \frac{G}{A}. \tag{3.79}$$

–   *Single-acting hydraulic cylinder* (plunger cylinder according to Figure 3.11(a))
    Applying Eq. (3.71) and Eq. (3.74), we obtain

$$F = p \cdot A_K, \quad A_K = \frac{d^2 \cdot \pi}{4},$$
$$v_K = \frac{Q}{A_K}, \tag{3.80}$$
$$P = F \cdot v_K = Q \cdot p.$$

–   *Double-acting hydraulic cylinder* according to Figure 3.11(b)
    When the piston surface is subjected to A_{K_1} with the pressure p_1 and the flow rate Q_1, the force, piston speed and power are calculated as follows:

$$F_1 = p_1 \cdot A_{K_1}, \quad A_{K_1} = \frac{d_1^2 \cdot \pi}{4},$$
$$v_{K1} = \frac{Q}{A_{K_1}},$$
$$P_1 = F_1 \cdot v_{K_1} = Q_1 \cdot p_1.$$

When the piston surface is subjected to A_{K_2} with p_2 and Q_2 to

$$F_2 = p_2 \cdot A_{K_2}, \quad A_{K_2} = \frac{\pi}{4} \cdot (d_1^2 - d_2^2),$$
$$v_{K_2} = \frac{Q}{A_{K_2}}, \tag{3.81}$$
$$P_2 = F_2 \cdot v_{K_2} = Q_2 \cdot p_2.$$

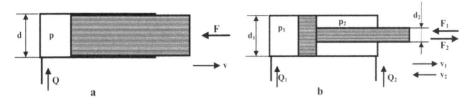

Figure 3.11: Hydraulic cylinder: (a) single-acting, (b) double-acting.

To cover the stroke h, the time [s] is used:

$$t_1 = \frac{h}{v_{K_1}} = \frac{h \cdot A_{K_1}}{Q_1},$$

$$t_2 = \frac{h}{v_{K_2}} = \frac{h \cdot A_{K_2}}{Q_2}.$$

In the case $p_1 = p_2$ and $Q_1 = Q_2$, follows $F_1 > F_2$, $v_{K_1} < v_{K_2}$ and $t_1 > t_2$.
(Frictional resistance in the piston guide and seal sleeves by a cylinder can be taken into account with a efficiency $\eta_z \approx 0.95$.)

Rotary equipment

With a delivery or displacement volume per revolution V [m³/U, l/U]:
– *Pump* driven with the speed n_P,

$$Q \,[\text{m}^3/\text{s}] = V_P \,[\text{m}^3/\text{U}] \cdot n_P \,[\text{U}/\text{s}],$$
$$Q \,[l/\text{min}] = V_P \,[l/\text{U}] \cdot n_P^* \,[\text{U}/\text{min}].$$

(3.82)

If the leakage oil loss of the pump is given by η_{vP}, the effective flow rate of the pump will be $\overline{Q} = Q \cdot \eta_{vP}$ and, to convey Q, the pump must be driven instead of n_P with $\overline{n}_P = \frac{n_P}{\eta_{vP}}$.
– *Hydraulic motor* loaded with volume flow Q:

$$n_M \left[\frac{U}{s}\right] = \frac{Q\,[\frac{\text{m}^3}{\text{s}}]}{V_M\,[\frac{\text{m}^3}{U}]}, \quad \omega_M \left[\frac{1}{s}\right] = 2\pi \cdot n_M \left[\frac{U}{s}\right] = 2\pi \frac{Q\,[\frac{\text{m}^3}{\text{s}}]}{V_M\,[\frac{\text{m}^3}{U}]},$$

$$n_M^* \left[\frac{U}{\text{min}}\right] = \frac{Q\,[\frac{l}{\text{min}}]}{V_M\,[\frac{l}{U}]}.$$

(3.83)

If the leakage oil loss of the motor is given by η_{vM}, then the effective speed of the hydraulic motor is

$$\overline{n}_M = \frac{Q}{V_M} \cdot \eta_{vM}.$$

The torque delivered by a hydraulic motor is calculated by applying Eqs. (3.27), (3.77) and (3.83) to obtain

$$M_M = \frac{P}{\omega} = \frac{Q \cdot \Delta p}{\omega} = \frac{V_M \cdot n_M \cdot \Delta p}{2 \cdot \pi \cdot n_M} = \frac{V_M \cdot \Delta p}{2 \cdot \pi},$$

$$M_M \ [\text{Nm}] = \frac{1}{2 \cdot \pi} \cdot V_M \left[\frac{\text{m}^3}{\text{U}} \right] \cdot \Delta p \left[\frac{\text{N}}{\text{m}^2} \right], \qquad (3.84)$$

$$M_M \ [\text{Nm}] = \frac{100}{2 \cdot \pi} \cdot V_M \left[\frac{1}{\text{U}} \right] \cdot \Delta p \ [\text{bar}] = 15.9 \cdot V_M \left[\frac{1}{\text{U}} \right] \cdot \Delta p \ [\text{bar}],$$

where

n	revolution speed [rps]
n^*	revolution speed [rpm]
M	torque [Nm]
V	displacement, absorption volume [m³/U] = [m³]
Index P	pump
Index M	motor

Taking into account a mechanical efficiency of the hydromotor η_{mM},

$$\overline{M}_M = M_M \cdot \eta_{vM},$$

and taking into account a volumetric efficiency of the hydromotor η_{vM},

$$\overline{\omega}_M = \omega_M \cdot \eta_{vM},$$

and the power of the hydraulic motor loaded with Q and p (see Eqs. (3.27), (3.83), (3.84)) is

$$P_{\text{eff}} = \overline{M}_M \cdot \overline{\omega}_M = \left(\frac{1}{2 \cdot \pi} \cdot V_M \cdot \Delta p \cdot \eta_{mM} \right) \cdot (\omega_M \cdot \eta_{vM})$$

$$= \frac{1}{2 \cdot \pi} \cdot V_M \cdot \Delta p \cdot \eta_{mM} \cdot 2\pi \cdot \frac{Q}{V_M} \cdot \eta_{vM}, \qquad (3.85)$$

$$P_{\text{eff}} = Q \cdot \Delta p \cdot \eta_{ges}.$$

3.8.3 Hydraulic accumulator system

Hydraulic accumulators take up a certain volume of fluid under pressure and can release this volume under pressure back into a hydraulic circuit when needed.

Construction types
Basically, a distinction can be made between two types of storage (see Figure 2.12):

- Accumulators in which the pressure remains constant when the liquid is discharged; this is the case with weight accumulators, since the same force is always applied (Figure 2.12(a)). In practice, such accumulators do not occur, because necessary weights are much too big.
- Accumulators in which the pressure decreases as the liquid is discharged; this is the case with spring accumulators, regardless of whether the spring is a mechanical spring or, a gas spring (Figure 2.12(b)–(e)).

In the following, the conditions for gas accumulators will be examined in more detail, as these are used in a *central hydraulic pressure stations* of stage technology systems. To avoid corrosion and for safety reasons, nitrogen is generally used as the filling gas. (At higher temperatures and under pressure, an oil–oxygen mixture ignites – diesel effect.)

Dimensioning of hydraulic accumulators

It is desired that the accumulator can deliver the amount of fluid ΔV while allowing the pressure to drop from $p_{max} = p_1$ to $p_{min} = p_2$.

On the gas side, the following theorem generally applies to a polytropic change:

$$p \cdot V^n = \text{const.},$$

and for states 1 and 2,

$$p_1 \cdot V_1^n = p_2 \cdot V_2^n, \quad V_1 = V_2 \cdot \sqrt[n]{\frac{p_2}{p_1}}. \tag{3.86}$$

With $\Delta V = V_2 - V_1$ follows

$$\Delta V = V_2 \cdot \left(1 - \sqrt[n]{\frac{p_2}{p_1}}\right), \tag{3.87}$$

where

$p_{1,2}$ pressure in filled or deflated state [N/m^2] ($p_1 > p_2$)
$V_{1,2}$ gas volume in filled or deflated state [m^3] ($V_1 < V_2$)
V_0 gas volume at p_0 (storage tank size including any connected gas cylinders)
ΔV hydraulic fluid output volume of the accumulator (utilizable volume)
n polytropic exponent ($n = 1$ to 1.4)

If we want to refer to a certain preload condition characterized by the variables V_0 and p_0, it results in the following formulas:

$$p_1 \cdot V_1^n = p_2 \cdot V_2^n = p_0 \cdot V_0^n,$$

$$V_1^n = \frac{p_0}{p_1} \cdot V_0^n,$$

$$V_2^n = \frac{p_0}{p_2} \cdot V_0^n.$$

If the condition $\Delta V = V_2 - V_1$ is used, it follows that

$$\Delta V = V_0 \cdot \left(\sqrt[n]{\frac{p_0}{p_2}} - \sqrt[n]{\frac{p_0}{p_1}} \right),$$

$$V_0 = \frac{\Delta V}{\sqrt[n]{\frac{p_0}{p_2}} - \sqrt[n]{\frac{p_0}{p_1}}}.$$

(3.88)

If the change from state 1 to state 2 occurs very slowly so that a temperature equalization with the environment can always take place, an *isothermal change* is present. In this case $n = 1$ can be set.

At high filling or withdrawal rates, i. e., with rapid changes of the state no heat exchange can take place and an *adiabatic change* is present. In this case, $n = \kappa = 1.4$ should be expected. In fact, n will lie between the values 1 and 1.4.

To keep the pressure drop in the accumulator as low as possible when hydraulic fluid is withdrawn, a gas volume $V_G \geq 5 \cdot V_F$ or even more is assigned to a piston accumulator for the fluid volume V_F in gas cylinders in central hydraulic pressure stations of stage buildings.

3.8.4 Piping

In order to keep the flow losses or the heating of the hydraulic fluid and the sound emission low, the following flow velocities should not be exceeded in hydrostatic systems of the stage:

in *pressure lines*   3 m/s
in *suction lines*   0.7 m/s
in *tank return lines*   2 m/s

(Higher velocities can also be permitted for pressure lines in the pressure center.) According to Eq. (3.74), the required nominal cross-section can be calculated from this.

3.9 Hydraulic fluids

In hydrostatic drives, the hydraulic fluid is an energy carrier for transmitting power effects, but it must also dissipate the heat energy released by losses. Furthermore, the hydraulic fluid takes over tasks of lubrication, corrosion protection, as well as the removal of wear material (abrasion) and the collection of solid particles in filters.

Hydraulic fluids should not have a too high viscosity to ensure good suction through the pump and to keep flow losses low; on the other hand, they should not have a too low viscosity to keep leakage oil losses at sealing gaps low.

In principle, various fluids can be used as hydraulic media, but their very different properties must be taken into account: mineral oils, vegetable oils, oil–water mixtures, water–glycol solutions (polyglycols), synthetic anhydrous liquids (esters).

According to DIN 51 524, *hydraulic oils based on mineral oil* are divided into the following categories:

- HL oils with active ingredients to increase aging resistance and corrosion protection, suitable up to fluid pressures of about 200 bar;
- HLP oils, which are provided with high-pressure additives to reduce wear at high pressures; and
- HVLP oils, in which active ingredients are also added to reduce the dependence of viscosity on pressure and temperature.

HLP oils are usually used in hydrostatic systems if operation does not take place at particularly large temperature differences. In stage technology, however, regulations often require the use of flame-retardant fluids with specific ignition and burning behavior, which are designated HFA-HFD according to ISO. *Flame-retardant hydraulic fluids* include:

- HFA fluids: oil-in-water emulsions with a water content of 80–98 %, i. e., combustible content maximum 20 %.
- HFB fluids: water-in-oil emulsions with a water content of more than 40 %, i. e., combustible content maximum 60 %.
- HFC liquids: aqueous solutions, e. g., water–glycol with a water content of 35–55 %.
- HFD fluids: anhydrous synthetic fluids such as phosphate esters, chlorinated hydrocarbons and organic esters.

In stage technology, systems with water hydraulics (HFA and HFC fluids) were often used in old stage systems for fire protection reasons. When using HFA fluids, mechanical wear, erosion and cavitation erosion, as well as microbial infestation, yield particular technological problems.

Nowadays, HFD fluids, namely *organic esters*, are frequently used as hydraulic media. In the case of synthetic fluids, it should be borne in mind that chemical impurities may possibly trigger electrochemical corrosion processes. Therefore, these fluids should be checked temporarily with regard to their composition in order to be able to take suitable countermeasures in good time, if necessary.

Among these organic esters, a product under the trademark "Quintolubric N 822" is very frequently used in stage technology, which is to be classified as an HFD-U fluid and is based on polyol ester. This hydraulic fluid will therefore be described in more detail with regard to its properties:

- The *auto-ignition temperature is* 460 °C, i. e., below this temperature auto-ignition is impossible, the *flash point is* 350 °C, i. e., below this temperature the liquid will not burn if an open flame is used as an ignition source. If the ignition source is removed, the liquid shows a *self-extinguishing effect.*
- Compatibility with all metals, all mineral oil-based hydraulic fluids and the most common phosphate esters is given. Furthermore, almost all commonly used standard seals can be used in the devices.
- The specific gravity is 0.92 kg/dm³ and allows floating in the presence of water impurities, which enables their removal. It is important to keep the liquid free of water, because it is immiscible with water and the water would cause adverse operating behavior in the form of water droplets.
- The liquid is biodegradable and nontoxic.
- The lubricating properties are very good over a wide temperature range.

Various characteristic values of hydraulic fluids

Specific mass (mass density)
The density depends on the product, but also on the temperature and pressure. Rough guide values can be taken from Table 3.5.

Table 3.5: Specific mass of hydraulic fluids.

	kg/dm³	kg/m³
Mineral oils	0.90	900
HFA	0.99	990
HFB	0,95	950
HFC	1.04–1.09	1040–1090
HFD, Quintolubric	0.92–1.45	920–1450

Viscosity
The kinematic viscosity is usually given in $[mm^2/s]$. It is generally strongly dependent on temperature and pressure; viscosity decreases with increasing temperature. The permissible viscosity range depends above all on the pump and motor design used in the system and can be found in the manufacturer's specifications. At 50 °C, the viscosity of oils is about $20 \div 150\ mm^2/s$, that of water is $0.6\ mm^2/s$.

Compressibility
In most considerations, hydraulic media can be considered incompressible, but sometimes the slight compressibility does need to be taken into account. The values defined as the *coefficient of compressibility*

$$\beta_K = \frac{\Delta V}{V_0} \cdot \frac{1}{\Delta p} \left[\frac{m^2}{N} \right], \left[\frac{1}{bar} \right],$$

$$\Delta V = V_0 - V_1, \tag{3.89}$$

$$\Delta p = p_1 - p_0,$$

$$\Delta V = \beta_K \cdot V_0 \cdot \Delta p,$$

and the reciprocal value is referred to as the *compression modulus*,

$$K = \frac{1}{\beta_K} = \frac{V_0 \cdot \Delta p}{\Delta V} = E_H \left[\frac{N}{m^2} \right], [bar],$$

$$\Delta V = \frac{V_0 \cdot \Delta p}{K}. \tag{3.90}$$

The *compressibility of* a fluid depends primarily on the pressure level, but also on the temperature: The higher the pressure, the lower the β_K or the greater the K. Air inclusions in the hydraulic fluid increase compressibility. At very high pressures (>350 bar), however, air inclusions hardly affect compressibility any more.

The higher the temperature, the larger the β_K or the lower the K. However, the dependence on temperature is negligible in the usual temperature range.

For *mineral oil*, depending on the pressure level, the following applies on average:

$$\beta_K = 6.7 \cdot 10^{-5}\, 1/bar = 6.7 \cdot 10^{-6}\, cm^2/N = 6.7 \cdot 10^{-10}\, m^2/N,$$

$$K = E_H = 1.5 \cdot 10^4\, bar = 1.5 \cdot 10^5\, N/cm^2 = 1.5 \cdot 10^9\, N/m^2 = 1.5\, kN/mm^2,$$

$$(K = 1.6\, kN/mm^2 \text{ at normal atmospheric pressure,}$$

$$K = 1.4\, kN/mm^2 \text{ at a pressure of 100 to 300 bar).}$$

For *water*, on average one has

$$\beta_K = 4.8 \cdot 10^{-5}\, 1/bar,$$

$$K = 2.1 \cdot 10^4\, bar = 2.1 \cdot 10^9\, N/m^2 = 2.1\, GPa.$$

This means that water has a lower compressibility than oil.

The *compressibility modulus K* of a liquid corresponds to the *modulus of elasticity* of a solid and is therefore also referred to as E_H. Comparatively, the *modulus of elasticity* of steel is

$$E_{Steel} = 206\, [kN/mm^2] = 206 \cdot 10^9\, [N/m^2].$$

Oil is thus $206/1.5 \approx 140$ times and water $206/2.1 \approx 100$ times more elastic than steel.

Therefore, a lifting podium supported by spindles or racks is much more rigidly supported than a podium supported on hydraulic cylinders. The much higher compressibility of the fluid has the effect of a greater change in position of the podium when the load changes, but above all of a greater "susceptibility to vibration" in the unlocked state, since the natural frequency of the system is much lower (see Section 3.10).

Thus, if the internal pressure in a cylinder is increased by Δp by increasing the load, the change ΔV in fluid volume can be calculated using Eq. (3.89) or Eq. (3.90). The piston shifts by

$$l_p = \frac{\Delta V}{A}, \tag{3.91}$$

where

V_0 volume of hydraulic fluid in the cylinder [m³]
ΔV volume change of hydraulic fluid due to pressure change [m³]
A piston area [m²]
l_p displacement of the piston due to pressure change [m]
β_K compressibility coefficient [m²/N, 1/bar, …]
$K = E_H$ compression modulus [N/m² = Pa, bar, …]

Strictly speaking, this consideration would also have to take into account the expansion of the cylinder tube when the pressure is increased. However, this is generally negligible.

Volume change with temperature change
The *volumetric coefficient of thermal expansion = coefficient of thermal expansion* is

$$\beta_\vartheta = \frac{\Delta V}{V_0} \cdot \frac{1}{\Delta\vartheta} \left[\frac{1}{\text{grd}} \right],$$
$$\Delta V = V_1 - V_0, \tag{3.92}$$
$$\Delta\vartheta = \vartheta_1 - \vartheta_0,$$
$$\Delta V = \beta_\vartheta \cdot V_0 \cdot \Delta\vartheta;$$

for *hydraulic oil* approx. $\beta_\vartheta = 7.0 \cdot 10^{-4}$ [1/grd]
for *water* $\beta_\vartheta = 1.8 \cdot 10^{-4}$ [1/grd]

With this, the ratio

$$l_\vartheta = \frac{\Delta V}{A}, \tag{3.93}$$

where
l_ϑ displacement of the piston due to the temperature change [m]
$\Delta\vartheta$ hydraulic fluid temperature change [grd]

can therefore be used to calculate the change in volume with a change in temperature and the resulting displacement of the piston of a cylinder.

Strictly speaking, the change in diameter of the cylinder tube with temperature change would have to be taken into account. However, this is negligibly small.

Specific heat

The specific heat c is the quantity of heat Q required to heat 1 kg of a substance by 1 grd. The following relationship therefore applies:

$$Q = c \cdot m \cdot \Delta\vartheta, \tag{3.94}$$

where
Q amount of heat supplied or dissipated = energy = work [Nm = J]
m mass of the substance [kg]
c specific heat [J/(kg·grd)]
$\Delta\vartheta$ temperature difference (heating or cooling)

Table 3.6 shows the specific heat of hydraulic fluids in new and old units.

Table 3.6: Specific heat of hydraulic fluids.

	kJ/(kg·grd)	kcal/(kg·grd)
Mineral oil	1.9–2.1	0.45–0.5
Water, HFA fluids	4.18	1.0
HFB, HFC fluids	3.3	0.8
HFD fluids, Quintolubric	1.26	0.3

The specific heat of mineral oil is therefore only about half that of water, and that of quintolubric only about one-third that of water. Therefore, if the same amount of heat is applied to a weight of mineral oil, it will heat up about twice as much as the same weight of water, and quintolubric will heat up about three times as much. Therefore, a plant with water as the operating medium may not require a cooling system, but the same plant with oil or quintolubric as the hydraulic medium must be equipped with a cooling system.

For hydraulic fluids, particular attention should be paid to the following during equipment maintenance:

– The water content in HLP mineral oils and HFD-U fluids should not exceed 0.1 % for extended periods.
– Since the liquids are subject to an aging process due to oxidation, hydrolysis, degradation of additives, etc., the aging condition should be monitored by temporarily determining the "neutralization number" and checking the air separation capacity or foaming behavior.

3.10 Vibrations

When dimensioning components, it must be ensured that no disturbing vibrations occur during operation. In stage technology, this criterion is of particular importance. For example, it would not be tolerable if a podium with set-up scenery would be stimulated to oscillate with clearly perceptible vibrations by ballet dancers or drives of the stage machinery.

In general, vibrations are more or less regularly occurring fluctuations of state variables. In the problem addressed here, we are dealing with mechanical vibrations in which a component periodically oscillates around a rest position. Figure 3.12(a) shows an oscillation with periodic change of a state value x (e. g., position coordinate, velocity, acceleration, force or angular position, angular velocity, angular acceleration, moment) over time t, whereby the following applies:

$$x(t) = x(t + T). \tag{3.95}$$

A characteristic value of a variable is its *effective value* defined as a *root mean square value (RMS)*:

$$x_{\text{eff}} = \sqrt{\frac{1}{T} \int x^2(t)dt}, \tag{3.96}$$

where
x condition value
t time [s]
T period of oscillation = duration of one period [s]
x_{eff} effective value (root mean square value – RMS) of the state variable

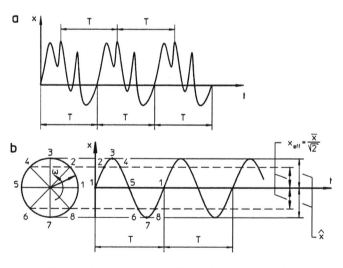

Figure 3.12: Oscillations: (a) periodic oscillation, (b) harmonic oscillation.

Most of the oscillations occurring in nature and technology are sinusoidal oscillations of the form (Figure 3.12(b))

$$\bar{x} = x \cdot \sin(\omega \cdot t);$$

they are also called *harmonic oscillations*. The further considerations refer to this type of oscillations. The following applies to them in particular

$$f = \frac{1}{T},$$

$$\omega = 2 \cdot \pi \cdot f,$$

$$f = \frac{\omega}{2 \cdot \pi},\tag{3.97}$$

$$\bar{x} = \frac{1}{2} \cdot (x_{max} - x_{min}),\tag{3.98}$$

$$\bar{x}_{eff} = \sqrt{\frac{1}{T} \int (\bar{x} \cdot \sin \omega t)^2 dt} = \frac{x}{\sqrt{2}},\tag{3.99}$$

where
f *frequency* = number of oscillations per time unit [1/s]
ω angular frequency = number of oscillations in 2π seconds [1/s]
$\bar{x}$ amplitude = half value of the total oscillation amplitude [dimension of the value x]
T period of oscillation [s]

A sinusoidal curve for x over t can be obtained by rotating a point on the circumference of a circle with radius x around the center of the circle with angular velocity ω and plotting the respective x-coordinate of the point over time t. Therefore, this angular velocity ω is called *angular frequency*. Therewith, the above formula relationships can also be explained:

At an angular velocity ω, the point runs along the circumference of the circle of length $u = 2 \cdot \bar{x} \cdot \pi$ with the circumferential velocity $v = \bar{x} \cdot \omega$.

The circulation time T is therefore

$$T = \frac{u}{v} = \frac{2\pi}{\omega} = \frac{1}{f}.$$

The following explanations are not intended to teach vibration theory in the sense of a textbook, but to provide a collection of formulas relevant to problems in stage engineering.

3.10.1 Single-mass oscillator

First of all, oscillators are considered in which an imaginary mass (mass point) concentrated at one location executes an oscillating motion. If such a *single-mass oscillator* is excited once, e. g., by a deflection, it oscillates with a very specific system-specific frequency, the so-called *natural frequency* f_{nat} or ω_{nat}.

In the case of an *undamped natural oscillation*, this oscillation would theoretically continue indefinitely. In fact, damping influences are always present, so that the oscillation decays sooner or later depending on the type and magnitude of the damping (*damped natural oscillation*).

However, it is also easy to imagine that a system which is constantly excited in time with its natural frequency or similar to its natural frequency builds up to oscillations of particularly large amplitudes. If *natural frequency* and *excitation frequency* coincide, this is called *resonance (resonance frequency)*. In a completely undamped system, the amplitude of oscillation in this case would theoretically become infinitely large. Again, it depends on the type and size of the damping how large the oscillation amplitudes actually become.

Natural frequency of a spring–mass system
If the mass of the spring is small compared to the mass m, the spring can be assumed to be massless for simplification. For the *spring–mass oscillator with translatory motion* of the mass according to Figure 3.13(a), one has:

$$\omega_{eig} = \sqrt{\frac{c}{m}},\qquad (3.100)$$

$$c = \frac{F}{\delta},\qquad (3.101)$$

where
m vibrating mass [kg]
c spring constant of the spring [N/m]
F force [N]
δ displacement [m] due to force F [N]

Here F is the force to be applied at the position of the mass in the direction of the oscillating motion, which is required to make a deflection by a distance δ.

If two springs are connected in parallel according to Figure 3.13(b, c), the resulting spring constant is

$$c = c_1 + c_2,$$

or for more springs,

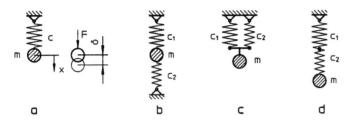

Figure 3.13: Spring–mass oscillator: (a) single spring, (b, c) parallel connection of springs, (d) series connection of springs.

$$c = \sum c_i. \tag{3.102}$$

If two springs are connected in series according to Figure 3.13(d), then

$$\frac{1}{c} = \frac{1}{c_1} + \frac{1}{c_2},$$

or for more springs,

$$\frac{1}{c} = \sum \frac{1}{c_i}. \tag{3.103}$$

For a *spring–mass oscillator with rotating motion* of the mass according to Figure 3.18(a), the following applies in an analogous manner (massless torsion spring):

$$\omega_{eig} = \sqrt{\frac{c}{I_m}}, \tag{3.104}$$

$$c = \frac{M}{\varphi}, \tag{3.105}$$

where
I_m mass moment of inertia of the vibrating mass [kg m^2]
c torsional spring constant of the torsion spring [N/rad]
M torque [Nm]
φ angle of rotation [rad] as a result of torque M [Nm]

Thus, in this case, the mass m is replaced by the mass moment of inertia I_m of the mass m, and the spring constant refers to a required torque M to achieve a torsional angle φ.
    Some specific examples are discussed in more detail below.

Elastic tension rod (Figure 3.14(a))
If the spring is a rod assumed to be massless with cross-sectional area A and made of a material with modulus of elasticity E, then using Hooke's law of strength theory, the displacement δ under a force action F is:

$$\frac{\delta}{l} = \frac{F}{A \cdot E},$$

$$c = \frac{F}{\delta} = \frac{A \cdot E}{l},$$

(3.106)

where

A   cross-sectional area of the bar [m²]
E   modulus of elasticity of the bar [N/m²]
l   length of the elastic rod [m]

The case of a load falling into an elastic rope discussed in Section 3.7 can also be interpreted as a vibration phenomenon, although damping is not taken into account here.

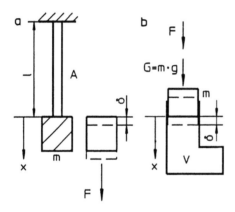

Figure 3.14: Rod and liquid column as spring: (a) tension rod, (b) liquid column.

The vibration equations for the initial conditions $x(t = 0) = 0$ and $\dot{x}(t = 0) = v_0$ are

$$x(t) = \frac{v_0}{\omega} \cdot \sin(\omega \cdot t),$$

$$\overline{x} = \frac{v_0}{\omega},$$

$$\dot{x}(t) = v_0 \cdot \cos(\omega \cdot t),$$

$$|v_{\max}| = v_0,$$

(3.107)

$$\ddot{x}(t) = -v_0 \cdot \omega \cdot \sin(\omega \cdot t),$$

$$|a_{\max}| = v_0 \cdot \omega,$$

$$\omega = \sqrt{\frac{c}{m}},$$

where

$x(t)$ evolution of the deflection from the rest position over time
$\dot{x}(t)$ evolution of velocity over time
$\ddot{x}(t)$ evolution of acceleration over time

The rest position is determined by the elongation of the rope due to the weight of mass m by

$$\Delta l_0 = \frac{m \cdot g}{c},$$

the amplitude of oscillation is

$$\overline{x} = \Delta l - \Delta l_0.$$

According to Eq. (3.69),

$$\Delta l = \frac{m \cdot g}{c} \cdot \left[1 + \sqrt{1 + \frac{2 \cdot h \cdot c}{m \cdot g}} \right],$$

$$\overline{x} = \frac{m \cdot g}{c} \cdot \left[1 + \sqrt{1 + \frac{2 \cdot h \cdot c}{m \cdot g}} \right] - \frac{m \cdot g}{c} = \frac{m \cdot g}{c} \cdot \sqrt{1 + \frac{2 \cdot h \cdot c}{m \cdot g}}.$$

However, according to Eq. (3.107), the amplitude can also be calculated as $\overline{x} = \frac{v_0}{\omega}$ where v_0 can be calculated as follows:

When the distance A is covered from the zero position, the kinetic energy given by v_0 and the potential energy released are converted into the rope's elongation work:

$$\frac{m \cdot v_0^2}{2} + m \cdot g \cdot \overline{x} = \frac{1}{2} \cdot F \cdot \overline{x}, \quad v_0 = \sqrt{\frac{2}{m} \left(\frac{1}{2} F \cdot \overline{x} - m \cdot g \cdot \overline{x} \right)}, \tag{3.108}$$

$$a_{max} = v_0 \cdot \omega. \tag{3.109}$$

Liquid column (Figure 3.14(b))

A column of fluid also has resilient properties as a result of the compressibility of a fluid. The compression modulus K of the hydraulic fluid corresponds to the elasticity modulus E_H (see Section 3.9, Eq. (3.90)).

Therefore, if a mass m is supported by a piston of area A according to Figure 3.14(b) on a fluid volume V_{ab}, the spring stiffness and natural frequency of the fluid volume can be calculated as follows (see Eqs. (3.28), (3.90), (3.91) and (3.101)):

$$c = \frac{F}{\delta} = \frac{\Delta p \cdot A}{\frac{\Delta V}{A}} = \frac{\Delta p}{\Delta V} \cdot A^2 = \frac{\Delta p \cdot A^2 \cdot E_H}{V_0 \cdot \Delta p} = \frac{A^2}{V_0} E_H, \tag{3.110}$$

$$\omega_{eig} = \sqrt{\frac{c}{m}} = \sqrt{\frac{A^2 \cdot E_H}{V_0 \cdot m}}, \tag{3.111}$$

where
A piston area [m^2]

V_0 elastic volume of hydraulic fluid [m³]

E_H modulus of elasticity of the hydraulic fluid = coefficient of compressibility K [N/m²]
(see Section 3.9).

Synchronous cylinder (Figure 3.15(a))
If the mass m is clamped between two liquid columns connected in parallel, we get

$$c = c_1 + c_2 = E_H \cdot A^2 \cdot \left(\frac{1}{V_1} + \frac{1}{V_2} \right).$$

(3.112)

If we set $V_1 = A - x$ and $V_2 = a - (h - x)$, we get

$$c = E_H \cdot A \cdot \left(\frac{1}{x} + \frac{1}{h - x} \right).$$

The piston position x_0, at which the spring stiffness and thus the natural frequency are minimal, is obtained by differentiating and setting to zero ($dc/dx = 0$):

$$x_0 = \frac{h}{2},$$

$$c_{\min} = \frac{4 \cdot A \cdot E_H}{h}.$$

The natural frequency in this piston position is therefore

$$\omega_{\text{eig}} = \sqrt{\frac{c_{\min}}{m}} = \sqrt{\frac{4 \cdot A \cdot E_H}{h \cdot m}}.$$

(3.113)

Differential cylinder (Figure 3.15(b))
A differential cylinder can be treated in an analogous manner. For further generalization, additional volumes V_I and V_{II} are added to the liquid volumes on both sides of the piston to take into account pipe sections up to the next shut-off device,

$$c = c_1 + c_2 = \frac{A_1^2 \cdot E_H}{A_1 \cdot x + V_I} + \frac{A_2^2 \cdot E_H}{A_2 \cdot (h - x) + V_{II}}.$$

(3.114)

From the condition $dc/dx = 0$ analogously as before one has

$$x_0 = \frac{\dfrac{A_2 \cdot h}{\sqrt{A_2^3}} + \dfrac{V_{II}}{\sqrt{A_2^3}} - \dfrac{V_I}{\sqrt{A_1^3}}}{\dfrac{1}{\sqrt{A_2}} + \dfrac{1}{\sqrt{A_1}}},$$

$$c_{\min} = \frac{A_1^2 \cdot E_H}{V_1} + \frac{A_2^2 \cdot E_H}{V_2},$$

(3.115)

$$V_1 = A_1 \cdot x_0 + V_I,$$

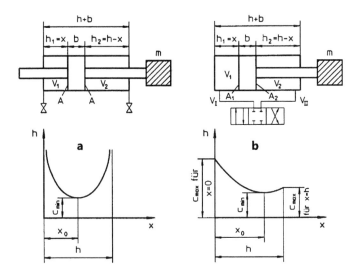

Figure 3.15: Double-acting cylinders: (a) synchronous cylinder (volume V_1 and V_2), (b) differential cylinder (volume $V_1 + V_I$ and $V_2 + V$).

$$V_2 = A_2 \cdot (h - x_0) + V_{II},$$

$$\omega_{\text{eig}} = \sqrt{\frac{c_{\min}}{m}}.$$

Bending oscillator (Figure 3.16)

Applying Eq. (3.100), the following applies to a cantilever beam according to Figure 3.16(a):

$$\delta = \frac{F}{3 \cdot E \cdot J} \cdot l^3,$$

$$c = \frac{F}{\delta} = \frac{3 \cdot E \cdot J}{l^3},$$

(3.116)

and, for an articulated beam on two supports according to Figure 3.16(b),

$$\delta = \frac{F}{3 \cdot E \cdot J} \cdot \frac{a^2 \cdot b^2}{l},$$

$$c = \frac{F}{\delta} = \frac{3 \cdot E \cdot J \cdot l}{a^2 \cdot b^2},$$

(3.117)

where
E modulus of elasticity of the beam [N/m²]
J area moment of inertia of the beam cross-section [m⁴]
a, b, l lengths according to Figure 3.16 [m]

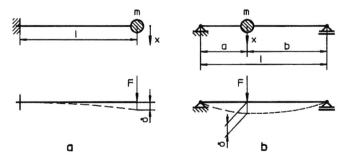

Figure 3.16: Bending oscillator with single mass: (a) cantilever beam, (b) beam on two supports.

Beams supported differently can be treated analogously using the appropriate formulas for deflection.

Transversely vibrating string with central mass (Figure 3.17)
For a (massless) string tensioned with prestressing force S, the following applies in application of equation (3.101) according to the laws of rope statics:

$$\delta = \frac{F}{S} \cdot \frac{a \cdot b}{l},$$
$$c = \frac{F}{\delta} = \frac{S \cdot l}{a \cdot b}.$$

(3.118)

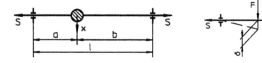

Figure 3.17: Transversely vibrating string with single mass.

Rotary oscillator (Figure 3.18(a))
Analogous to the spring index, we can define

$$c = \frac{M}{\varphi},$$
$$\omega_{\text{eig}} = \sqrt{\frac{c}{I}},$$

(3.119)

where
M torque [Nm]
φ torsion angle [rad]
I moment of inertia of the oscillating mass [kg m^2]

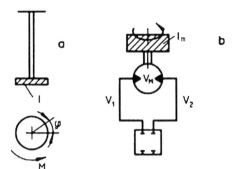

Figure 3.18: Rotary transducer: (a) torsion bar with mass, (b) hydraulic motor.

If the torsion spring is a torsion bar with a circular cross-section whose mass inertia is negligible, then

$$\varphi = \frac{M \cdot l}{G \cdot I_p},$$

$$c = \frac{M}{\varphi} = \frac{G \cdot I_p}{l},$$

where

I_p polar area moment of inertia [m^4]
G shear modulus of the material of the torsion bar [N/m^2]
l length of the torsion bar [m]
φ angle of twist [rad]

Hydromotor (Figure 3.18(b))

A mass m moved by a hydraulic motor also represents a vibratory system of this type. If the hydraulic motor has the displacement V_M (see Sections 2.3.2, 3.8.2), it can be assumed that the clamped volume in the motor is $V_M/2$ on each side. If a volume V_I is connected to one side with piping, and a volume V_{II} is connected to the other side, the clamped volumes are as follows:

$$V_1 = \frac{V_M}{2} + V_I \quad \text{and} \quad V_2 = \frac{V_M}{2} + V_{II}. \tag{3.120}$$

The natural frequency of this system can also be determined by Eqs. (3.104), (3.105). If the amount of liquid corresponding to its displacement V_M is supplied to the hydromotor, the rotor rotates by the angle 2π; therefore, a change in volume ΔV corresponds to an angle of rotation

$$\varphi = 2\pi \cdot \frac{\Delta V}{V_M}.$$

Therefore, applying Eqs. (3.105), (3.84), and (3.90),

$$c = \frac{M}{\varphi} = \frac{\frac{1}{2\pi} \cdot V_M \cdot \Delta p}{2\pi \cdot \frac{\Delta V}{V_M}} = \left(\frac{V_M}{2\pi}\right)^2 \cdot \frac{\Delta p}{\Delta V} = \left(\frac{V_M}{2\pi}\right)^2 \cdot \frac{E_H}{V_0}. \tag{3.121}$$

If we now substitute the volumes for V_0 according to Eq. (3.119), we obtain

$$c = c_1 + c_2 = \left(\frac{V_M}{2\pi}\right)^2 \cdot E_H \cdot \left(\frac{1}{V_1} + \frac{1}{V_2}\right) \tag{3.122}$$

and, according to Eq. (3.104),

$$\omega_{eig} = \frac{V_M}{2\pi} \cdot \sqrt{\frac{E_H}{I_m} \cdot \left(\frac{1}{V_1} + \frac{1}{V_2}\right)}. \tag{3.123}$$

For the special case $V_1 = V_2 = V$, we get

$$\omega_{eig} = \frac{V_M}{2\pi} \cdot \sqrt{\frac{2E_H}{I_m \cdot V}}. \tag{3.124}$$

Natural frequency of the mathematical pendulum
Finally, the so-called *mathematical pendulum* in Figure 3.19(a) should be mentioned, in which a mass m suspended from a massless thread oscillates under the effect of gravity. For this oscillating system, the following applies in the case of small deflections:

$$\omega_{eig} = \sqrt{\frac{g}{l}}, \tag{3.125}$$

where
g acceleration due to gravity $g = 9.81\,\text{m/s}^2$
l pendulum length [m]

i. e., in this case the natural frequency is independent of the size of the vibrating mass.
Concerning the *nonmathematical pendulum (compound pendulum)* in Figure 3.19(b), see Eq. (3.134).

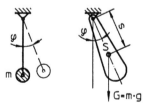

Figure 3.19: Pendulums: (a) mathematical, (b) notmathematical.

3.10.2 Oscillating continuum

The term oscillating continuum refers to an oscillating component in which no distinction can be made between a massless spring and an oscillating mass; a mass-containing component oscillates as a result of its inherent elasticity. Such an oscillating continuum has an infinite number of natural frequencies: a lowest natural frequency and many superimposed harmonics of higher frequency. Resonance occurs again when the excitation frequency coincides with one of these natural frequencies. The lowest natural frequency is usually of particular interest.

Natural frequency of a vibrating continuum
Only the formulas for the longitudinal and transverse vibrations of a beam with constant cross-section are given.

Longitudinal vibration of a girder

$$\omega_{eig} = K_n \sqrt{\frac{c}{m}}. \tag{3.126}$$

The factor K_n depends on the type of bearing (the boundary conditions); c is calculated according to Eq. (3.106).
- For completely free longitudinal movement or fixed support (clamping) at both ends, set as boundary condition

$$K_n = \pi \cdot n, \quad n = 1, 2, 3, \ldots.$$

Applying Eq. (3.122) to a *column of liquid* (cross-sectional area A) confined in a tube of length l, Eq. (3.107) results in

$$c = \frac{A^2}{V_0} \cdot E_H = \frac{A^2}{A \cdot l} \cdot E_H = \frac{A}{l} \cdot E_H,$$
$$m = V \cdot \rho = A \cdot l \cdot \rho,$$
$$c = \frac{m \cdot E_H}{l^2 \cdot \rho},$$
$$\omega_{eig} = \pi \cdot n \cdot \sqrt{\frac{c}{m}} = \frac{\pi \cdot n}{l} \cdot \sqrt{\frac{E_H}{\rho}} = \frac{\pi \cdot n}{l} \cdot v_s, \tag{3.127}$$
$$v_s = \sqrt{\frac{E_H}{\rho}},$$
$$f_{eig} = \frac{\omega_{eig}}{2\pi} = \frac{n \cdot v_s}{2 \cdot l} \quad \text{(according to Eq. (3.97)).}$$

The lowest frequency ω_0 or f_0 is obtained for $n = 1$.

The wavelength is calculated from this (Eq. (3.138)) to be

$$\lambda = \frac{v_s}{f} = \frac{2 \cdot l}{n}.$$

Resonance is present when the tube length l meets this condition:

$$l = \frac{n \cdot \lambda}{2}, \quad l = \frac{\lambda}{2}, \lambda, \frac{3\lambda}{2}, 2\lambda, \ldots \quad (3.128)$$

- For the boundary conditions with one end free, the other fixed end satisfies:

$$K_n = \pi \cdot \frac{2n - 1}{2},$$

$$\omega_{eig} = \pi \cdot \frac{2n - 1}{2 \cdot l} \cdot v_s,$$

$$f_{eig} = \frac{2n - 1}{4 \cdot l} \cdot v_s, \quad (3.129)$$

$$\lambda = \frac{v_s}{f} = \frac{4 \cdot l}{2n - 1}.$$

Resonance is available for

$$l = \frac{2n - 1}{4} \cdot \lambda, \quad \lambda = \frac{\lambda}{4}, \frac{3\lambda}{4}, \frac{5\lambda}{4}, \ldots, \quad (3.130)$$

where
E_H compression modulus of the liquid [N/m²]
v_s speed of sound in the liquid [m/s], $v_s = \sqrt{\frac{E_H}{\rho}}$
K_n boundary condition factor [–]
L girder length [m]
n natural number, namely $1, 2, 3, \ldots$
λ wavelength [m]
ρ specific mass of the liquid [kg/m³]

Bending vibration of a beam (Figure 3.20)
For the bending vibration of a beam with constant material occupancy over the beam length l, the natural frequencies are to be calculated according to the following formula:

$$\omega_{eig} = K_n^2 \cdot \sqrt{\frac{E}{\rho}} \cdot \sqrt{\frac{J}{A}}, \quad (3.131)$$

where
A cross-sectional area of the girder [m²]
E modulus of elasticity [N/m²]
ρ density [kg/m³]

J area moment of inertia of the beam [m⁴]

v_s speed of sound in the girder material [m/s], $v_s = \sqrt{\dfrac{E}{\rho}}$

The values for K_n can be found in Table 3.7.

Table 3.7: Values for K_n in Eq. (3.131).

a	$K_1 \approx 4.73/l$	$K_2 \approx 7.85/l$	$K_3 \approx 11.00/l$	$K_n \approx (2n+1) \cdot \pi/(2 \cdot l)$
b	$K_1 \approx \pi/l$	$K_2 \approx 2\pi/l$	$K_3 \approx 3\pi/l$	$K_n \approx n - \pi/l$
c	$K_1 \approx 1.88/l$	$K_2 \approx 4.69/l$	$K_3 \approx 7.86/l$	$K_n \approx (2n-1) \cdot \pi/(2 \cdot l)$
d	$K_1 \approx 3.93/l$	$K_2 \approx 7.07/l$	$K_3 \approx 10.21/l$	$K_n \approx (4n+1) \cdot \pi/(4 \cdot l)$

Boundary conditions are as in Figure 3.20.

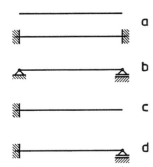

Figure 3.20: Natural frequencies of the transversely vibrating beam – boundary conditions: (a) free–free, clamped–clamped, (b) hinged–jointed, (c) clamped–free, (d) clamped–jointed.

The natural frequency for bending vibrations is also relevant for a rotating shaft and results in the *critical angular velocity* ω_k or, with Eq. (3.9), the *critical rotational speed*

$$n_k = \frac{1}{2\pi} \cdot \omega_k.$$

As an example, the bending-critical speed of the first degree (vibration loop between the bearings) is given by K_n for $n = 1$ of a tubular shaft articulated on both sides (case b). Applying Eq. (3.131), the following is obtained:

$$\omega_k = \left(\frac{\pi}{l}\right)^2 \cdot \sqrt{\frac{E}{\rho}} \cdot \sqrt{\frac{J}{A}}. \tag{3.132}$$

From this follows for the tubular shaft (e. g., also for a cardan shaft) with

$$A = \frac{\pi}{4} \cdot (D^2 - d^2),$$

$$J = \frac{\pi}{64} \cdot (D^4 - d^4),$$

$$\omega_k = \frac{\pi^2}{4} \cdot \sqrt{\frac{E}{\rho}} \cdot \frac{\sqrt{D^2 + d^2}}{l^2},$$

(3.133)

where

J area moment of inertia [m^4]

A pipe cross-section [m^2]

ρ density of the pipe material [kg/m^3]

E modulus of elasticity of the pipe material [N/m^2]

D outer diameter of the pipe shaft [m]

d inner diameter of the pipe shaft [m]

l length of pipe shaft from joint to joint [m]

If one sets for steel $\sqrt{\frac{E}{\rho}} = 5100$ m/s, one obtains with Eq. (3.17)

$$n_k = 120 \cdot 10^3 \cdot \frac{\sqrt{D^2 + d^2}}{l^2} \ [\text{rpm}],$$

$$D, d, l \ [\text{m}].$$

For a solid shaft, $d = 0$ is to be set and one obtains

$$\omega_k = \frac{\pi^2}{4} \cdot \sqrt{\frac{E}{\rho}} \cdot \frac{D}{l^2},$$

$$n_k = 120 \cdot 10^3 \cdot \frac{D}{l^2} \ [\text{rpm}],$$

$$D, l \ [\text{m}].$$

Natural frequency of the compound pendulum (nonmathematical pendulum)
In contrast to the string pendulum explained for the single-mass oscillators, the following applies to the compound pendulum (also called physical pendulum) in Figure 3.19(b) for small deflections:

$$\omega_{\text{eig}} = \sqrt{\frac{m \cdot g \cdot s}{I_m}},$$

(3.134)

where

m pendulum mass [kg]

I_m moment of inertia of mass m around the center of rotation [kg m^2]

g gravitational acceleration [m/s^2], $g = 9.81$ m/s^2

s center of gravity – distance from the center of rotation [m]

3.10.3 Vibration excitation

The possibilities to excite vibrations are, of course, very diverse. Here, only a few cases are given as examples, which can be of relevance especially in stage technology.

Vibration excitation by rotating components

Rotating mass

The excitation frequency f_{err} is

$$f_{err} = n = \frac{n^*}{60} \ [1/s], \tag{3.135}$$

where
f frequency [1/s]
n speed [rps]
n^* speed [rpm]

Cardan shaft

The mass of the intermediate shaft of a cardan shaft (see Section 4.5) runs with a nonuniformity of twice the rotational frequency of the input and output shaft

$$f_{err} = 2 \cdot n = 2 \cdot \frac{n^*}{60}. \tag{3.136}$$

Gear dives (tooth drives)

Likewise, due to the constantly changing tooth meshing in the pairing with another gear, a gear wheel emits interference pulses of the frequency

$$f_{err} = z \cdot n = z \cdot \frac{n^*}{60}. \tag{3.137}$$

Chain gear

As a result of the polygon effect (see Section 4.2.2), also a sprocket driven with uniform angular velocity causes a nonuniform longitudinal movement of the chain and – apart from special geometrical cases – also a nonuniform rotation of a chain wheel driven by this chain. The frequency of the nonuniformity of the longitudinal movement of the chain is to be calculated according to Eq. (3.137), where z is the number of teeth of the sprocket.

Hydraulic pump, motor

A rotating hydraulic unit generates a pulsating delivery flow with the frequency f_{err} depending on the number of delivery or displacer elements. The frequency f_{err} is to be calculated according to Eq. (3.137), where z is the number of displacement elements (the number of pistons in a piston pump, the teeth in a gear pump, the vanes in a vane cell pump).

Frictional vibrations

Periodic excitations can also result from the so-called *slip–stick effect*. If a mass m, driven by an elastic power transmission, is to be moved in a sliding manner on a surface, and if the coefficient of friction of the static friction μ_S is greater than that of the motion μ_D, the following phenomenon can occur at low relative velocity.

As a result of the high static coefficient of friction, no sliding movement initially occurs despite the force effect; however, the components transmitting the displacement force are elastically deformed and tensioned like a spring. At a certain level of tension, the frictional connection is broken and sliding motion begins. The tension is released abruptly. However, this again leads to a situation where the relative motion of the two bodies sliding against each other becomes zero. Thus, static friction is again present, and this process is repeated anew. By improving the lubrication conditions, the slip–stick effect can usually be made to disappear.

Actors movements

Finally, in stage events, vibration excitation can also occur through rhythmic movements of performers, in marching groups of people, by ballet dancers, etc.

3.10.4 Perception of vibrations

Vibrations in certain frequency ranges are perceptible to humans and, if they are not intentional, are usually disturbing, whether one sees or feels a vibrational movement. Humans find vibrations in the range of about 1 to 10 Hz particularly unpleasant. The *average sensibility limit of* humans is at an effective vibration velocity of about v_{eff} = 0.11 mm/s (see Eqs. (3.96), (3.99) with v for the value x). Vibrations of the air in the frequency range from about 16 (20) to 16,000 (20,000) Hz are perceived as sound (see Section 3.11).

When dimensioning stage technology equipment, it must be taken into account that, e. g.:
–   Components that can be walked on do not have too low natural frequency.
–   Rapidly rotating masses are balanced particularly well.
–   Machines are set up on vibration isolation metals and largely decoupled from the rest of the structure in terms of vibration.

3.11 Acoustics

Acoustics is the study of sound and includes sound generation, propagation and reception.

With regard to the acoustic design of a stage and auditorium, care must be taken, on the one hand, to ensure that the speech and singing of the performers and the sound of the orchestra are perceived by all spectators in as good a quality as possible; on the other hand, care must be taken to ensure that sound emissions from technical equipment required for the performance hardly reach the spectators or are not perceived by them as disturbing.

In this book, which deals with stage equipment, only the problems of sound emission and propagation from stage drives will therefore be dealt with. Permissible tolerance values for this are very low, since even relatively quiet noises are perceived as disturbing during alteration works on an open stage.

3.11.1 Sound and hearing sensation

The nature of sound and its propagation
Sound is defined as mechanical vibrations in solid, liquid or gaseous media in the frequency range of human auditory perception. Sound conduction is of particular importance:
–   In air, referred to as *airborne sound*, and in solids, referred to as *structure-borne sound*. Sound transmission in liquids is – considering stage technology – only of interest in connection with hydraulic systems.
–   The propagation of sound in gases and liquids takes place in the form of longitudinal waves as periodic pressure fluctuations (compression and rarefaction) in the medium. If these pressure fluctuations arrive at the eardrum of the ear, they are perceived as sound. In solid bodies, sound is also transmitted as a transverse wave, since shear stresses can also act in solids.
–   If, for example, sound is emitted by an electric motor, this means that vibrations of components of this motor and air flow noise excite the surrounding air to vibrate, so that small pressure fluctuations are superimposed on the atmospheric air pressure. These pressure waves spread out in all directions and finally reach the human ear, which then perceives these pressure fluctuations as sound. Vibrations of engine components can also be transmitted via the steel structure on which the engine is mounted, so that components excited to vibrations via structure-borne sound can become further sources of airborne sound propagation.
–   However, the airborne sound not only arrives at the ear and causes the eardrum to vibrate; it can also excite other components to vibrate. In particular, this occurs when the frequency of the sound matches the natural frequency of a component (e. g., a window pane clanging as a car passes by). Natural frequency is the number

of oscillations per second at which a component vibrates on its own when briefly stimulated by a knock to vibrate, (like a tuning fork, gong, guitar string, etc.). If the *natural* and *excitation frequencies* are equal, this is called *resonance*.

–   When a sound wave hits a wall, part of the sound is reflected (reflection), the remaining part is absorbed by the wall. Of this absorbed part, a certain percentage is reradiated through the wall (transmission), while another part is converted into heat in the wall, i. e., destroyed as sound energy (dissipation). In silencers, as much sound energy as possible has to be converted into heat.

Frequency of the sound and magnitude of the sound pressure

The *frequency of the sound* is the number of pressure fluctuations per second. In perception, it determines the pitch of the tone. The range of audible sound in young healthy people extends from about 6 to 16 kHz (20 kHz). In older people, this range is greatly reduced. Other ranges apply for animals, for example dogs can perceive far higher tones than humans (dog whistle).

The fundamental of a bass voice is about 85 Hz, that of a soprano voice reaches about 1400 Hz. The concert pitch is 440 Hz, in Austrian and German orchestras 443 Hz, in Switzerland 442 Hz.

Frequency and speed of sound determine the wavelength according to the formula

$$\lambda = \frac{v_s}{f}, \tag{3.138}$$

where

v_s speed of sound [m/s] in temperature ranges relevant to stage technology:
  in air $\approx$ 335–345 m/s
  in water $\approx$ 1400–1500 m/s
  in mineral oil $\approx$ 1200–1300 m/s
  in steel $\approx$ 5100 m/s
f frequency [1/s = Hz]
λ wavelength [m]

The greater the *sound pressure* arriving at the ear or the pressure fluctuations effective at the ear, the louder the sound perceived by the human being. The human *hearing threshold* is at a pressure of approx. $20 \cdot 10^{-6}$ Pa = 20 µPa, the *pain threshold* at approx. 100 Pa = $100 \cdot 10^6$ µPa. The sound pressure range of interest in acoustics thus varies between a minimum and maximum value in a ratio of approx. 1 to 1 million.

With regard to human hearing sensation, another physiological fact must be taken into account. As already mentioned, humans perceive a low sound pressure as quiet and a high sound pressure as loud. However, sound waves of different frequencies do not cause the same loudness perception in humans, even if the sound pressure objectively determined with a measuring device is the same. Our ear is particularly sensitive to frequencies in the range of 500 to 5000 Hz. Therefore, the measured sound pressure value must still be corrected with regard to the actual perception situation for humans by means of reduction factors that depend on the level of the frequency. In most cases, including stage technology, an *A-weighting* of the sound pressure level specified in the standard is used. A sound with a frequency of 1000 Hz is assigned a value of 1. Measured sound pressure values for tones of lower and higher frequencies are reduced according to the curve shown in Figure 3.21. On the ordinate of the diagram, the sound pressure is plotted as a related quantity (see Section 3.11.2). Therefore, due to the definition of the level values with the logarithm, reduction amounts are given in dB. The A-weighted result is specified as dB (A) value.

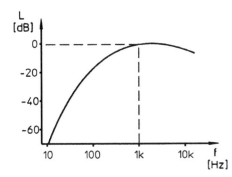

Figure 3.21: A-weighting of the sound level.

Consequences for stage technology

For example, an electric motor attached to a lifting podium as a climbing drive represents a sound source. This motor will emit a certain *sound power*, i. e., a certain sound energy per time unit, to its surroundings. Since the motor is attached to the podium, the sound is transmitted to other components of the podium by means of structure-borne sound conduction and possibly also sets the wooden stage floors at the podium in vibration. Airborne sound waves excited in this way spread out spherically in all directions as pressure fluctuations, are partially reflected at walls, partially absorbed, perhaps also excite other components to resonance vibrations; ultimately, pressure fluctuations of a certain magnitude arrive in the auditorium and are perceived as noise.

The magnitude of the *sound pressure* perceived by the audience will therefore depend not only on the magnitude of the *sound power* emitted by the actual sound generator, but also on the distance from the sound source and the room conditions that influence the way the sound propagates. It will depend on how much sound is reflected and absorbed in the stage and auditorium space (e. g., what decorative material is on stage, how many textiles are suspended in the upstage area) and how much sound power may be emitted through openings to another room. The sound pressure will therefore also not be the same at every point in the auditorium.

From this presentation, it is very easy to recognize the problem for the stage technician. For the audience, the only thing that is ultimately relevant is the sound pressure that reaches them. For this reason, most technical specifications for stage equipment specify a maximum permissible sound pressure value measured at a seat in the middle of the first row of the audience.

From this maximum permissible sound pressure value in the auditorium, however, it is hardly possible to conclude what sound power a sound source in the stage may actually emit in order not to exceed this limit value. The conditions with regard to sound propagation are very difficult to estimate and depend, as explained, on very many influencing factors. In addition, it must be taken into account that several sound sources can be activated simultaneously during stage operation.

At this point, it should be noted that extremely low values are often specified in technical specifications that cannot be met in reality or require special measures that are hardly economically justifiable. Sometimes, specifications even demand values that are lower than the basic noise level in the auditorium.

However, it should be noted here, as in many other places, that the choice of construction method for stage equipment, its structural design and, in particular, the choice of drive concepts can have a great deal of influence on noise behavior. Some concrete hints are given in Section 3.11.3.

3.11.2 Sound field parameters

After the introductory remarks in Section 3.11.1, this section will explain the physical terms in a little more detail.

Terms

In the explanation of the term *"sound pressure"*, it has already been pointed out that the auditory sensation of humans from the *threshold of hearing* to the *threshold of pain* covers a very large pressure range in the ratio of one to one million. It is therefore customary in acoustics not to work with absolute values but with relative values using a logarithmic scale.

A logarithmic scale is also appropriate because our hearing perceives according to a logarithmic principle. An increase in sound power by a factor of ten is perceived approximately as a doubling of loudness. Furthermore, the human ear achieves a higher resolution at lower sound pressures.

In this section, the most important *sound field parameters* used in acoustics will be explained. In addition to the dimensional absolute values, the logarithmically scaled relative values are also given as so-called *sound level values*. In the formulas written down in the following "lg" stands for the decimal logarithm, $\log_{10}$.

The decimal logarithm of the ratio of two similar sound field quantities is called "Bel" [B] or "Decibel" [dB],

$$L_x = \lg \frac{x_1}{x_2} \ [\text{B}] = 10 \cdot \lg \frac{x_1}{x_2} \ [\text{dB}].$$

Sound power P [W]
The *sound power* is the power radiated per unit of time by a sound source. It is an analogous quantity to the radiated heat power of a radiator (= radiated heat energy per unit of time). The sound power is therefore a characteristic quantity for the sound source and completely independent of the ambient situation and is used for the noise evaluation of machines.

Sound power level L_p [dB]

$$L_p = 10 \cdot \lg\left(\frac{P}{P_0}\right),$$

$$P = P_0 \cdot 10^{L_p/10},$$

(3.139)

with $P_0 = 10^{-12}$ W = 1 pW (picowatt).
Table 3.8 gives guide values for sound power and sound power level.

Table 3.8: Sound power and sound level of some noise sources.

	P [W]		L_p [dB]
Turbojet	10,000	$= 10^4$	160
Propeller plane	1,000	$= 10^3$	150
Pain threshold	100	$= 10^2$	140
Large orchestra	10	$= 10^1$	130
Car horn	1	$= 10^0$	120
Loud radio	0.1	$= 10^{-1}$	110
Car on highway	0.01	$= 10^{-2}$	100
Subway interior noise	0.001	$= 10^{-3}$	90
Loud conversation	0.0001	$= 10^{-4}$	80
Normal conversation	0.00001	$= 10^{-5}$	70
Office	0.000001	$= 10^{-6}$	60
Silent conversation	0.0000001	$= 10^{-7}$	50
Whisper noise	0.00000001	$= 10^{-8}$	40

Sound intensity I [W/m²]

The *sound intensity* is the amount of energy flowing in a certain direction through a certain area in the unit of time, i. e., the sound power transmitted per unit area. Sound intensity measurement can be used primarily for the localization and assessment of noise sources.

Sound intensity level (sound power level) L_I [dB]

$$L_I = 10 \cdot \lg\left(\frac{I}{I_0}\right),$$
$$I = I_0 \cdot 10^{L_I/10},$$

(3.140)

with $I_0 = 10^{-12}$ W/m² $= 1\,\text{pW/m}^2$.

Sound pressure p [N/m² = Pa]

The *sound pressure* is the pressure variation superimposed on the static air pressure.

The static air pressure is approximately 10^5 Pa $= 1\,\text{bar}$. The pressure fluctuations at normal speech volume are of the order of microbar (µbar). The pressure p is therefore a variable over time and the characteristic value for p is the RMS-value p_{eff} (see Section 3.10) of the pressure fluctuation.

The measurement of sound pressure is used to assess the impact of sound on humans and records the magnitude of perceptible pressure variations at a given location, resulting from sound emissions, reflections and absorptions in the environment.

Sound pressure level L_p [dB]

$$L_p = 10 \cdot \lg\left(\frac{p}{p_0}\right)^2 = 20 \cdot \lg\left(\frac{p}{p_0}\right),$$
$$p = p_0 \cdot 10^{L_p/20},$$

(3.141)

with $p_0 = 20 \cdot 10^{-6}$ Pa $= 20\,\mu\text{Pa}$ (hearing threshold).

Sound power P, sound intensity I and sound pressure p are related in spherical sound propagation in the far field according to the following relations:

$$I = \frac{P}{4\pi \cdot r^2} = \frac{p^2}{\rho \cdot c}.$$

Thus, the following proportionalities exist:

$$I \approx P \approx p^2 \approx \frac{1}{r^2},$$

(3.142)

where

r distance from the sound source [m]

ρ density of air $\rho \approx 1.3 \,[\text{kg/m}^3]$

c speed of sound [m/s]:

  $c \approx 332 \,\text{m/s at } 0\,°\text{C}$

  $\approx 343 \,\text{m/s at } 20\,°\text{C}$

  $\approx 349 \,\text{m/s at } 30\,°\text{C}$

Under open-field conditions, this means, for example:

–   The sound intensity decreases with the square of the distance.

–   If the distance to the sound source is doubled, the sound intensity is reduced to 1/4 and the sound pressure is reduced to 1/2 (for $r_2 = 2r_1$, $I_2 = I_1/4$ and $p_2 = p_1/2$), the sound intensity level and the sound pressure level are reduced by 6 dB.

Sound velocity v [m/s]

Sound velocity is the speed of oscillation of the particles transmitting the sound.

Sound velocity level L_v [dB]

$$L_v = 10 \cdot \lg\left(\frac{v}{v_0}\right)^2 = 20 \cdot \lg\left(\frac{v}{v_0}\right),$$

$$v = v_0 \cdot 10^{L_v/20},$$

(3.143)

with $v_0 = 5 \cdot 10^{-8}\,\text{m/s} = 50\,\text{nm/s}$ (nanometer/second).

Equivalent continuous sound level $L_{\text{päqu}}$

If an oscillating state variable $x(t)$ changes its magnitude with the period T, its effective value RMS is defined as a root mean square value according to Eq. (3.96). Measuring instruments measure this RMS-value, e. g., the sound pressure p_{eff} or the sound pressure level $L_{p_{\text{eff}}}$ (referred to as p or L_p), over very short time intervals.

  If this RMS value does not remain constant over a longer period of time, but changes its value, an averaging can be performed in an analogous manner over a time interval Δt. This average value is called the *equivalent continuous sound level* according to

$$L_{\text{päqu}} = 10 \cdot \lg \frac{1}{\Delta t} \int_{\Delta t} \left(\frac{p(t)}{p_0}\right)^2 dt,$$

(3.144)

or, with Eq. (3.141), according to $\left(\frac{p_i}{p_0}\right)^2 = 10^{L_{p_i}/10}$,

$$L_{\text{päqu}} = 10 \cdot \lg \frac{1}{\Delta t} \int\limits_{\Delta t} 10^{L(t)/10} \, dt. \tag{3.145}$$

It is a measure of the energy content of the signal during the measurement period.

Very often a time interval of $\Delta t = 60$ seconds is chosen and the sound pressure A-weighted is taken into account.

If sound levels L_p act at time intervals Δt_i, where $\sum_i \Delta t_i = \Delta t$ then Eq. (3.144) can be expressed in the form

$$L_{\text{päqu}} = 10 \cdot \lg \frac{1}{\Delta t} \sum_i \left(\left(\frac{p_i}{p_0} \right)^2 \cdot \Delta t_i \right),$$

$$L_{\text{päqu}} = 10 \cdot \lg \frac{1}{\Delta t} \sum_i (10^{L p_i/10} \cdot \Delta t_i), \tag{3.146}$$

where

Δt_i time interval in which L_{pi} acts [s]

Δt total time interval [s]

L_p sound pressure level in time interval $\sum_i \Delta t_i = \Delta t$ [–]

$L_{\text{päqu}}$ equivalent (average) sound pressure level in the time interval Δt [–]

Reflection and absorption of sound

Sound hitting a wall is partly reflected, partly absorbed. The *reflectance* and *absorption coefficient* can be used to quantitatively describe the relationships with respect to sound power:

$$P_{\text{ges}} = P_{\text{ref}} + P_{\text{abs}},$$

$$r = \frac{P_{\text{ref}}}{P_{\text{ges}}}, \quad a = \frac{P_{\text{abs}}}{P_{\text{ges}}}, \tag{3.147}$$

where

r reflectance coefficient [–]

a absorption coefficient [–]

P_{ges} sound power hitting the wall [W]

P_{ref} sound power reflected from the wall [W]

P_{abs} sound power absorbed by the wall [W]

Thus $r + a = 1$, i. e., for complete reflection, $r = 1$ and $a = 0$; for complete absorption, $r = 0$ and $a = 1$.

Reverberation time T_N [s]

The *reverberation time* is the time required for a suddenly switched-off sound source to decay and for the sound pressure to drop to 1/1000 of its value. If the sound pressure is initially p_1, the time until the value $p_2 = p_1/1000$ is measured.

Converted into the related quantity of the sound pressure level, this means that if the sound pressure level is initially $L_p = 20 \cdot \lg(p_1/p_0)$, the sound pressure level of the pressure is p_2, namely

$$L_{p_2} = 20 \cdot \lg\left(\frac{p_2}{p_0}\right) = 20 \cdot \lg\left(\frac{1}{1000} \cdot \frac{p_1}{p_0}\right) = 20 \cdot \lg\left(\frac{p_1}{p_0}\right) + 20 \cdot \lg\frac{1}{1000}$$

$$= 20 \cdot \lg\left(\frac{p_1}{p_0}\right) + 20 \cdot (-3) = L_{p_1} - 60 \text{ dB}. \tag{3.148}$$

This means that the reverberation time T_N is the time period in which the sound pressure level drops by 60 dB when the sound source is suddenly switched off.

The reverberation time T_N is a decisive parameter for room acoustics. Just as the sound sensation of a note struck on a piano behaves differently depending on whether it is played with or without a pedal, the reverberation time is also important for the hearing sensation in an event room.

As a guideline, one can state that in speech theaters, the reverberation time should be about $0.8 \div 1.4$ s, in music theaters $1.1 \div 1.7$ s, and in concert halls $1.5 \div 2.5$ s.

According to Sabine, the reverberation time can be calculated using the following empirical formula:

$$T_N = 0.163 \cdot \frac{V}{\sum_i A_i \cdot a_i} = 0.163 \cdot \frac{V}{A_{\text{äqu}}},$$

$$\sum_i A_i \cdot a_i = A_{\text{äqu}}, \tag{3.149}$$

where
V volume of the room [m³]
A_i absorbing surfaces [m²], these are room boundaries, fixtures, people
a_i absorption coefficient of surfaces A_i [–]
$A_{\text{äqu}}$ equivalent sound absorption area [m²]; this is a fictitious area with the absorption coefficient $a = 1$, which would swallow the same proportion of sound energy as the total surface of the room and the objects and people in it

Sound reduction index R [dB]

The *sound reduction index* indicates what proportion of the sound power arriving at a wall passes through the wall; it is defined by the following formula:

$$R = 10 \cdot \lg \frac{P_a}{P_d} \text{ [dB]}, \tag{3.150}$$

where

P_a incident sound power [W]

P_d penetrating sound power [W]

From this an application of Eq. (3.139), due to $\lg 10 = 1$, gives

$$R = 10 \cdot \lg \frac{P_a}{P_d} = 10 \cdot \lg \frac{10^{L_{P_a}/10}}{10^{L_{P_d}/10}} = \lg \frac{10^{L_{P_a}}}{10^{L_{P_d}}} = (L_{P_a} - L_{P_d}) \cdot \lg 10 = (L_{P_a} - L_{P_d}) \cdot 1, \tag{3.151}$$

$$R = L_{P_a} - L_{P_d} \text{ [dB]}.$$

This sound insulation value is a decisive characteristic value for building acoustic measures to restrain or reduce sound propagation. If, for example, the backstage or a side stage can be separated from the main stage by a fire door, such a door is usually also required to have a sound-insulating effect in order to be able to carry out decoration preparations in the side stage during the performance on the main stage without causing noise nuisance for performers and spectators. The specification then stipulates a certain minimum sound reduction index R, often $R = 30$ dB.

Basic noise level

The *basic noise level* is the lowest A-weighted sound level in dB measured at a location during a certain period of time, which is caused by distant sounds and at the impact of which quietness is still perceived.

Guideline values for the basic noise level are given, for example, for residential areas, etc. There is also a certain basic noise level in the auditorium of a theater. On the one hand, there is always a certain noise situation due to the people present in the room; on the other hand, street noise from outside penetrates into some theaters. Taking this aspect into account, specifications – as already mentioned – often make exaggerated stipulations regarding the maximum permissible sound pressure level value in the first row of the audience.

Calculating with level values

Level addition of sound sources

If n individual sound sources at a measurement location cause the same sound intensity I, then the intensity at the measurement location with the effect of all n sound sources is in sum $I_{ges} = I \cdot n$. The sound intensity level of all n sound sources is therefore

$$L_{I_{ges}} = 10 \cdot \lg \frac{I \cdot n}{I_0} = 10 \cdot \lg \frac{I}{I_0} + 10 \cdot \lg n.$$

The sound pressure level yields because of $I \approx p^2$ (Eq. (3.142))

$$L_{p_{ges}} = 10 \cdot \lg \frac{p^2 \cdot n}{p_0^2} = 10 \cdot \lg \left(\frac{p}{p_0}\right)^2 + 10 \cdot \lg n,$$

or, in general,

$$L_{ges} = L + 10 \cdot \lg n. \tag{3.152}$$

For two sound sources ($n = 2$), because of $10 \lg 2 = 3$, this means that two equally loud sound sources increase the sound level by 3 dB if they act simultaneously; ten sound sources increase it by 10 dB.

If n sound sources at the measurement location have different sound intensity I_i, then an analogous approach with Eq. (3.140) for L_I and L_p gives

$$L_{I_{ges}} = 10 \cdot \lg \frac{\sum_i I_i}{I_0} = 10 \cdot \lg \sum_i \frac{I_i}{I_0} = 10 \cdot \lg \sum_i 10^{L_{Ii}/10},$$

or, in general,

$$L_{ges} = 10 \cdot \lg \sum_i 10^{L_i/10}.$$

Level subtraction

If the sound sources 1 and 2 cause a sound level L_{1+2} at the measuring location and the sound source 2 alone a sound level L_2, then the sound level L_1 originating only from the sound source 1 is calculated from

$$I_1 = I_{1+2} - I_2,$$
$$p_1^2 = p_{1+2}^2 - p_2^2,$$
$$L_{I_1} = 10 \cdot \lg \frac{I_1}{I_0} = 10 \cdot \lg \left(\frac{I_{1+2}}{I_0} - \frac{I_2}{I_0}\right).$$

The same applies to L_p, so that the following can be written in general:

$$L_1 = 10 \cdot \lg(10^{L_{1+2}/10} - 10^{L_2/10}). \tag{3.153}$$

This formula can be used, for example, if the influence of the background noise is to be eliminated during a sound measurement. Index 1 then refers to the sound level of the sound source without background noise, index 2 to that of the background noise.

3.11.3 Measures for noise reduction

Measures at the sound receiver

A reduction of the sound pressure level acting on humans by means of measures at the sound receiver is irrelevant in stage technology applications, unless special measures are involved, e. g., in the pressure station of a hydrostatic system. For example, in the case of very high sound pressure levels, people can be protected from excessive exposure by wearing hearing protection or by staying in sound isolated rooms.

Measures at the sound source

Use of components with low noise emission
Of course, we will endeavor to use *components* with the lowest possible *noise emission* as a matter of principle:
-   The noise emission of a gearbox can be influenced by the design of the gearbox, the type and quality of the gearing and the selection of the transmission ratio.
-   Using *cardan shafts*, high speeds must be avoided, or particular importance must be attached to good balancing.
-   In the case of *electric motors*, the installation of a cooling fan can be dispensed with in some cases for short-time operation, as this causes the most clearly perceptible noise. However, the noise component caused by the magnetic field, in particular vibrations of laminations, can also be clearly audible.
-   *Rolling bearings* operating at high speeds can emit clearly perceptible noise. Therefore, in some cases, sliding bearing and guides may be more favorable.
-   *Hydropumps* generally cause pressure pulsations. (Screw pumps behave favorably, since they generate almost no geometrically induced volume flow pulsations.)

Placement of noise sources in suitable room areas
If possible, unavoidable *sound sources* should be relocated to rooms or areas of rooms from which sound propagation into the auditorium does not occur or can be prevented:
-   Thus, *electric motors* of larger rated power for podiums will be placed in the basement area of the lower stage, if possible, and will not travel with the podium as a climbing drive.
-   *Hydraulic pumps* will be housed in a separate hydraulic pressure station.
-   Smaller *hydraulic power units* can be designed in submerged oil or submersible pump arrangements, i. e., the pump and motor are mounted on the reservoir in a vibration-isolated manner and are partially or completely submerged in the hydraulic medium so that sound is transmitted as fluid sound, but the reservoir walls have an insulating effect.

– If possible, *winch drives* with higher sound emission in the upper stage will be separated from the general stage area and the ropes will only pass through small openings.

Avoidance of the transmission of structure-borne sound
In addition, an attempt must be made to *prevent the transmission of structure-borne sound* as far as possible. The sound generator, e. g., the motor or the motor with the entire base frame of a drive unit, must be elastically connected to the rest of the structure via *vibration metals* so that vibration decoupling is achieved at least for critical frequency ranges.

Encapsulation of noise sources with sound absorbing hoods
If a sound source has to be located in a very unfavorable location in terms of space and if it is not possible to reduce the emitted sound power at the device, the expense of air-borne sound insulation by encapsulation must be accepted by fitting a sound insulation hood. This modification of a noise source initially causes an increase in level within the enclosure, but the level is then reduced by the enclosure wall.

Sound insulation is achieved by arranging sound partitions which, due to their mass and natural frequency, allow only a small amount of sound power to pass through. *Sound damping* takes advantage of frictional effects of flowing air to convert sound energy into heat, as is the case with curtains, upholstery, carpets, sound insulation mats, etc.

3.11.4 Measures to influence the room acoustics

Influencing the room acoustics is possible mainly through three measures:
– Suspended sound-reflecting surfaces above the podiums in the form of height-adjustable individual panels.
– Possibility of coupling sound chambers via heavy gate closures.
– Curtains to change the area of the sound absorbing materials.

These room acoustic measures were applied, for example, in the Palace of Arts in Budapest (BTR 3/2006).

4 Project planning and design information on stage equipment components

In this chapter, some supplementary notes are given on important components of the stage technology, in particular on components of the drive technology.

4.1 Ropes and rope drives

Rope drives are used in the upper stage mainly for load bar and point hoists, in the lower stage for lifting platforms, but also for moving stage wagons or turntables.

4.1.1 Wire ropes, rope sheaves and rope drums

Types of wire ropes

In a wire rope, several individual wires take over the total tensile force. Depending on how the wires are bundled, a distinction is made between several types. Apart from braided ropes, which are hardly ever used, wires or strands consisting of wires are helically wound around a core in a rope.

If one or more layers of wire are stranded around a core wire, this results in a *spiral rope* (Figure 4.1(a)) or a *strand* as the construction element of a *stranded rope*. If several strands are wound in one or more layers around a hemp or plastic core, the result is a *round strand rope* according to Figure 4.1(b)–(e); if they are wound around a core strand, the result is a *stranded spiral rope* according to Figure 4.1(f).

If six stranded ropes are helically wound again around a core, the result is a *cable laid rope* as shown in Figure 4.1(h). This type of construction results in particularly flexible ropes, but is not used as wire ropes in stage technology. They are used, for example, as hawsers for mooring ships to bollards on the quay wall. This type of rope is nevertheless mentioned here because it can also be used as a natural fiber rope in stage technology (see Section 4.1.6).

Spiral ropes in standard design according to Figure 4.1(a) are very resistant to bending and are not used as moving ropes running over sheaves or for drum winches.

Multistrand spiral ropes as shown in Figure 4.1(f) are used in particular for single-strand suspended loads, i. e., for point hoists, since they can be manufactured in a rotation-free (or low-rotation) design and therefore do not tend to untwist under load.

Figure 4.1(g, i) also shows cross-sections of special ropes with so-called *compacted strands*, which are also characterized to be rotation-free, but also by a high fill factor (see Eq. (4.2)). The rope shown in Figure 4.1(i) with the designation "Quadrolift" is frequently used for point hoists, but also for prospect hoists for reasons of reducing the rope types used in theaters.

https://doi.org/10.1515/9783111366968-004

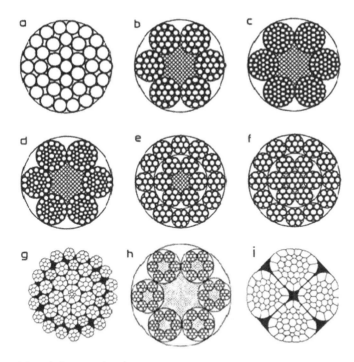

(a)   spiral rope or strand
(b)   round strand rope with hemp core according to DIN 3060, 6 × 19 standard
(c)   round strand rope with hemp core according to DIN 3066, 6 × 37 standard
(d)   round strand rope with hemp core according to DIN 3064, 6 × 36 Warrington-Seale
(e)   spiral round strand rope (strand spiral rope) according to DIN 3069, 18 × 7 rotation resistant
(f)   spiral round strand rope with core strand
(g)   special rope with compacted strands, rotation resistant, "Casar Powerlift"
(h)   cable laid rope
(i)   special rope with compacted strands, rotation resistant, "Casar Quadrolift"

Figure 4.1: Types of wire ropes – rope cross-sections. Photo credits for (g), (i): Drahtseilwerk Saar GmbH (D-Kirkel-Limbach).

In scenic use, galvanized ropes interfered with some lighting effects due to reflection. Black plastic covers were used, but these made it impossible to check the rope for wire breakage unless the covers could be moved.

For some years now, wire ropes with a black surface coating have also been commercially available, as well as accessories such as thimbles, shackles, turnbuckles, wire rope clamps, etc.

Breaking force of a rope

If wires of nominal strength σ_B are used, the *"calculated breaking force"* of the total rope can be calculated as the product of this breaking stress times the metallic rope cross-section:

$$S_{B,c} = A_m \cdot \sigma_B = A_s \cdot f \cdot \sigma_B, \tag{4.1}$$

$$f = \frac{A_m}{A_s}. \tag{4.2}$$

However, the complete rope breaks at a force that is already about 15 % lower, since in the helically laid and mutually contacting wires there are in reality more complex stress conditions than pure tensile stresses, as is assumed in the calculation of $S_{B,c}$. According to the standard, the smallest permissible real breaking force is referred to as the *minimum breaking force*, and the reduction factor k as the "*stranding factor*":

$$S_B = k \cdot S_{B,c}, \tag{4.3}$$

where
d_R diameter of the rope [mm]
A_s area of rope enveloping circle [mm²], $A_s = \frac{d_R^2 \pi}{4}$
A_m metallic rope cross-section [mm²] (sum of all wire cross-sections)
σ_B nominal strength of single wire [N/mm²]
$S_{B,c}$ calculated breaking force of the rope [N]
S_B minimum breaking force [N]
f fill factor [–], which takes into account the fact that only part of the rope cross-sectional area consists of load-bearing wires
k stranding factor [–], which takes into account the fact that the rope wires, due to their helical position in the rope, are stressed not only on tension but also on torsion, bending and compression, i. e., there is not a uniaxial but a multiaxial stress state

In rope catalogs, both the calculated and the minimum breaking load used to be specified. For example, a round strand rope often used for *load bar hoists* with the designation "6 × 19 Standard" according to DIN 3060 (6 strands of 19 wires each) with d_R = 5 mm (A_m ≈ 8.93 mm²) has a calculated breaking force of $S_{B,c}$ = 15.8 kN and a minimum breaking force of S_B = 13.6 kN at a nominal wire strength of 1770 N/mm². In this case f = 0.46 and k = 0.86.

Correctly dimensioning a rope means determining the actual tensile force acting in the rope for the most unfavorable load case and checking whether this existing rope force is sufficiently far below the breaking force.

The quotient of the rope breaking force and the existing rope force represents the safety factor. Depending on which value is used in the numerator, this safety factor refers to the calculated breaking force or the minimum breaking force. It is therefore important to note the value to which safety factors required in regulations refer:

$$\begin{aligned} v_c &= \frac{S_{B,c}}{S} \geq v_{c\,requ}, \\ v &= \frac{S_B}{S} \geq v_{requ}, \end{aligned} \tag{4.4}$$

where

S tensile force actually present in the rope from the load [N]
v_c safety factor in relation to the calculated breaking force [–]
v safety factor in relation to the minimum breaking force [–]

In the older standards, the minimum safety values were based on the calculated breaking force. Now, the minimum breaking force has been used in most applications, also in EN 17206. According to this standard, it is postulated that $v \geq 10$. This is twice the value required in the EU Machinery Directive (EU-Maschinenverordnung).

(The reason why people used to refer to the calculated breaking load was that in order to experimentally prove the strength of a rope, in this case only the actual nominal strength of the rope wires had to be proven, and very small wire testing machines were sufficient for this purpose. If the minimum breaking load of the entire rope is to be verified experimentally, the entire rope must be broken in a traction machine, which can require testing equipment for very large rope forces in the case of larger rope diameters.)

Furthermore, in the older stage technology standards, the safety factors were set in such a way that only static load values had to be taken into account for the rope force. Now, dynamic additional forces, i. e., mass forces occurring due to inertia during acceleration or deceleration, must also be taken into account (see Section 5.1). Therefore, if a mass m is suspended from the rope, the force in the rope $S = m \cdot (g + a)$, with "a" as the operationally intended nominal acceleration, must be taken into account.

When a rope is wrapped around a sheave or wound onto a drum, the rope is also subjected to bending. These additional stresses caused by the bending could be calculated and added to the tensile stresses to determine the total stress. In the same way as in crane construction, however, these stresses are not actually calculated in stage technology either, but this local stress increase is taken into account by specifying the smallest permissible bending radii. This is done by specifying a minimum value for the ratio "diameter of the rope sheave (or diameter of the rope drum) to the diameter of the rope." Thus, for example:

$$D_P/d_R \geq 20 \text{ for rope pulleys,} \quad D_P \geq 20 \cdot d_R,$$
$$D_D/d_R \geq 18 \text{ for rope drums,} \quad D_D \geq 18 \cdot d_R, \tag{4.5}$$

where

d_R diameter of the rope (of the enveloping circle)
D_P pulley diameter
D_D rope drum diameter
  (D_P and D_D are measured from rope center to rope center)

Rope elasticity

Under tensile stress, a rope stretches in a similar way to a tension spring; the greater the force, the greater the elongation of the rope. The elasticity of the rope can therefore be characterized by specifying the force required to produce an elongation of one unit length; this is called the *spring rate* or *spring constant c*:

$$c = \frac{\Delta F}{\Delta l},$$ (4.6)

where
ΔF change in tensile force [N]
Δl change in length [m, mm]
c spring rate, spring constant [N/m, N/mm]

For a rope, specified by diameter, metallic cross-section and design, the spring rate c changes with the length of the rope, because the longer the spring, the more it stretches for the same force effect. A characteristic value for the elongation behavior of the rope that is independent of the rope length is the *elastic modulus E_s* of the rope. As a result of the helical position of the wires and the flexibility of the core, the overall structure of the rope is much more elastic than the steel of a single wire. Applying Hooke's law, Eq. (4.6) gives the following relationship:

$$\varepsilon = \frac{\Delta l}{l} = \frac{\Delta \sigma}{E} \quad \text{with } \Delta \sigma = \frac{\Delta F}{A},$$

where
ε elongation [–]
$\Delta \sigma$ change of tensile stress with change of load by ΔF
A cross-sectional area
E modulus of elasticity [N/mm^2]

From this, for a rope one has:

$$\varepsilon = \frac{\Delta l}{l} = \frac{\Delta \sigma}{E} = \frac{\Delta F}{A_m \cdot E_s},$$ (4.7)
$$c = \frac{A_m \cdot E_s}{l},$$

where
L rope length [mm]
c spring rate of the rope of length l [N/mm]
A_m metallic cross-section of the rope [mm^2]
E_s modulus of elasticity of the rope [N/mm^2]
  $E_s \approx 80$–110 kN/mm^2 for a round strand rope with hemp core,
      100–130 for a round strand rope with steel core,
      120–140 for a stranded spiral rope,
      110   for a Quadrolift,
      130–160 for a spiral rope,
  for comparison $E_{Stahl} = 206$ kN/mm^2.

Rope pulleys – rope drums

Rope pulleys/Rope sheaves are fabricated from cast iron or steel, welded from steel or made from plastic. Multigrooved rope sheaves are also used to guide several ropes in the case of multiple suspended load bars. Rope sheave designs are shown as examples in Figure 4.2.

To avoid chafing of the rope on the groove edge and to prevent the rope from jumping out of the sheave, the *deflection angle a* of the rope from the sheave plane must not be greater than 4° (Figure 4.3). Particular attention must be paid to this when ropes with *offset pulleys* (*misalignment sheaves*) for point hoists are installed on the grid.

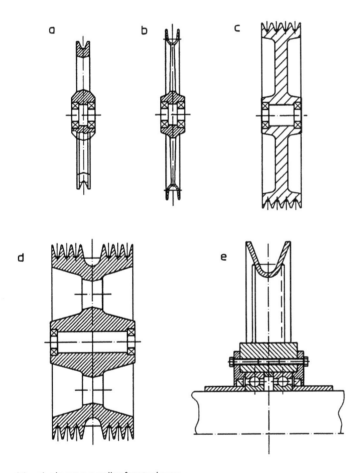

(a)   single-groove pulley for steel rope
(b)   single-groove pulley for hemp rope
(c)   multigroove pulley for steel rope
(d)   multigroove pulley for steel rope and a hemp rope
(e)   single-groove pulley in welded design

Figure 4.2: Rope sheaves according to DIN 56 919 "Rope sheaves for prospect hoists".

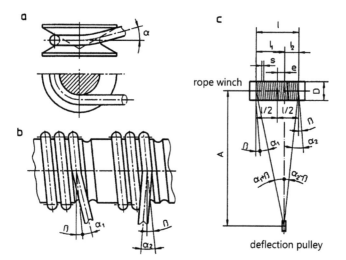

Figure 4.3: Rope deflection: (a) on a rope pulley, (b) on a rope drum, (c) shortest distance A of a deflection pulley.

For a rope drum, similar to a rope pulley, it must be noted that certain deflection angles must not be exceeded (Figure 4.3(b)). In the case of a rope drum, the angle of inclination β of the grooves on the drum shell must be taken into account. The deflection of the rope from this plane inclined by β must again be only about $\alpha_1 \approx \alpha_2 \approx 4°$. Thus, a deflection angle of $(\beta + \alpha)$ is allowed in one direction and $(\beta - \alpha)$ in the other. As a consequence, a minimum distance A must be maintained for a deflection pulley arranged as close as possible to the rope drum as shown in Figure 4.3(c), whereby the pulley must be located somewhat eccentrically to the center of the drum. With $\alpha_1 \approx \alpha_2$, it follows from $\tan(\alpha + \beta) = l_1/A$ and $\tan(\alpha - \beta) = l_2/A$ that

$$l = l_1 + l_2 = A[\tan(\alpha + \beta) + \tan(\alpha - \beta)],$$

$$\frac{l_1}{l_2} = \frac{\tan(\alpha + \beta)}{\tan(\alpha - \beta)},$$

$$l_1 \leq A \cdot \tan(\alpha + \rho), \tag{4.8}$$

$$l_2 \leq A \cdot \tan(\alpha - \rho),$$

$$A \geq \frac{l}{\tan(\alpha + \rho) + \tan(\alpha - \rho)},$$

where
A distance of the drum axis to the axis of the idler pulley
l wound drum length
l_1, l_2 winding sections

This criterion of rope deflection must be taken into account when setting offset pulleys for point hoists on the grid. More exact values for the permissible deviation, taking into account actually slightly different values of a_1 and a_2, could be found in the technical literature.

4.1.2 Block and tackle systems

By arranging fixed and movable sheaves (Figure 4.4), a force/displacement transmission similar to that of a lever can be achieved. While a fixed sheave according to Figure 4.4(a, b) only causes a rope deflection, loose moveable sheaves result in a pulley block (see Figure 4.4(c)–(g)).

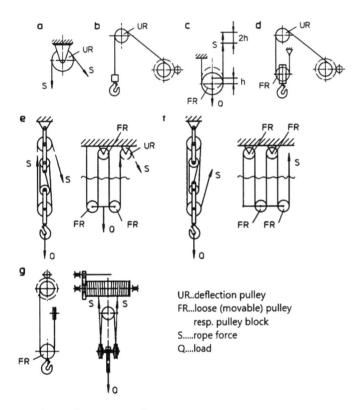

UR..deflection pulley
FR...loose (movable) pulley
        resp. pulley block
S.....rope force
Q....load

(a) fixed pulley as return pulley
(b) rope winch with return pulley
(c) pulley block with one loose pulley ($i = 2, z = 2$)
(d) rope winch with pulley block ($i = 2, z = 2$)
(e) rope pulley block ($i = 4, z = 4$)
(f) rope pulley block ($i = 5, z = 5$)
(g) rope winch with twin drum – winding of both rope ends ($i = 2, z = 4$), z number of carrying rope strands, i transmission ratio of the pulley block

Figure 4.4: Rope pulley block.

For simple pulley blocks according to Figure 4.4(c)–(f), where the load is pulled at one end of the rope, the pulley block transmission ratio i and the number of load-bearing rope strands z from which the load is suspended are equal, $i = z$. For a twin pulley block in Figure 4.5(g), $i = z/2$.

If we consider a movable pulley according to Figure 4.4(c, d), the load Q is suspended from two rope strands and the rope force is $S = Q/2$. If the load is to be lifted by the distance h, the rope must be pulled by the distance $2h$.

Generalizing one gets:

- If there is a *simple pulley block* and the number of supporting rope strands is z, the transmission ratio of the pulley block is $i = z$.

To hold the load Q in position, the necessary force required at the rope is

$$S = \frac{Q}{z} = \frac{Q}{i}. \tag{4.9}$$

However, if the load is to be moved by the distance h_Q, the rope must be moved by the distance

$$h_R = h_Q \cdot i. \tag{4.10}$$

If the movement of the load is to take place at the speed v_Q, then the required rope speed is

$$v_R = v_Q \cdot i. \tag{4.11}$$

If we disregard small losses in the system, the work done (Eq. (3.23)) and the power (Eq. (3.26)) considered at the load and at the pulled rope are, of course, equal, namely:

$$W = Q \cdot h_Q = S \cdot h_R = \frac{Q}{i} \cdot h_Q \cdot i = Q \cdot h_Q,$$

$$P = Q \cdot v_Q = S \cdot v_R = \frac{Q}{i} \cdot v_Q \cdot i = Q \cdot v_Q,$$

where

Q	load
S	rope force
h_Q, h_R	path of the load or rope
v_Q, v_R	speed of the load or rope
i	ratio of the block and tackle system
z	number of load-bearing rope strands

For the pulley block shown in Figure 4.4(c), $z = i = 2$.

- If the load is suspended from a twin block and tackle according to Figure 4.4(g) with z rope strands, the transmission ratio is

$$i = \frac{z}{2}.$$

To hold the load Q, therefore, a force of

$$S = \frac{Q}{\frac{z}{2}} = \frac{Q}{i} \tag{4.12}$$

to be applied. For a load path h_Q, both ends of the ropes must be moved a distance of

$$h_R = h_Q \cdot i. \tag{4.13}$$

And, for a speed v_Q, both ropes must be moved with a speed

$$v_R = v_Q \cdot i. \tag{4.14}$$

In this case, too, work and power are equal when considering the load Q and both rope strands on the driving and driven side.

In the case of the pulley block shown in Figure 4.4(g), $z = 4$ and $i = 2$, and the following are valid for this case:

$$W = Q \cdot h_Q = 2 \cdot S \cdot h_R = 2 \cdot \frac{Q}{2i} \cdot h_Q \cdot i = Q \cdot h_Q,$$

$$P = Q \cdot v_Q = 2 \cdot S \cdot v_R = 2 \cdot \frac{Q}{2i} \cdot v_Q \cdot i = Q \cdot v_Q.$$

In the previous considerations, frictional losses in the rope due to bending deformation at the sheaves and due to frictional losses in the rope sheave bearings were neglected because they are very small. These losses could be taken into account as sheave efficiency ($\eta_P \approx 0.96$ for slide bearings and $\eta_P \approx 0.98$ for roller bearings) and used to calculate a lifting and lowering efficiency of the rope pulley; see Section 3.5 or technical literature.

In most cases, pulley blocks are used to apply large forces with small driving forces due to the transmission ratio. However, pulley blocks can also serve the opposite purpose. If, for example, a curtain blade weighing 120 kN is to be lifted 8 m with a hydraulic cylinder, it is better to install a cylinder with a stroke of only $8/4 = 2$ m in an inverted pulley arrangement with $z = i = 4$ (Figure 1.262(b)), however, a force of $120 \cdot 4 = 480$ kN must then be applied. Bar hoists with hydraulic cylinder drives (see Section 1.8.3 and Figure 1.201(a)) and also hydraulic elevators in houses operate according to the same principle. With hydraulic drives, the higher forces are not problematic.

4.1.3 Winch drive

In a *winch drive*, the rope is wound onto a *rope drum*. The surface of the rope drum must generally be provided with helically cut grooves. When winding the rope drum, the type of rope guide must ensure that the rope is laid properly, i. e., turn by turn in the grooves provided for this purpose. If there is a risk of slack rope formation, the winch must be switched off immediately by a monitoring device in this case. When the winch is completely unwound, it must be ensured that at least two reserve windings remain on the rope drum; two windings are required in order to ensure that the rope end is bound in the drum with sufficient safety by means of its frictional connection (see Section 4.1.4), since the usually clamping alone would not be sufficient. Furthermore, it must be ensured that the rope must not unwind and rewind completely in opposite directions.

4.1.4 Traction sheave drive

In a winch drive, the tractive force is transmitted to the rope by positive locking. In a *traction sheave drive*, the transmission takes place at a sheave by friction. It must be ensured that the frictional connection between the rope and the sheave is actually sufficient to transmit the required circumferential force. The rope must therefore not slip. The check that no "slippage" occurs can be carried out as follows:

If a rope wraps around a traction sheave with the center angle a (calculated in radians [rad]), the friction coefficient μ acts between rope and sheave and the rope force on one side is S_1 and on the other side S_2 (see Figure 4.5(a, b)), where S_1 is the larger and S_2 the smaller rope force, the condition according to the theorem of Eytelwein says:

$$\frac{S_1}{S_2} \leq e^{\mu a} \quad \text{with } (S_1 > S_2). \tag{4.15}$$

For the circumferential force, one has an equivalent condition

$$U = S_1 - S_2 \leq S_2 \cdot (e^{\mu \cdot a} - 1), \tag{4.16}$$

where
S_1 greater rope force [N]
S_2 smaller rope force [N]
U circumferential (tangential) force [N]
μ friction value between rope and sheave [–]:
　　steel, cast iron　　$\mu = 0.12$
　　rubber with fabric　$\mu = 0.22$
　　light metal　　　　$\mu = 0.25$
a wrap angle [rad]
e notation for a number that occurs frequently in mathematics;
　　$e = 2.718282$ (similar to the notation $\pi = 3.141593\ldots$)

If this condition is fulfilled, then no slippage of the rope occurs; the circumferential force can be transmitted. If the equality sign applies, then the traction capacity is fully utilized and there is no "safety reserve" against slippage. Of course, the condition can only be fulfilled if $S_2 > 0$.

If $S_1 \neq S_2$, however, a so-called "*elongation slip*" inevitably occurs between the rope and the sheave, since the elastic elongation of the rope during rotation at the sheave must change, depending on the direction of rotation, from the elongation state under the larger force S_1 to the smaller elongation under the force S_2 or the elongation state under the smaller force S_2 to the larger elongation under the force S_1. Therefore, the rope speed cannot be determined exactly from the rotational speed. This has to be considered when the rope speed has to be measured.

A possible safety factor against slippage can be defined in many ways in the case of a traction sheave drive. Corresponding regulations must therefore specify the way in which a specific safety factor against slippage is taken into account. For example, it may be required that slipping must not occur even if the force S_1 is 25 % greater, i. e., the safety factor $v = 1.25$ is defined in this case as

$$v_{S1} = \frac{S_{1\,\text{perm}}}{S_{1\,\text{exis}}} \quad \text{with } S_{1\,\text{perm}} = S_2 \cdot e^{\mu\alpha},$$

and one should have

$$\frac{S_{1\,\text{perm}}}{S_{1\,\text{exis}}} \geq 1.25.$$

The following definition leads to the same result:

$$v_{S2} = \frac{S_{2\,\text{exis}}}{S_{2\,\text{requ}}} \quad \text{with } S_{2\,\text{requ}} = \frac{S_1}{e^{\mu\alpha}}.$$

So with both approaches one gets with $S_{1\,\text{exis}} = S_1$ and $S_{2\,\text{exis}} = S_2$, that

$$v_{S_1} = v_{S_2} = \frac{S_2}{S_1} \cdot e^{\mu\alpha}. \tag{4.17}$$

Other values for the safety result if the quotient calculation is related to the circumferential force U, to the coefficient of friction μ or to the wrap angle α:

$$v_U = \frac{U_{\text{perm}}}{U_{\text{exis}}} \quad \text{with } U_{\text{perm}} = S_2 \cdot (e^{\mu\alpha} - 1),$$

$$v_U = \frac{S_2}{S_1 - S_2} \cdot (e^{\mu\alpha} - 1), \tag{4.18}$$

$$v_\mu = \frac{\mu_{\text{exis}}}{\mu_{\text{requ}}} \quad \text{with } \mu_{\text{requ}} = \frac{1}{\alpha} \cdot \ln\frac{S_1}{S_2},$$

$$v_\alpha = \frac{\alpha_{\text{exis}}}{\alpha_{\text{requ}}} \quad \text{with } \alpha_{\text{requ}} = \frac{1}{\mu} \cdot \ln\frac{S_1}{S_2},$$

$$v_\mu = v_a \cdot \frac{\mu \cdot a}{\ln \frac{S_1}{S_2}}.$$ (4.19)

S_{exis} existing value of force S
S_{requ} required value of force S
S_{perm} permissible value of force S
v safety factor

The size of the safety value thus depends on the definition. This can be clearly seen if we consider the extreme case $S_1 = S_2$, for which, after all, the largest value of the safety factor must result. In this case, $v_1 = v_2 = e^{\mu a}$, but $v_U = v_a = v_\mu \to \infty$.

The verification of the frictional connection with Eq. (4.15) or Eq. (4.16) must, of course, be carried out taking into account all load cases. If one rope end is loaded with an always existing dead load E and a variable working load $Q = 0$ to Q_{max} and the other rope end is loaded with a counterweight $G = E + Q_{max}/2$, the more critical load case is given for $Q = 0$. This size of counterweight is almost always chosen because it minimizes the power required from the drive motor. (The maximum circumferential force for $Q = 0$ and $Q = Q_{max}$ is then $Q/2$; see Section 3.6.)

As can be seen from Eqs. (4.15) and (4.16), respectively, the following measures can be taken to improve the driving capability if sufficient frictional connection cannot initially be reached in a given situation:

– *Increase of the wrap angle a* by a pulley (Figure 4.5(b)), by multiple wrap on a drive drum (Figure 4.5(c) – scarcely executed, as the rope moves axially during rotation and limits the stroke) or a *capstan winch* (Figure 4.5(d)) or connection of two traction sheaves in series, e. g., in the form of a *double-grooved traction sheave* (Figure 4.5(e)) – applied when driving revolving platforms (turntables) via a rope drive.
– *Improving the friction conditions* by selecting a different *friction pairing*. For example, the coefficient of friction between a steel rope and a steel sheave is lower than that between a steel rope and a sheave lined with a rubber insert or special synthetic material insert.
– Increase of the frictional force by increasing the compressive force between rope and sheave which determines the frictional connection. This can be illustrated by the wedge groove (see Figure 4.5(h)). In this case, the frictional engagement is no longer given by a normal force N with the value $N \cdot \mu$, but by normal forces $2N'$ with the frictional engagement value $2N'\mu$. In Eqs. (4.15) and (4.16), this can be accounted for by a fictitious friction value, denoted μ_f, following the formula

$$\mu_f = \frac{1}{\sin \frac{\delta}{2}} \cdot \mu.$$ (4.20)

Similar considerations result in the following for a *semicircular groove* in Figure 4.5(f):

$$\mu_f = \frac{4}{\pi} \cdot \mu,$$ (4.21)

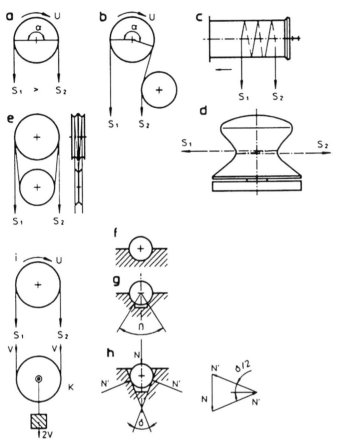

(a) principle sketch for the Eytelwein condition
(b) increase in traction capacity by increasing the wrap angle α with a pulley
(c) increase in traction capacity by multiple wrap of a drive drum
(d) increase in traction capacity by multiple wrap of a capstan winch
(e) increase in traction capacity with a double-grooved traction sheave
(f) increasing the traction capacity by increasing the compressive forces acting between the rope and the traction sheave by means of a semicircular groove
(g) increasing the traction capacity by means of a semicircular groove with undercut
(h) increasing the traction capacity by means of a wedge groove
(i) increasing the traction capacity with a tensioning pulley with tensioning weight

Figure 4.5: Traction sheave drive.

or for a semicircular gouge with undercut in Figure 4.5(g),

$$\mu_f = \frac{1 - \sin\frac{\beta}{2}}{\pi - \beta - \sin\beta}\mu. \tag{4.22}$$

–   *Increasing the span force S_2 (and thus also the span force S_1), e. g., by pretensioning* via a tensioning pulley with a tensioning weight (Figure 4.5(i)), a tensioning spring, hydraulically, etc.

Most elevator systems operate with traction sheaves. The elevator cabin is suspended on one end of the rope and a counterweight on the other. Cable cars and elevators are also driven by traction sheaves. In stage technology, traction sheave drives are seldom used for lifting equipment, but they are used, for example, to drive turntables and as drives for horizontal travel movements.

4.1.5 Clamping drive

In the case of a clamping drive (Figure 4.6), the tensile force on the rope is also transmitted by friction. However, in this case, the rope is held by friction – as if you were to clasp and hold it with your hands – by clamping forces acting radially on the rope from compression springs.

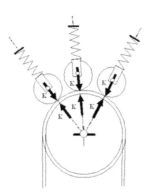

Figure 4.6: Clamping drive.

This principle is used, e. g., for the clamping drive according to Figure 1.236, since, in contrast to the traction sheave drive in Section 4.1.4, the rope force $S_2 = 0$ can be zero and the free rope end can be wound into a storage drum.

4.1.6 Fiber ropes

All previous explanations have referred to wire ropes. However, there are also fiber ropes made of natural fibers (hemp, manila, sisal) or man-made fibers (polyamide, polyester, polypropylene, polyethylene).

Natural fiber ropes are still used in stage engineering, but note that their use as load-bearing elements is only permitted to a very limited extent; see Chapter 5.

Like wire ropes, natural fiber ropes are also manufactured by multiple stranding. The starting elements for the production of wire ropes are wires and, in the case of natural fiber ropes, yarns spun from threads. Just as strands are produced from wires by simple stranding, stranded ropes by repeated stranding and cable laid ropes by again repeated stranding, yarns are also twisted into strands, *hawser lay ropes* (with three or four strands) are produced by repeated stranding and *cable laid ropes* (three three-strand hawsers to form a nine-strand cable laid rope) are produced by repeated stranding as shown in Figure 4.7.

 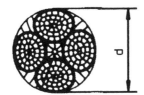 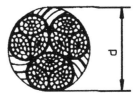

Figure 4.7: Fiber ropes: (a) form A: hawser lay, 3 strands, (b) form B: hawser lay, 4 strands, (c) form C: cable lay consisting of three three-strand hawsers (9 strands in total).

Depending on the natural product used, these *natural fiber ropes* have different strength values, but they do not differ much from each other:

Hemp (mulberry plant from Inner Asia) has the highest strength – hemp ropes should be marked with green identification stripes or threads. *Manila* (banana plant from the Philippines) has almost the same strength – manila ropes should be marked with black identification stripes or threads. *Sisal* (agave plant from Mexico, Brazil, East Africa) has lower strength – sisal ropes should be marked with red identification stripes or threads.

The strength values for a rope with a diameter of 22 mm are given as an example. In the standards, the minimum breaking load is given in daN, therefore this unit is also chosen in Table 4.1. It should also be noted that, unlike wire ropes, only the minimum breaking load is specified for fiber ropes, and not a "calculated breaking load" as for wire ropes.

Table 4.1: Minimum breaking loads in daN according to EN 1261 and EN 698.

Material	three-stranded	four-stranded
Hemp	3600	3240
Manila	3590	3230
Sisal	3340	3010

4.2 Chains and chain drives

Chains can also be used instead of ropes to transmit tensile forces. In stage technology, chains are used for chain hoists, podium drives and stage wagon drives.

4.2.1 Chains

Chain types

Round steel chains (Figure 4.8(a))
Round steel (link) chains are used in chain hoists, chain conveyors and as chains sling for loads. Toothless chain rollers for chain deflection are shown in Figure 4.8(b). Toothed sprockets are required to transmit forces as shown in Figure 4.8(c). In this case, chains with close tolerances for the chain pitch must be used.

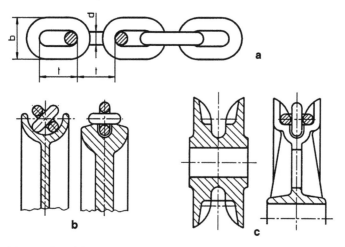

(a) round steel chain
(b) toothless chain sprocket
(c) toothed chain sprocket

Figure 4.8: Round steel chain and sprockets.

Steel link chains (Figure 4.9)
These chains consist of pins and plates and, except for special designs, are articulated in one plane only. Depending on the design, a distinction is made between:
– *Gall chains* (Figure 4.9(a));
– *Bush chains* (Figure 4.9(b)); and
– *Roller chains* (Figure 4.9(c, d)).

The bush chain can be used for higher operating speeds, since better sliding conditions are provided at the pin. Roller chains still have a roller mounted on the bush.

Roller chains are very frequently used in conveyor technology, in stage engineering applications, for example, as lifting chains for podiums. However, they are also used as transport chains in chain conveyors (carrying-chain conveyor), as shown in

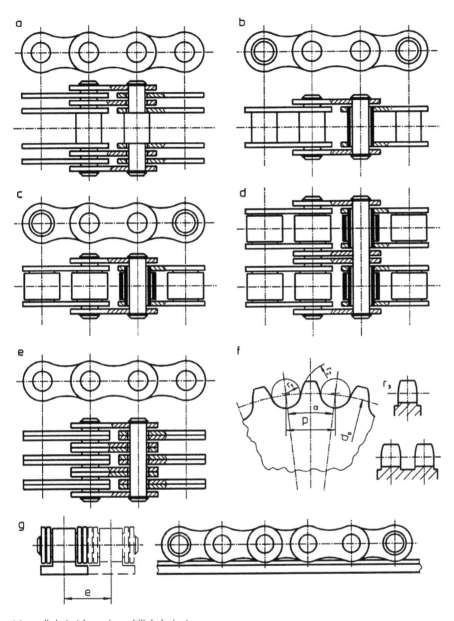

(a) gall chain (shown in multilink design)
(b) bush chain
(c) single roller chain
(d) double roller chain
(e) Fleyer chain
(f) sprocket
(g) roller chain as conveyor chain

Figure 4.9: Steel link chains.

Figure 4.9(g), and are therefore also used for transporting containers for decorative material in storage technology for theaters and operas.

The *Fleyer chain* shown in Figure 4.9(e) can be used to transmit particularly large forces; however, this type of chain cannot run over toothed wheels. (It is used, for example, on forklifts to move the lifting frame).

The sprocket teeth for steel link chains are shown in Figure 4.9(f). Such sprockets should always be used if the sprocket is actually wrapped by the chain. If a steel link chain serves only as a substitute for a pin gearing (see Section 4.4), a drive sprocket formed according to the laws of toothing theory should actually be used instead of a commercially available sprocket, except for subordinate applications.

Special steel link chains – push chains

Normally, a chain is only suitable for absorbing tensile forces. The term "support chain" is familiar from elevator construction, e. g., in paternoster elevators. By guiding the chain in a chain channel, the chain can support the elevator cabins in the event of chain breakage and prevent a fall; the zig-zag position of the chain links in the chain channel blocks any further pushing movement.

So-called *push chains* behave completely differently.

As a special design, which is also used in particular in stage technology, the push chains of the Serapid company should be mentioned, which are mainly offered in two systems also for stage technology applications.

The *LinkLift* system is designed for vertical handling of high loads, as is the case with lifting platforms. It is a push chain whose links consist of block-shaped elements that are stacked on top of each other in the upward movement to form a tower. The lifting column thus achieves particularly high stability and flexural rigidity. Figures 4.10, 4.11 and 4.12 show the LinkLift.

Depending on the size, loads of up to 150 kN can be moved and static loads of up to 200 kN can be accommodated. However, these large loads are only for smaller lifting heights (5 m for 150 kN dynamic or 3.5 m for 200 kN static). For smaller loads, up to 8 m stroke is possible. The maximum lifting speed is 0.2 m/s (with long guides on the drive housing up to 0.3 m/s). However, the load must be accurately guided in slide or roller guides or by shears. To ensure stability, some mounting instructions must be observed (see Figure 4.11).

For applications in stage technology as a drive for lifting platforms, see Section 1.7.1.

As an alternative to the LinkLift, the *ChainLift* is also offered. It is a product from the SERAPID industrial series. This system, shown in Figure 4.13, is designed for significantly higher load cycles and dynamics, but only allows lower loads and lifting heights below 2 m.

Push chains as shown in Figure 4.14 can be used for horizontal pulling and pushing movements.

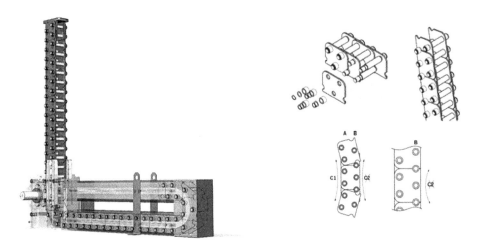

Figure 4.10: LinkLift. Image credit: SERAPID.

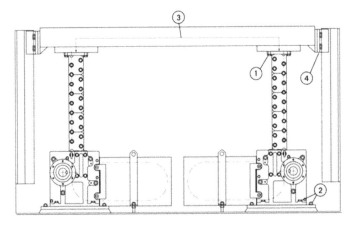

Figure 4.11: LinkLift. 1. The connection plate must be fixed to the platform without clearance. 2. The drive housing must be fixed to the base without play and aligned parallel to the connection plate. 3. The load must move absolutely vertically and parallel to the center of gravity axes of the lifting columns. 4. The platform must be guided parallel to the center of gravity axis of the lifting columns so that any horizontal movement is blocked. Picture credits: SERAPID.

The Silent ChainTrack system (SCT) was specially developed for the horizontal movement of stage wagons (see Figure 4.15). The chain has polymer rollers which run in specially designed aluminum channels and thus roll very quietly. With the SCT system, crossings of side and backstage wagons are possible (see Section 1.7.2). It is available in two basic sizes (pitches 40 and 60 mm). A special coupling carriage on the chain head is used for stage wagon connection. The chain is stored flexibly on the wall and ceiling or under the platform in a specially designed magazine (see Figures 1.138 and 1.139). The maximum travel speed is specified as 1 m/s.

Figure 4.12: LinkLift. Photo credit: SERAPID.

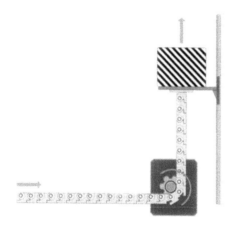

Figure 4.13: Chainlift. Picture credits: SERAPID.

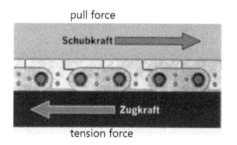

Figure 4.14: Horizontal push chain. Picture credits: SERAPID.

Figure 4.15: Silent Chain – chain guide. Picture credits: SERAPID.

Dimensioning of a chain

In standards and catalogs of manufacturers, the minimum breaking force is specified for each type and dimension of chain. It corresponds to the minimum breaking force of the rope. In addition, a test force is usually specified, the application of which does not cause any permanent deformation. In general, the relevant regulations EN 17206 require at least an 8-fold safety against breakage, $v \geq 8$. This is twice the value required in the EU Machinery Directive (EU-Maschinenverordnung), like it is the case with wire ropes:

$$v = \frac{F_B}{F} \geq v_{\text{requ}}, \tag{4.23}$$

where
F tensile force present in the chain from the load [N], [kN]
F_B breaking force of the chain [N], [kN]
v safety factor
v_{requ} required safety factor

Elasticity of chains

Similar to a rope, a chain also stretches under load. The chain links deform depending on their design, as elongation of the link plates and bending of the pins. Up to an operating force of about 40 % of the minimum breaking load, purely elastic deformation can be assumed for roller and bush chains.

A value corresponding to the modulus of elasticity E of a bar or a rope cannot be defined for a chain, since no constant cross-sectional area A is given over the chain length. If a tensile test is carried out on a chain and the chain force F is plotted on a diagram as shown in Figure 4.16 in relation to the chain breaking force F_B over the elongation ε (extension Δl in relation to the original length l_0), an approximately linear relationship is obtained. The slope of the straight line in the diagram is often referred to as the *relative spring rate* c_{rel} and is specified by manufacturers:

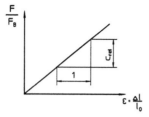

Figure 4.16: Relative spring rate of a chain.

$$\frac{\Delta F}{F_B} = c_{rel} \cdot \frac{\Delta l}{l_0} = c_{rel} \cdot \varepsilon,$$

$$\varepsilon = \frac{\Delta l}{l_0} = \frac{\Delta F}{c_{rel} \cdot F_B}.$$

(4.24)

With this, the spring rate c, which depends on the chain length, can be calculated as follows:

$$c = \frac{\Delta F}{\Delta l} = \frac{\Delta F/l_0}{\Delta l/l_0} = c_{rel} \cdot \frac{F_B}{l_0},$$

(4.25)

where
ΔF increase of tensile force in the chain [N]
F_B minimum breaking force of the chain [N]
l_0 original length of the chain [mm]
Δl lengthening of the chain due to ΔF [mm]
c spring rate [N/mm]
c_{rel} relative spring rate [–], $c_{rel} \approx 50$
ε elongation [–]

4.2.2 Chain drive

In contrast to a rope drive, the following effect occurs with a chain drive:
The pitch circle diameter of a sprocket in Figure 4.9(f) or Figure 4.17(a) is as follows:

$$d_0 = \frac{p}{\sin \frac{\alpha}{2}},$$

$$\alpha \text{ [rad]} = \frac{\pi}{z}, \quad \text{resp. } \alpha \text{ [°]} = \frac{180}{z},$$

(4.26)

where
d_0 pitch circle diameter of the sprocket [mm]
p chain pitch [mm]
z number of teeth of the sprocket
α half pitch angle of the sprocket [rad], [°]

If a sprocket wheel rotates with angular velocity ω, then the chain velocity v at an effective sprocket wheel radius r (Eq. (3.9)) is

$$v = r \cdot \omega.$$

When a chain runs around a sprocket, it should be noted that, due to the geometry of the sprocket, the effective chain radius, i. e., the normal distance between the chain strand and the center of the sprocket, always varies between a maximum value $r_{max} = r_0 = d_0/2$ and a minimum value $r_{min} = r_0 \cdot \cos \alpha$ during the rotation of the sprocket (see Figure 4.17(a)), i. e., even when driving a sprocket with constant angular velocity ω, the tangential velocity transmitted to the chain as circumferential velocity varies between a maximum value $v_{max} = \omega \cdot r_{max}$ and a minimum value $v_{min} = \omega \cdot r_{min}$.

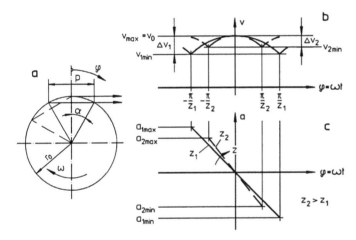

Figure 4.17: Polygon effect at the chain wheel: (a) velocity vectors at the chain wheel, (b) evolution of the transversal velocity, (c) evolution of the transversal acceleration of the chain.

This so-called *polygon effect*, as it can be easily seen from the formulas written below, has a greater effect the smaller the number of teeth of the sprocket, because the greater the difference between r_{max} and r_{min} then becomes.
    The variable chain velocity v is

$$v = \omega \cdot r = \omega \cdot r_0 \cdot \cos \varphi \quad \text{with } \varphi = \omega \cdot t. \qquad (4.27)$$

For $\varphi = 0$, the maximum velocity is given by

$$v_{max} = \omega \cdot r_{max} = \omega \cdot r_0 = v_0 \qquad (4.28)$$

and, for $\varphi = \pm\alpha$, the minimum velocity

$$v_{min} = \omega \cdot r_{min} = \omega \cdot r_0 \cdot \cos \alpha = v_0 \cdot \cos \alpha. \tag{4.29}$$

The following applies to the chain acceleration a:

$$a = \frac{dv}{dt} = -\omega^2 \cdot r_0 \cdot \sin \alpha \tag{4.30}$$

and, for $\varphi = \pm \alpha$, as extreme value with $\sin \alpha = \frac{p/2}{r_0}$,

$$a_{max} = -\omega^2 \cdot r_0 \cdot \sin(\pm\alpha) = \pm \frac{p \cdot \omega^2}{2}, \tag{4.31}$$

where
r effective radius [m]
r_0 nominal radius (radius of the chain polygon) [m], $r_0 = d_0/2$
p chain pitch [m]
v chain speed [m/s]
a chain acceleration [m/s^2]
α half pitch angle of sprocket [rad], [°]
φ angle of rotation [rad] varying with time t
ω angular speed of the sprocket [1/s]
t time [s]
z number of teeth [–]

The evolution of the velocity and acceleration can be seen in Figure 4.17(b, c).

Vibration excitations resulting from the polygon effect can become problematic especially if the excitation frequency coincides with a natural frequency of the system and resonance phenomena occur as a result. In any case, the abrupt acceleration changes lead to sudden excitations. However, if sufficiently large numbers of teeth are selected and the speeds are not too high, this effect hardly has a disturbing effect.

In addition to the described longitudinal vibrations in the direction of chain travel, the polygon effect also results in transverse vibrations of the chain.

4.3 Wedge and spindle drive

4.3.1 Wedge drive

Forces and velocities
According to Figure 4.18, the following relationship applies to the velocities for the wedge drive:

$$v_1 = v_2 \tan \alpha. \tag{4.32}$$

In an inclined plane at the angle of inclination α, a force F_H is required to lift a load F_Q and a force F_L is required as a restraining force to prevent a sinking movement due to a load F_Q. The magnitudes of these forces depend on the angle α and the friction conditions at the contact surface in the inclined plane. In the following considerations, it is assumed that there is no friction on other surfaces.

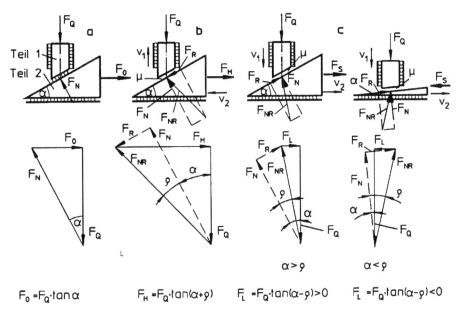

$$F_0 = F_Q \cdot \tan\alpha \qquad F_H = F_Q \cdot \tan(\alpha+\rho) \qquad F_L = F_Q \cdot \tan(\alpha-\rho)>0 \qquad F_L = F_Q \cdot \tan(\alpha-\rho)<0$$

Figure 4.18: Force ratios at the wedge (forces acting on part 1): (a) without friction, especially also at the wedge surface between part 1 and 2, (b)–(d) with friction in the wedge surface; (b) movement in the stroke direction, (c) movement in the sink direction – $\alpha > \rho$ – no self-locking, (d) movement in the sink direction – $\alpha \leq \rho$ – self-locking.

Without friction at the contact surface, the following applies according to Figure 4.18(a):

$$F_0 = F_H = F_L = F_0 \cdot \tan \alpha. \tag{4.33}$$

With the friction effect, determined by the friction coefficient μ or the friction angle ρ ($\tan \rho = \mu$), according to Figure 4.18(b, c) for a movement in the direction to hoist one has

$$F_H = F_Q \cdot \tan(\alpha + \rho), \tag{4.34}$$

and for a movement in the direction to lower (sink sense)

$$F_S = F_Q \cdot \tan(\alpha - \rho), \tag{4.35}$$

where

F_Q load on part 1 [N]
$F_{H,L}$ displacement force on part 2 [N]
a pitch angle [°]
ρ friction angle [°]

In the sink sense, in case $a > \rho$ as equilibrium force, the load F_Q requires a positive force F_L as a restraining force. If $a < \rho$, then, according to Eq. (4.35), $\tan(a - \rho) < 0$ and thus the force F_L becomes negative. In this case, no restraining force is required for lowering, but a driving force F_L is needed.

In the case $a > \rho$, the wedge can be displaced by the action of a force F_Q if no sufficient force F_L acts as a restraining force, and the load F_Q can be lowered. In the case $a \leq \rho$, this lowering motion cannot be caused even by a force F_Q, no matter how large. In fact, a force F_L must be applied for lowering. This condition is called *self-locking* (see also Section 3.5).

Efficiency

Using the formulas for the efficiency derived in Section 3.5, the following applies when hoisting (*hoisting efficiency*):

$$\eta_H = \frac{P_0}{P_H} = \frac{F_0 \cdot v_2}{F_H \cdot v_2} = \frac{\tan a}{\tan(a + \rho)}, \tag{4.36}$$

and when lowering (*lowering efficiency*)

$$\eta_L = \frac{P_L}{P_0} = \frac{F_L \cdot v_2}{F_0 \cdot v_2} = \frac{\tan(a - \rho)}{\tan a}. \tag{4.37}$$

Thus, in the case of self-locking, because of $a \leq \rho$, the lowering efficiency $\eta_L \leq 0$. In the limiting case $a = \rho$, this means $\eta_L = 0$ and, using the formula $\tan 2\rho = \frac{2\tan\rho}{1-(\tan\rho)^2}$,

$$\eta_H = \frac{\tan\rho}{\tan 2\rho} = \frac{1 - (\tan\rho)^2}{2} \approx \frac{1}{2}.$$

Only if $\tan 2\rho$ is negligibly small, i. e., $\tan 2\rho \approx 0$, then $\eta_H = 0.5$, i. e., the statement that self-locking is present if the hoisting efficiency is less than 50 % is only approximately valid, exactly it must mean that the lowering efficiency must be less than/equal to zero.

Equation (3.48) does not apply exactly because the power losses during lifting and lowering are different. These can be calculated from Eqs. (4.36) and (4.37):

$$P_{VH} = P_H - P_0 = P_0 \cdot \left(\frac{1}{\eta_H} - 1\right) = P_0 \left[\frac{\tan(a + \rho)}{\tan a} - 1\right], \tag{4.38}$$

$$P_{VL} = P_0 - P_S = P_0 \cdot (1 - \eta_S) = P_0 \left[1 - \frac{\tan(a - \rho)}{\tan a}\right], \tag{4.39}$$

with

$$P_0 = F_0 \cdot v_2 = \frac{F_Q \cdot \tan\alpha \cdot v_1}{\tan\alpha} = F_Q \cdot v_1,$$ (4.40)

where
P power [W]
η efficiency [–]
Index: 0 lossless, H hoisting, L lowering, V loss share

In stage technology, wedge drives are suitable for levelling platforms, as small lifting distances are generally sufficient.

4.3.2 Spindle drive

A *spindle drive* consists of a pairing of screw and nut, where, depending on the mode of operation (see Section 1.7.1 – Lifting platforms), the screw rotates with ω and the nut moves with v, or the screw is stationary and the nut performs the movements with v and ω.

Since a thread can be considered as a wedge in screw form, all relationships derived for the wedge gear are also applicable to the screw drive.

First of all, we will discuss the classic *sliding spindle* drive with sliding friction on the wedge surface. However, there are also screw drives in which the sliding motion on the wedge surface is replaced by a rolling motion; in this case, it is referred to as *rolling spindle drive*.

Sliding spindle drive
For such motion gears, trapezoidal threads according to ISO DIN 103 with the standard designation "Tr $d \times p$" are used (Figure 4.19(a)). Here d indicates the outer diameter of the spindle thread and p the pitch of the thread. For threads with a large pitch, several threads can be arranged in parallel next to each other as a multi-start thread with the pitch p_n:

$$p_n = n \cdot p.$$ (4.41)

According to Figure 4.19(a), the following geometric parameters are defined:
d outer diameter of the spindle thread [mm]
d_K core diameter of the spindle thread [mm]
d_F screw thread diameter (effective diameter) of the spindle thread [mm]; this is the diameter of a cylinder concentric to the spindle axis, which intersects the thread profile in such a way that tooth and gap are equal

p profile pitch of the thread [mm]; this is the distance between two similar profile points in an axial section. For a single-start spindle, this corresponds to the pitch of the thread

p_n pitch of the thread with n threads [mm]; this is the axial path difference that results when passing through a helix of the thread profile after one revolution. If, for example, the nut is rotated by 360° while the spindle is held stationary, it moves axially by the distance p_n

a helix angle of the thread [°]

If the shell of a cylinder with the pitch diameter d_F is unwound into a plane, the helix with the diameter d_F appears as a straight line according to Figure 4.19(b). The helix angle a can therefore be calculated with reference to the diameter d_F as follows:

$$\tan \alpha = \frac{p_n}{d_F \cdot \pi} \quad \text{(for } n = 1 \text{ one has } p_n = p\text{)}.$$

Figure 4.19(b) shows that – as with a spiral staircase – the pitch angle is larger at the core diameter and smaller at the outer diameter.

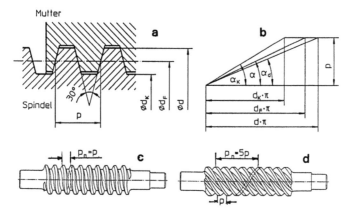

Figure 4.19: Sliding spindle drive: (a) trapezoidal thread according to ISO standard, (b) pitch of a trapezoidal thread, (c) single-start, (d) multistart.

To calculate the forces or torques, you can apply the formulas for the wedge gear by letting the forces F_H and F_L act on the radius $d_F/2$ as torques M_H and M_L:

$$M_0 = \frac{d_F}{2} \cdot F_Q \cdot \tan \alpha, \tag{4.42}$$

$$M_H = \frac{d_F}{2} \cdot F_Q \cdot \tan(\alpha + \rho), \tag{4.43}$$

$$M_L = \frac{d_F}{2} \cdot F_Q \cdot \tan(\alpha - \rho). \tag{4.44}$$

Therefore, the following applies to the efficiencies in an analogous manner:

$$\eta_H = \frac{P_0}{P_H} = \frac{M_0 \cdot \omega}{M_H \cdot \omega} = \frac{\tan \alpha}{\tan(\alpha + \rho)}, \tag{4.45}$$

and when lowering

$$\eta_L = \frac{P_L}{P_0} = \frac{M_L \cdot \omega}{M_0 \cdot \omega} = \frac{\tan(\alpha - \rho)}{\tan \alpha}. \tag{4.46}$$

Again, the lifting and lowering power losses are slightly different and can be calculated using Eqs. (4.38) and (4.39).

Applying Eq. (4.32), for the angular velocity ω of the spindle or nut rotation and the longitudinal movement of the spindle or nut with velocity v, the following relations hold:

$$v = \frac{d_F}{2} \cdot \omega \cdot \tan \alpha, \tag{4.47}$$

$$P_0 = M_0 \cdot \omega = \frac{d_F}{2} \cdot F_Q \cdot \tan \alpha \cdot \omega = F_Q \cdot v. \tag{4.48}$$

The power losses at a spindle drive with self-locking are relatively high. They also lead to high thermal loads, so that only relatively low power is transmitted with spindle drives and only short duty cycles of the drive are permissible (see Section 1.7.1). In stage technology, spindle platforms are therefore mainly used for orchestra and balancing platforms.

Rolling spindle drives
In this case, sliding friction is replaced by rolling friction. This results in high efficiencies and the thermal loads are very low. The self-locking effect sometimes desired with spindle drives cannot, of course, be achieved.

In *ball spindle drives*, balls act as rolling elements, as shown in Figure 4.20(a). The balls roll through the threads of the nut and must be returned inside or outside the nut body, depending on the design. Particularly large forces can be transmitted with *planetary spindle drives*. In the design shown in Figure 4.20(b), small thread rolls mounted in a cage are installed in the nut and roll against the spindle and nut thread in a planetary motion.

In the case of rolling thread spindles, the nominal thread diameter d_0 (according to the manufacturer's specification) should be used instead of the pitch diameter d_F in application of Eq. (4.41). Equations (4.36) and (4.37) can be used again to determine the efficiencies. In the following table, guideline values for the friction ratios for sliding and rolling screw drives are given for comparison:

Sliding spindle drive $\rho = 6°$ $\tan \rho = \mu = 0.1$
Ball spindle drive $\rho = 0.34°$ $\mu = 0.006$
Planetary spindle driver $\rho = 0.46°$ $\mu = 0.008$

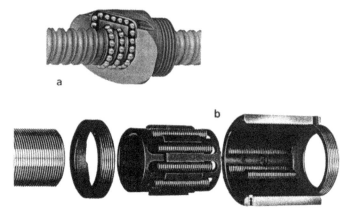

Figure 4.20: Rolling spindle drives: (top) ball spindle dive, (bottom) planetary spindle drive. Picture credits: SKF.

4.4 Gear drives

Toothed drives can be used to transmit motion under force effects by positively coupling toothed components. This book is not intended to teach the theory and calculation of gearing; only some basic terms are explained and notes on the specific subject matter are given.

4.4.1 Toothing

Involute gearing

From the condition that a uniform movement at the input should also result in a uniform movement at the output via positive locking, a tooth form of the second element can be constructed for each selected tooth form of the one element. In general mechanical engineering – and thus also in stage technology –, the so-called *involute gearing* according to Figure 4.21(a, b) is used almost exclusively. As the name expresses, the tooth flanks have a curved shape, which is called an involute: If a tensioned thread is wound from a pulley according to Figure 4.21(c), the end of the thread describes such a curve.

The size of the tooth is characterized by a ratio called *modulus m*. If there are to be z teeth on the circumference u of a circle, the following geometric relationship exists:

$$p = \frac{\pi \cdot d}{z},$$ (4.49)

$$d = \frac{p}{\pi} \cdot z = m \cdot z \quad \text{with } m = \frac{p}{\pi}.$$ (4.50)

If two cylindrical gears with the numbers of teeth z_a and z_b mesh according to Figure 4.21(a), two conditions must be fulfilled:

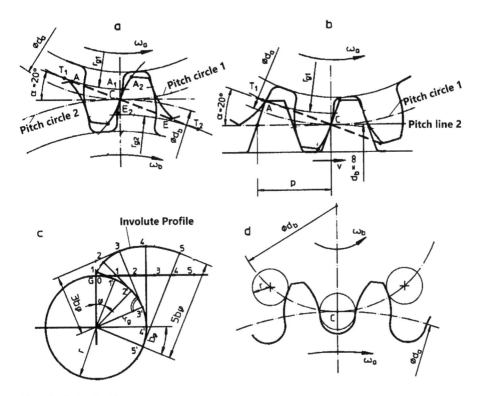

(a) wheel–wheel pairing
(b) wheel–rack pairing
(c) geometry of the involute
(d) pinion toothing

Figure 4.21: Involute gearing.

– The circumferential speeds at the pitch circle of both gears must be the same:

$$v = \frac{d_a}{2} \cdot \omega_a = \frac{d_b}{2} \cdot \omega_b = \frac{m \cdot z_a}{2} \cdot \omega_a = \frac{m \cdot z_b}{2} \cdot \omega_b.$$

From this follows

$$i = \frac{d_a}{d_b} = \frac{z_b}{z_a} = \frac{\omega_a}{\omega_b} = \frac{n_a}{n_b}. \tag{4.51}$$

– The power must be the same at the input and output sides assuming lossless transmission:

$$P_a = M_a \cdot \omega_a = P_b = M_b \cdot \omega_b.$$

From this follows

$$i = \frac{\omega_a}{\omega_b} = \frac{M_b}{M_a}. \tag{4.52}$$

If the efficiency of the gear stage $\eta < 1$ is to be taken into account, the output torque for drive with M_a is

$$M_b = M_a \cdot i \cdot \eta.$$

If the pitch diameter of a gear becomes infinitely large, a gear becomes a rack with trapezoidal teeth according to Figure 4.21(b). Thus, a rotating motion can be converted into a linear motion, and vice versa.

If a pinion with the number of teeth z drives a rack according to Figure 4.21(b), the following applies:

$$v = \frac{d}{2} \cdot \omega = \frac{m \cdot z}{2} \cdot \omega, \tag{4.53}$$

and, because of $P = M \cdot \omega = F \cdot v$,

$$\frac{M}{F} = \frac{v}{\omega}, \tag{4.54}$$

where

F longitudinal force on the rack and circumferential force on the pinion [N]
M torque [Nm]
P power [W]
d pitch diameter [m]
m module [m]
p pitch [m], which is the distance measured as the arc length between two successive right or left flanks
z number of teeth [–]
i translation ratio [–]
v circumferential speed at the pitch circle [m/s]
n speed [rps], $n°$ [rpm]
ω angular velocity [1/s]
η efficiency [–]

Pinion toothing
In addition to involute gearing, so-called *pin gearing* is sometimes used in special cases. In the case of a pin gearing, the teeth are replaced by cylindrical pins as shown in Figure 4.21(d). The teeth of the pinion must be specially shaped according to the laws of gear geometry and do not correspond to the involute shape, but to a cycloid. The driving pinion toothing is a special form of cycloid toothing, which is rarely used in the general form because of the more complex production (e. g., in the watch industry).

As described in Section 1.7.2, tensioned steel link chains are often used as drive rods for stage wagon drives. If normal commercial chain sprockets are used as pinions for the drive, the result is that the tooth shape of these sprockets does not correspond to the cycloid shape required by the exact toothing geometry, because the sprockets are formed in the tooth base from circular flanks with connecting tangents with an opening angle of 30° and a rounded tooth tip. This can possibly have a negative effect on the running behavior, but only in special critical cases.

Worm and worm wheel

A form of gearing frequently used in stage technology is the *worm gear*. In this case, a worm shaft (worm) and a worm wheel are in mesh (Figure 4.22).

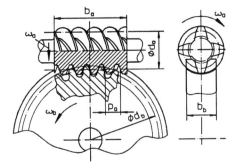

Figure 4.22: Worm gear.

Equation (4.51) also applies to a worm gear. The mean helix angle of the worm with z_a teeth and the modulus m is calculated as follows:

$$\tan \alpha = \frac{m \cdot z_a}{d_a}, \quad m = \frac{p_a}{\pi}, \tag{4.55}$$

where

m axis modulus [m] (on the worm shaft in the axis section and on the worm wheel in the center face section)

p_a axial pitch [m]

z_a number of teeth of the worm shaft = number of gears of the worm (multiple thread) (z_a = 1 to 4) [–]

z_b number of teeth on worm wheel [–]

The efficiency of a worm gear unit must not be neglected under any circumstances, since sliding takes place in the meshing as with a threaded spindle and nonnegligible losses occur. Therefore, the formulas given for a screw drive in Section 4.3 are also applicable.

For a *driving worm* (gear transmission into slow), the following is valid:

$$\eta_H = \frac{\tan\alpha}{\tan(\alpha + \rho)}, \quad M_b = M_a \cdot i \cdot \eta_H, \tag{4.56}$$

while for a *driving worm wheel* (gear transmission into fast),

$$\eta_L = \frac{\tan(\alpha - \rho)}{\tan\alpha}, \quad M_a = M_b \cdot \frac{1}{i} \cdot \eta_L, \tag{4.57}$$

where

i	worm gear ratio [–]
M_a	torque at the worm shaft [Nm]
M_b	torque at the worm gear shaft [Nm]
α	mean pitch angle of the screw [°]
ρ	friction angle of the worm–worm wheel pairing [°] (tan $\rho = \mu$)
$\eta_{H,L}$	efficiency

If $\alpha \leq \rho$ or $\eta_L \leq 0$, self-locking occurs.

4.4.2 Gearbox

To achieve larger transmission ratios, several gear stages can also be connected in series:

$$i_{b/a} = \frac{z_2}{z_1} \cdot \frac{z_4}{z_3} \cdot \frac{z_6}{z_5} \cdots = i_{2/1} \cdot i_{4/3} \cdot i_{6/5} \cdots, \tag{4.58}$$

$$\eta_{b/a} = \eta_{2/1} \cdot \eta_{4/3} \cdot \eta_{6/5} \cdots, \tag{4.59}$$

where

$i_{b/a}$	gear ratio
$i_{2/1}$	gear ratio of the first gear stage
$\eta_{b/a}$	efficiency of the gearbox
$\eta_{1/2}$	efficiency of the first gear stage

Since electric drive motors generally run at a high nominal speed, in most technical applications gear units have to be integrated into the slow speed section of the power transmission. Depending on the transmission ratios, the axis positions of the input and output shafts and the operating characteristics required, spur/helical, bevel, helical-bevel or worm gear units or combinations can be used (see technical literature). In the case of drives with slow-running hydraulic motors, the interposition of a transmission gearbox can sometimes be omitted.

However, gear units can be used not only for the transmission of speed and torque, but also for power split. If, for example, several spindles have to be driven by one drive

motor for a spindle drive, its rotary motion must be transmitted to all spindles via intermediate gears (Figure 1.108). Another type of power split is provided by planetary gear
units, but their mode of operation is not discussed here.

4.5 Cardan shafts

In stage applications, cardan shafts can cause disturbing vibration excitations and sound
emissions. For this reason, the kinematic properties of a cardan shaft are explained in
more detail in this section.

Kinematic conditions at a universal joint
A cardan shaft consists of two universal joints connected by an intermediate shaft
(Figure 4.23). Therefore, to investigate the motion behavior, we first consider the sit-
uation at one universal joint (cardan joint) with the diffraction angle β of two shafts
(Figure 4.24(a)).

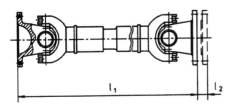

Figure 4.23: Cardan shaft (diffraction angle $\beta = 0$).

The following relationship exists between the rotation angles δ_1 and δ_2:

$$\frac{\tan \delta_2}{\tan \delta_1} = \frac{1}{\cos \beta},$$

$$\delta_2 - \delta_1 = \arctan\left(\frac{\tan \delta_1}{\cos \beta}\right) - \delta_1. \tag{4.60}$$

In Figure 4.24(b) the angle difference $(\delta_2 - \delta_1)$ is plotted against δ_1. This shows that first
shaft 2 leads and then shaft 1.

The ratio of the angular velocities of shafts 1 and 2 is obtained by differentiation
giving

$$\frac{\omega_2}{\omega_1} = \frac{\cos \beta}{1 - (\cos \delta_1)^2 \cdot (\sin \beta)^2}. \tag{4.61}$$

Thus, if shaft 1 is driven at constant angular velocity, the angular velocity of shaft 2 will
still vary between a maximum value for $\delta_1 = 0°$ and a minimum value for $\delta_1 = 90°$ as a
function of the angular position δ_1 as shown in Figure 4.24(c). The extreme values are:

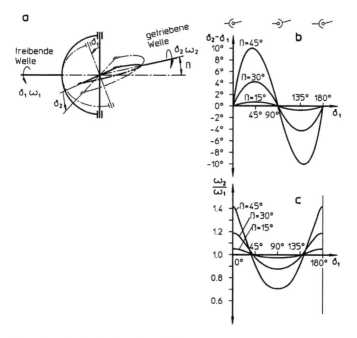

(a) schematic of a simple universal joint
(b) difference angles $\delta_2 - \delta_1$ over δ_1
(c) ratio value of angular velocities ω_2/ω_1 over δ_1 for different diffraction angles β

Figure 4.24: Simple universal joint (cardan joint).

$$\max \omega_2 = \frac{1}{\cos \beta} \cdot \omega_1, \qquad (4.62)$$

$$\min \omega_2 = \cos \beta \cdot \omega_1. \qquad (4.63)$$

Kinetic conditions at a universal joint

Since the power flow must remain constant – apart from minor frictional losses – the torque M_{t1} effective in shaft 2 is also variable with the angular velocity ω_2. Applying Eqs. (3.27) and (4.61) for the condition gives:

$$P_1 = M_{t1} \cdot \omega_1 = P_2 = M_{t2} \cdot \omega_2,$$

$$M_{t2} = \frac{\omega_1}{\omega_2} \cdot M_{t1} = \frac{1 - (\cos \delta_1)^2 \cdot (\sin \beta)^2}{\cos \beta} \cdot M_{t1}. \qquad (4.64)$$

However, the redirection of the moment vector M_{t1} in the universal joint into the moment vector M_{t2} also results in the effect of bending moments M_{b1} in the bearing of shaft 1 and M_{b2} in shaft 2. From the triangles shown in Figure 4.25, the following extreme values can be derived for $\delta_1 = 0°$ and $\delta_2 = 90°$:

when $\delta_1 = 90°$,

$$w_2 = w_1 \cdot \cos\beta \quad (= \min w_2),$$
$$M_{t2} = M_{t1} \cdot \frac{1}{\cos\beta} \quad (= \max M_{t2}),$$
$$M_{b1} = M_{t1} \cdot \tan\beta \quad (= \max M_{b1}),$$
$$M_{b2} = 0 \quad (= \min M_{b2});$$

(4.65)

when $\delta_1 = 0°$,

$$w_2 = w_1 \cdot \frac{1}{\cos\beta} \quad (= \max w_2),$$
$$M_{t2} = M_{t1} \cdot \cos\beta \quad (= \min M_{t2}),$$
$$M_{b1} = 0 \quad (= \min M_{b1}),$$
$$M_{b2} = M_{t1} \cdot \sin\beta \quad (= \max M_{b2}).$$

(4.66)

These periodic variations in angular velocities and rotational and bending moments result in vibrational excitations that can sometimes be problematic in stage applications.

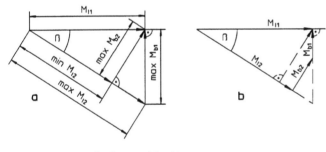

(a) extreme values for $\delta = 0$ and $\delta = 90$
(b) general case

Figure 4.25: Bending and torques in shafts 1 and 2.

Kinematics and effective torques in a cardan shaft

In a cardan shaft, two universal joints are connected in series. By observing certain geometric conditions, it can be achieved that the second universal joint completely compensates for the nonuniformity of the first universal joint. If then shaft 1 is driven with constant w_1, the intermediate shaft rotates nonuniformly with w_2, but shaft 3 moves again with constant $w_3 = w_1$.

Such behavior occurs when the following conditions are met:
-   The diffraction angle of both universal joints must be equal ($\beta_1 = \beta_2 = \beta$), in so-called Z or W arrangement (Figure 4.26).
-   Shafts 1 and 3 and intermediate shaft 2 must be in one plane.
-   The forks of the intermediate shaft must be in one plane.

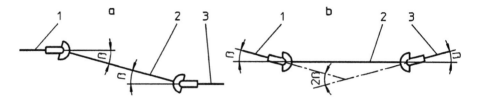

Figure 4.26: Compensation of the cyclic irregularity at the cardan shaft by connecting two universal joints in series in (a) Z-arrangement, (b) W-arrangement.

However, it must be remembered that by fulfilling these conditions, uniformity in the movement of shaft 3 is achieved with uniformity in the speed of shaft 1; however, the nonuniformity in the rotation of the intermediate shaft and the moment effects in the universal joints remain. The speed of rotation varies twice between the maximum and minimum values for each revolution of shafts 1 and 3. The intermediate shaft thus represents a vibration exciter with double the frequency of the rotary motion of the input and output shafts.

In general mechanical engineering, it is usually sufficient:

– To implement a Z or W arrangement in the geometry of the cardan shaft arrangement so that no nonuniformities occur in output speed and output torque, and
– Avoid operating speeds close to the critical speed in order to prevent large bending vibrations (see Section 3.10).

In some stage drives, however, fast-running cardan shafts can still lead to intolerable vibrations and sound excitation. Transmitted as structure-borne sound in the construction, disturbing noise can be generated by resonance excitation of corresponding construction elements. The wooden construction of a podium can act like the resonance box of a stringed instrument.

Accurate balancing of the cardan shaft can improve the situation; however, the most successful measure is to avoid high cardan shaft speeds. Therefore, it is generally expedient to reduce the speeds of cardan shafts somewhat by installing ahead a transmission gear, although for cost reasons one also endeavors not to set the speeds too low for the reason that the torques to be transmitted are as small as possible.

4.6 Special low-friction bearings

If two bodies come into contact and a relative movement is to take place, frictional forces have to be overcome, the magnitude of which depends primarily on the force effect at the contact surfaces and the coefficient of friction.

4.6.1 Hydrostatic bearing

The coefficient of friction can be decisively reduced if a liquid is introduced between the two solid surfaces. In oil-lubricated fast-running plain bearings, this lubricant film formation occurs due to hydrodynamic effects. However, fluid friction can also occur hydrostatically by injecting oil between the two friction surfaces. In stage engineering, this hydrostatic bearing principle has already been used, for example, for the pivot bearing of a large revolving stage (see Section 1.7.3 or Figure 1.46). However, this design has not prevailed.

4.6.2 Air cushion technique

Particularly low frictional resistance can also be achieved by creating an air gap between the two solid sliding surfaces.

If two glass plates are placed on top of each other, the upper plate slides almost smoothly on the lower plate until the air film has drained off. If the upper plate is pierced in the middle and air is blown through this opening, then constant sliding of the upper plate can be achieved. A stable air cushion can thus be created when an air cushion of sufficient overpressure can be formed between the opposing surfaces of two bodies by blowing in air.

Air cushion systems for internal transport are based on this principle. The mode of operation and in particular the design of the air cushion is shown in Figure 4.27: In the rest position, the load is supported on skids or support wheels to prevent damage of the rubber seal. By blowing in air, the annular rubber bellows are filled with air. Initially, the seal of the bellow to the ground will be maintained. Then, by applying sufficient force due to the excess air pressure in the buoyancy chamber, the load is lifted until air begins to escape through the gap between the bellows and the floor. The load then floats almost frictionlessly on the resulting air film.

The need for compressed air depends very much on the condition of the floor. On a porous floor, e. g., on a normal concrete floor, no air film can form because the air escapes through capillary channels. The smoother and the less porous the surface, the more favorable the load-bearing behavior (see Figure 4.28).

Such air cushion support modules are available on the market in various diameters. By using several modules, very large loads can be manipulated.

The air gap is approx. 0.05–0.25 mm, the air pressure in the buoyancy chamber approx. 2–4 bar. Depending on the size of the support module, considerable load-bearing forces can thus be achieved.

The load capacity can be calculated as follows:

$$F = A \cdot \Delta p,$$

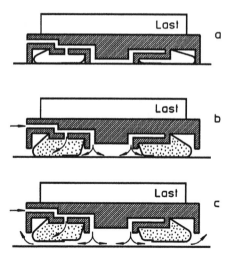

(a)   rest position, the load rests on supports or on wheels
(b)   the annular rubber bellows and the buoyancy chamber are filled with air
(c)   the pressure in the buoyancy chamber is sufficient to lift the load slightly and air around the bellows
      escapes. Thus, a floating condition can be established as a state of equilibrium

Figure 4.27: Mode of operation of a rubber air cushion. Picture credits: DELU-Luftkissentransport-
Gerätetechnik GmbH (D-Nuremberg), hereinafter referred to as DELU.

$$A = \frac{d^2 \pi}{4},$$ (4.67)

$$\Delta p = p_i - p_a,$$

where
F load capacity [N]
A area of the support module [mm^2]
d diameter of the support module [mm]
p_a external pressure [N/mm^2], $p_a \approx 1\,\text{bar} = 10\,\text{N/cm}^2 = 0.1\,\text{N/mm}^2$
p_i internal pressure in the buoyancy chamber [N/mm^2]
      (This gives, for example, a load capacity $F = 14.1\,\text{kN}$ for $p_i = 3\,\text{bar}$ and $d = 300\,\text{mm}$.)

The air cushion technology is therefore advantageously used for the assembly of large
loads and for internal transport systems. Also in stage technology there are always possible applications.

They may involve the transport and assembly of heavy decorative elements. This
is the case, for example, when the stage of the Vienna State Opera is equipped for the
Opera Ball with box units that correspond in appearance to those in the auditorium.
But it can also be used to manipulate grandstands and large-area pallets with mounted
rows of seats, so-called chair trolleys. Such applications have already been referred to
in Section 1.7.4.

Nm³/min

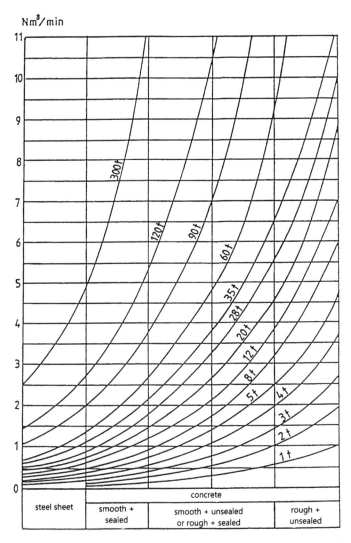

Figure 4.28: Compressed air consumption for different loads with different soil conditions. Image credit: DELU.

A disadvantage for use on a stage is the fact that there are gaps of about 10 mm in podiums and that the normal wooden stage floor does not provide a suitable surface. Plastic mats are then placed on the transport surfaces, if necessary, to create a floor suitable for the air cushion technique.

A special alternative for overriding air gaps would be to use a suitable control system to disable air cushions located above a gap, i. e., to disconnect them from the air supply for the dwell time in the gap area.

However, there are also differently designed air cushion elements on the market which, due to their design and mode of operation, lose only a very small part of their load-bearing capacity when driving over a gap. In this system, the air is blown out of a bellows with a very large number of holes (air nozzles), so that only a very small percentage of these nozzles become ineffective when a gap is crossed (see Figure 4.29). With such air cushions, floor gaps, but also level differences up to about 10 mm in height, can be traversed without any problems. The disadvantage of this air cushion design, however, is that its air consumption is somewhat higher. These air cushions operate at a maximum overpressure of 1 bar. The smallest air cushion of size 380 mm × 380 mm has a load capacity of 8 kN at an air consumption of approx. 300 Nl/min, while the largest air cushion of size 1220 mm × 1220 mm has a load capacity of 90 kN at an air consumption of approx. 900 Nl/min.

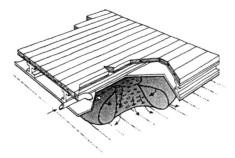

Figure 4.29: "Theater air cushion".

In Section 1.7.2 it was pointed out that stage trolleys working according to this system were built in the Muziektheater Amsterdam. For the conference center in Kuwait (Figure 1.37), the same principle was applied for chair trolleys.

4.7 Brakes

As a machine element, there exist very many brake-types in drive technology. Since controlled drives are generally used in stage technology, braking is performed by the drive motor and the mechanical brake only has a holding function as a *"holding brake,"* since it only engages when the stage unit has become to a standstill. Only in the case of an emergency stop function, which leads to an immediate disconnection of the power supply, or in the case of a failure of the power supply, the load must be brought to a standstill in this fault case with the mechanical friction brake from the movement, the brake must therefore only act as a *"stop brake"* in these exceptional cases.

Basically, all brakes in drive technology are designed in such a way that the braking force is applied by compression springs and the brake is always in braking-function when at rest. To enable movement, the brake must first be released. This is why we speak of *"passive brakes"* as opposed to *"active brakes,"* which are released when not in use and where braking force must first be applied. A passive brake is not released until the drive motor can hold or move the load using its motor torque. In the event of a power failure in the case of electric drives or a pressure drop in the case of hydraulic drives, the brake is immediately reapplied as a result of the energy stored in the compression springs and the compression spring automatically generates the required braking force.

In the case of hoist drives on the stage, it must be considered that persons are under suspended loads or persons are moved with these hoist drives – with podiums or a flybar system. Therefore, there is an increased safety requirement compared to normal hoist technology and it is prescribed, among other things, that braking must be redundant, i. e., two independently acting brakes must be available (see Section 5.1).

These two brakes usually act on the high-speed shaft on the motor side, i. e., they are ineffective in the event of a gearbox break. This is taken into account by overdimensioning the gear unit and all components up to the load. The relevant standards therefore require, for example, that when calculating the structural elements carrying the lifting load in the power flow between the brake and the load handling attachment, twice the nominal dynamic load must be applied; this also applies in particular to the gear unit. Furthermore, the verification of a finite life fatigue strength is required.

Another possibility is to have brake shoes acting directly on the rope drum, so that – assuming an appropriate sensor system – an application of this brake would also bring the load to a standstill in the event of a gear break.

According to EN 17206, release monitoring must generally be provided for brakes used in stage equipment in order to prevent impermissible operating states. This can be done via microswitches or contactless with proximity sensors.

When dimensioning brakes, consider the following:

- Of course, the brake must be designed in such a way that the test load can also be braked over an appropriate braking distance. The minimum braking torque (nominal braking torque) specified by the manufacturer must be taken into account.
  In this respect, EN 17206 states: "Load securing devices shall be capable of bringing the test load to a stop even if one of them fails. Load tests shall be carried out with a load of at least 1.10 ELL "Entertainment Load Limit" (ELL... it is the maximum load an item of lifting equipment is designed to raise, lower or sustain) ensuring the load capacity for the test load and checking the effective operation of the following elements: (a) brakes; (b) couplings; (c) components in hydraulic, pneumatic and electrical systems."
- It should also be borne in mind that, for example, in the event of a power failure (if the brake actually has to act as a stop brake), a delay time of around 120 ms occurs before the braking process starts, during which the lowering speed is increased, so that a higher speed must be expected rather than the maximum rated lowering speed. This also increases the sliding speed at the friction linings and the actual energy to be dissipated during the braking process, which is converted into thermal energy in the brake by friction. This thermal energy resulting from the friction work

heats up the brake. As a result of the enclosed design of the stage brakes, heat dissipation is very limited. Repeated braking during test procedures without cooling phases must therefore be avoided.

– When designing load safety devices, care must be taken to ensure that the reaction time for stopping is such that a load at the level of the permissible ELL cannot be accelerated to a speed greater than 1.5 times the nominal speed.

– According to EN 17206, the delayed application of a second brake is permissible; this must be taken into account for calculating the stopping distance in the event of a fault. Two redundant brakes are sometimes switched in such a way that the release of the first brake is switched off on the DC side, so that the brake delivers its braking torque relatively quickly and the release of the second brake is switched off on the AC side and thus only applied with a delay at a somewhat later time. The actual switching times depend on the temperature and the size of the air gap between the brake disk and the coil carrier (depending on the wear condition of the brake linings). However, the delay time for the application of the second brake must be so short that, in the event of failure of the first brake, this delayed application of the second brake still results in an acceptable braking distance. However, this requirement generally leads to a time interval in which both brakes actually act simultaneously and result in a large deceleration or high load values if both brakes are functional. Therefore, the effect of a delayed application is relatively small, so that it is not often used.

– The additional dynamic loads occurring during the braking process must, of course, be taken into account in the strength-related dimensioning of the components in the power flow. It must also be borne in mind that these values can actually be much higher than the nominal braking torque, since the braking torque can also be approx. 60 % higher than the specified minimum braking torque, i. e., 1.6 times higher, depending on the manufacturer's specifications.

– Abrupt braking during lifting can also result in the load being lifted further due to its inertia when the hoist drive stops and then falling into the slack rope (see Section 3.7).

– Since in stage technology – as mentioned – it is almost exclusively a matter of controlled drives in which speed reduction in normal operation takes place electrically (hydraulically) with defined deceleration, the brakes are only used as holding brakes. Only in the event of malfunctions, resp. in the event of a power failure, they have to act as stopping brakes. As it turned out, dust deposits on the brake linings can possibly lead to the fact that then, in an actual emergency case, the required braking torques guaranteed by the manufacturer are no longer achieved; in extreme cases with the consequence that a load can no longer be brought to a standstill from the lowering movement. For this reason, it should be ensured – whether by means of corresponding instructions in the operating and maintenance instructions or by means of special automated circuits provided for in operation – that brake linings are activated at appropriate intervals by briefly driving against the sliding brake in order to remove "lubricating dust particles."

In stage technology, it is particularly important that the brake is operated very quietly, i. e., "clacking" when shifting is avoided, because such noises are particularly disturbing.

Figure 4.30 shows a particularly low-noise disc brake *ROBA-stop®-silenzio* specially designed for stage applications. It is shown as a double brake, but is, of course, also available as a single brake.

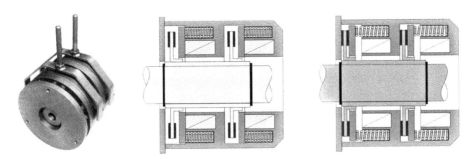

Figure 4.30: ROBA-stop®-silenzio dual-circuit brake: (left) view, (center) sectional view of brakes open, (right) sectional view of brakes closed. Picture credits: mayr® power transmission.

Since such disc brakes require very short switching distances, the brake is released electromagnetically. The picture also shows the levers for manual release, as required, for example, for emergency manual drive.

In the usual design with two identical redundant brakes, twice the braking torque acts after the second brake is applied – as explained earlier. If both brakes act as stop brakes, this results in very high additional dynamic forces and deceleration values, especially if – as explained earlier – braking torques that are possibly 60 % higher than the minimum braking torques guaranteed by the manufacturer come into effect. These force effects must of course be covered in terms of strength if the hoist is correctly designed.

However, the operator must also be made aware that in the event of a fault, decorations suspended in the hoist are subject to increased loads. For this reason, EN 17206 requires the manufacturer's documentation to contain appropriate information (see Chapter 5).

The situation is particularly problematic with performer flies, since the human body can only tolerate relatively low acceleration or deceleration values without damage to health, especially in a horizontal position.

In the case of nominal loads, the loads occurring in the event of a fault must be taken into account for the strength-related dimensioning; in the case of partial loads, smaller loads but even higher values of deceleration occur in the event of a fault, since the brakes must be designed for the nominal load case and are therefore overdimensioned for a partial load.

To avoid doubling the braking torque by simultaneous application of two brakes, a multicircuit brake with four independent brake circuits was developed as an alternative. Three brake circuits together provide the required nominal braking torque. The fourth circuit provides the necessary redundancy, since according to generally accepted safety philosophy, in principle only the failure of one component has to be assumed ("single-failure safety" according to EN 17206). This reduces the loads and makes the braking processes smoother. The fourth brake additionally brakes with only 1/3 of the braking torque delivered by the other three braking elements together, i. e., the maximum braking torque is not 2-fold but 1.33-fold. Figure 4.32 shows this brake with the designation *ROBA®-quadrostop*.

A special solution has also been developed in which a double brake delivers approximately the same braking torque, regardless of whether the compression springs of both brakes are acting or only the compression springs of one brake. This is the case with the *"ROBA-stop®-stage"* brake as shown in Figure 4.31. If both brakes are functional, both sets of compression springs press from the left and right with the spring force F_1. If one of the two sets of compression springs ("one of the brakes") fails, the other effective compression springs (of the "second brake") are slightly released via the air gaps and press against the other brake disk via the four friction surfaces and the movable intermediate disk with a braking force F_2. The force F_2 will be slightly smaller than F_1 because of the small relaxation in the air gaps. (In the sectional drawing shown, the air gaps are very large for clarity, but this does not correspond to reality). The ROBA-stop-stage is therefore designed according to the slightly lower force F_2.

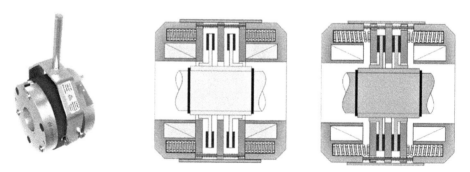

Figure 4.31: ROBA-stop®-stage dual-circuit brake: (left) view, (center) sectional view of brake open, (right) sectional view of brake closed. Picture credits: mayr® power transmission.

The requirement according to EN 17206 that two independently acting and independently controlled brakes must be provided, especially for hoists (provided that there is no self-locking from the movement), is usually met by attaching the two brakes to the high-speed motor shaft in order to be able to work with small braking torques.

Figure 4.32: ROBA®-quadrostop. Photo: mayr® power transmission.

However, it is also possible that brakes act on the rope drum of the hoisting winch. Since very high braking torques are required here, caliper brakes are generally used. Braking at the drum can also prevent the load from falling, e. g., in the event of a broken gearbox, if the damage is detected by an appropriate sensor system. In the event of a gearbox breakage, it must be remembered that until the brake is applied, the load is accelerated higher in the falling movement, since the inertia of the rotating masses in the motor-gearbox in the power transmission is eliminated.

If disc brakes are to be used to brake directly on the rope drum, the brake shown in Figure 4.33 with, e. g., six brake calipers is suitable. Five braking elements together again provide the required braking torque, the sixth element provides the required redundancy by additionally braking with 20 % of the braking torque of the other five elements, i. e., the maximum braking torque is only 1.2 times.

Figure 4.33: ROBA®-diskstop. Photo: mayr® power transmission.

Figure 4.34 shows brakes on the drive of bar hoist winches. In addition, a mechanically acting gear breakage protection system for a worm gear (see Figure 4.35) is presented. In addition to the worm wheel, it has an additional pilot wheel. The pilot and worm wheels are driven together by the worm. In the event of a break, the pilot wheel rotates and releases the movement locking safety pin.

Figure 4.34: Winches of load bar hoists with brakes. Photo: mayr® power transmission.

Figure 4.35: Gear fracture protection of a worm gearbox. Picture credits: Lightpower/ASM.

4.8 Notes on the standard-compliant dimensioning of hoists (see also Section 5.1)

4.8.1 Required verifications according to standard

In normal operation, a drive is stopped electrically (or hydraulically) with a maximum deceleration determined by the system. Existing mechanically acting brakes are only applied at standstill, i. e., they are only used as holding brakes and not as stopping brakes.

In case of an emergency stop due to power failure (emergency stop of category 0 according to EN ISO 13850 or EN 60204), the mechanical brakes engage as stop brakes.

According to EN 17206, power units must be designed in such a way that unintentional hazardous movements are excluded. This can be achieved, for example, by self-locking from motion (dynamic self-locking) or by two independently acting load-securing devices, e. g., brakes acting independently of each other in each operating state. Each brake acting alone must be suitable for bringing the winch to a standstill under nominal load from the lowering movement at nominal speed within an acceptable braking distance (stopping distance).

The standard specifies concrete dimensioning criteria for the operating case and for the accident case, which means that two strength verifications must be provided, with the more critical load case ultimately determining the dimensioning:

In *normal operation*, all components of the power unit between the load carrying device and the load safety device shall be designed in accordance with the standard so that *twice the rated load in service will not cause permanent deformation or failure of the component, using 400 hours of operation at rated speed as the basis for calculations; (unless a longer period of operation is appropriate* – see ISO 4301-1).

Nominal load is the sum of the maximum permissible payload – referred to in the standard as *"entertainment load limit...ELL"* plus the weight of the load handling attachment plus the dynamic mass force at the operationally specified maximum acceleration. This load is often referred to in the standard as the "nominal load in the operating case" or "nominal dynamic load".

In order to take account of the effects of *malfunctions* (stop category 0 – power failure – see Section 5.1), all power plant components between the load carrying device and the load safety device must be designed in accordance with the standard in such a way that the *simple load does not lead to permanent deformation or failure of the component in the event of a malfunction.*

While the load assumptions for the normal load case are easy to determine, this is not the case for the disturbance load case, since a winch with a hoisting load is, in a simple approximation, an oscillatory two-mass system coupled by the hoisting rope, consisting of the winch with rotating masses and the load suspended from an elastic rope, with the brakes applied in the event of a fault acting on the rotating masses of the winch.

EN 17206 also requires manufacturing companies to specify in their documentation, among other things, the "acceleration and deceleration values for operating and fault conditions." In particular, the note "This is especially relevant for performer fly systems" calls for the specification of the "highest possible deceleration values (low load, with all load safety devices operating at the highest possible efficiency)".

If one disregards special designs of brakes mentioned in Section 4.7, this means that both redundant brakes are applied so that the double braking torque is effective.

4.8.2 Computational investigation of the system behavior in the event of a fault

In order to investigate the system behavior in the event of a fault by calculation, a simulation program can be used and/or measurements can be taken on a real system.

A simplified dynamic model of the MATLAB-SIMULINK simulation program used for this purpose is based on two differential equations describing the drive unit, on the one hand, and the load unit, on the other; the rope is assumed to be a spring-damper system and forms the connection between the differential equations (see Figure 4.36):

Drive unit $\quad I_{red} - \ddot{\varphi} = F - r_T - M_{(t)}$

Load unit $\quad m - \ddot{x} = m - g - F$

Rope force $\quad F = c_{(\varphi)} - (x - r_T - \varphi + \Delta l_0) + d - (\dot{x} - r - \dot{\varphi})$

The reduced mass moment of inertia is used for I_{red}, the drum radius for r_T, the spring rate of the rope for $c_{(\varphi)}$, the rope elongation under dead load at the start time $t = 0$ for Δl_0 and the damping of the rope for d. However, the rope damping can be disregarded, since only the critical first oscillation maximum value is of interest for the rope force and deceleration and the damping has little influence on it.

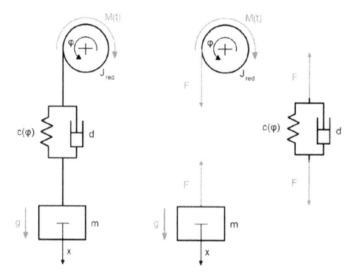

Figure 4.36: System of the mathematical model.

The results given here as an example refer to a *point hoist with a load capacity of 500 kg* (nominal load approx. 5 kN mass of load carrying device 20 kg, nominal speed 1.8 m/s, operational maximum acceleration 3 m/s²) and were calculated with the MATLAB-SIMULINK simulation program. In principle, the results apply generally to point and load bar hoists. How large the peak values for deceleration and load actually are for a specific winch depends on its technical data.

It was assumed that in the event of a malfunction, twice the braking torque would occur if both brakes were functional.

Furthermore, the following simplifying assumptions were made in the simulation:

- Both brakes apply at the same time, with the braking force increasing linearly from zero to the nominal value (see dot and dash line in Figure 4.37).
- Efficiencies are not considered in the example printed here.
- The change of rope length during the braking process is not taken into account.
- The dead weight of the rope is neglected.

Besides the spring constant (spring rate) $c = \frac{E_S \cdot A_m}{l}$ (see Eq. (3.66)), a damping constant can be assigned to the rope. However, since only the maximum value of the vibration amplitude is decisive for dimensioning, the influence of the damping with decrease of the vibration amplitude over time is of no further interest.

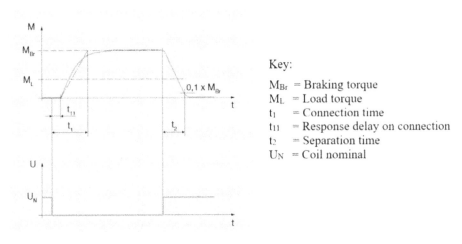

Key:

M_{Br} = Braking torque
M_L = Load torque
t_1 = Connection time
t_{11} = Response delay on connection
t_2 = Separation time
U_N = Coil nominal

Figure 4.37: Braking torque–time diagram. Picture credits: Mayr company.

Sudden braking in the event of a fault during the lowering movement of the nominal load

The "MATLAB-SIMULINK" curve pictured in Figure 4.38 shows the values of the deceleration $a_{stör}$ and the rope force $F_{stör}$ calculated as a function of the spring rate of the rope or the length of the rope. The curve shows that the maximum values for deceleration and rope force are given for rope lengths in a middle range. In this specific case, the maximum deceleration and rope force occur at a spring rate of $c \approx 200 \frac{kN}{m}$ and at a rope length of approx. 14 m, respectively, and amounts to $a_{stör} \approx 32.5\,\text{m/s}^2 \approx 3.3\,g$ and $F_{stör} \approx 22\,\text{kN}$.

The nominal static load is $F_{stat} = m_L \cdot g = 520 \cdot 9.81 = 5101\,\text{N} \approx 5.1\,\text{kN}$

The nominal dynamic load is $F_{dyn} = m_L \cdot (g + a) = 520 \cdot (9.81 + 3) = 6660\,\text{N} = 6.66\,\text{kN}$

In this case, it follows that the rope force in case of failure is $22/5.1 = 4.3$ times the static nominal load and $22/6.66 = 3.3$ times the dynamic nominal load in case of normal operation.

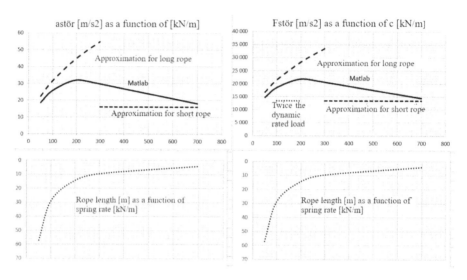

Figure 4.38: Evolution of the deceleration $a_{\text{stör}}$ and the rope force $F_{\text{stör}}$ in case of failure plotted against the spring rate c when *lowering the nominal load* mass of 500 kg of a 5 kN point hoist.

The diagram also shows the 2-fold dynamic nominal load to be applied in accordance with the standard $2 \times 6.66 = 13.32$ kN and it can be seen that the load in the event of a fault also exceeds this load by far.

For *very short rope lengths*, i. e., loads lifted almost up to the grid, the following limit consideration could be made as an approximation: The elasticity of the rope has hardly any effect and winch and load could be regarded as rigidly connected. If one also makes the simplifying assumption that the braking force after a reaction time of

$$t_{\text{reak}} = \frac{t_1 - t_{11}}{2} + t_{11} = \frac{t_1 + t_{11}}{2} \tag{4.68}$$

suddenly comes into full effect, the maximum deceleration values or the largest rope force can be calculated without complex simulation by applying the formulas given in Sections 3.2 and 3.3 involving

v	lifting or lowering speed of the load [m/s]
ω_M	angular velocity of the motor [1/s]
t_{reak}	response time = reaction time [s]
v_{reak}	speed of the load after the reaction time until the brakes are actually engaged) [m/s]
a_{reak}	acceleration (deceleration) during reaction time [m/s²]
a_{brems}	deceleration of the load during braking [m/s²]
$\varepsilon_{\text{brems}}$	deceleration of rotating masses during braking [1/s²]
m_L	load mass [kg] = mass of the payload m_Q plus mass of the load carrying device m_E

m_{red} reduced mass [kg] (fast rotating masses reduced to the translationally moved load mass)

I_{MBG} mass moment of inertia of rotating masses on the motor shaft (parts of motor, brake, fast running gear shaft...) [kg m^2]

I_T mass moment of inertia of the rope drum [kg m^2]

M_{brems} braking torque of a brake [Nm]

$a_{stör}$ deceleration at the load in the event of a fault [m/s^2]

$F_{stör}$ force effect on the load (in the rope) in the event of a fault [N]

During this reaction time, the lowering speed increases to

$$v_{reak} = v + a_{reak} \cdot t_{reak},$$

$$a_{reak} = \frac{m_L \cdot g}{m_L + m_{red}}, \quad m_{red} = I_{MBG} \cdot \left(\frac{\omega_M}{v}\right)^2 + I_T \cdot \left(\frac{\omega_T}{v}\right)^2, \tag{4.69}$$

$$s_{reak} = \frac{v + v_{reak}}{2} \cdot t_{reak}.$$

Deceleration during the braking process with the moments from braking and load reduced to the load movement:

$$a_{stör} = a_{brems} = \frac{(2 \cdot M_{brems} - M_L) \cdot \frac{\omega_M}{v}}{m_L + m_{red}}, \quad M_L = m_L \cdot g \cdot \frac{r_T}{i}, \tag{4.70}$$

or

$$\varepsilon_{brems} = \frac{2 \cdot M_{Brems} - M_L}{I_{MBG} + I_{T_{red}} + I_{L_{red}}}, \quad a_{brems} = \frac{v}{\omega_M} \cdot \varepsilon_{brems}, \tag{4.71}$$

$$t_{brems} = \frac{v_{reak}}{a_{brems}}, \quad s_{brems} = \frac{v_{reak}^2}{2 \cdot a_{brems}},$$

$$t_{bremsges} = t_{reak} + t_{brems}, \quad s_{bremsges} = s_{reak} + s_{brems}, \tag{4.72}$$

$$F_{stör} = m_L \cdot (a_{stör} + g).$$

The result is shown in Figure 4.38 as "Approximation for short rope."

For *very long rope lengths*, i.e., load lowered almost to the stage floor, it can be assumed as an approximation that the winch comes to a sudden stop and the load at the end of the rope is excited to oscillate with the initial velocity v_{reak}, circular frequency of the oscillation

$$\omega = \sqrt{\frac{c}{m_L}}, \tag{4.73}$$

amplitude of the oscillation $s_{max} = A = \frac{v_0}{\omega}$ and velocity at zero point v_{reak}.

Maximum acceleration (deceleration) is

$$a_{stör} = a_{max} = v_{reak} \cdot \omega = v_{reak} \cdot \sqrt{\frac{c}{m_L}}. \tag{4.74}$$

Maximum force in the rope is

$$F_{stör} = F_{stat} + F_{a\,max} = m_L \cdot (g + a_{max}) = m \cdot \left(g + v_{reak} \cdot \sqrt{\frac{c}{m_L}}\right). \tag{4.75}$$

The result is shown in Figure 4.38 as "Approximation for long rope." It can be seen in the diagram that higher spring rates (shorter rope) result in much too high values, since the influence of the braking effect on the load is greater the more rigidly the two systems winch and load are coupled and therefore the initial velocity of the load oscillation v_{reak} becomes smaller, the shorter the rope.

Both approximation approaches for very small and very large rope lengths thus do not yield the maximum values for deceleration and rope force occurring in the medium rope length range.

Sudden braking in the event of a fault during the lifting movement of the nominal load

Due to the sudden deceleration of the rope drum at the winch and the kinetic energy inherent in the load, the rope force is reduced or the rope is relieved, or even slack, and the load then falls into the relieved rope. The resulting rope oscillation results in a maximum value of deceleration or rope force, and these peak values are greater, the stiffer the rope is as a spring, or the greater the spring rate of the rope, or the shorter the rope is. The greatest load therefore occurs at the shortest rope length, i. e., at the greatest stroke height.

(It should be noted at this point that slack rope can also occur in the event of an emergency stop when the load is lowered, but this only occurs when the load swings back upwards and is therefore not from important relevance.)

Figure 4.39 again shows the results for the values of $a_{stör}$ and $F_{stör}$ as a function of spring rate and rope length, respectively, as a graph in the "MATLAB" curves.

In the diagrams of Figure 4.39, the values for $a_{stör}$ and $F_{stör}$ determined from this approximate calculation are again plotted as curves labeled "approximation." The diagrams show that the highest values are obtained at the highest spring rate of the rope and at the shortest rope length.

For a largest spring rate $c = 700\,kN/m$ or a shortest rope length of about 4 m, MATLAB calculations yield values of $a_{stör} \approx 39\,m/s^2 \approx 4\,g$ and $F_{stör} \approx 25.5\,kN$.

Furthermore, in the diagram $F_{stör}$ above the spring rate c, the value of the rope force at twice the nominal dynamic load is again recorded as a calculation value for the normal operating case according to the standard.

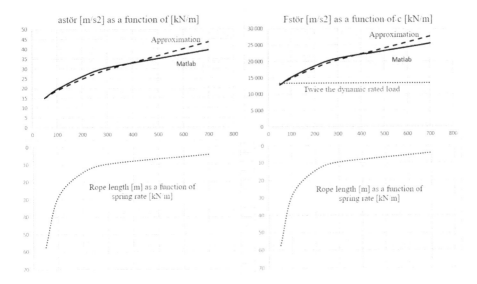

Figure 4.39: Evolution of the deceleration $a_{\text{stör}}$ and the rope force $F_{\text{stör}}$ in case of failure plotted against the spring rate c when *lifting the nominal load* mass of 500 kg of a 5 kN point hoist.

In this case, it follows that the load in the event of a fault is $25.5/5.1 = 5$ times the static nominal load and $25.5/6.66 = 3.83$ times the dynamic nominal load.

The diagram also shows the 2-fold dynamic nominal load to be applied in accordance with the standard 2 times $6.66 = 13.32$ kN and it can be seen that the load in the event of a fault also exceeds this load by far.

One could again make the following consideration as an approximate calculation:

The elasticity of the rope has during the first phase hardly any effect and winch and load could be considered as rigidly connected. If we again make the simplifying assumption that the braking force suddenly takes effect after a reaction time of $t_{\text{reak}} = \frac{t_1 - t_{11}}{2} + t_{11} = \frac{t_1 + t_{11}}{2}$ (see Eq. (4.68)). The maximum deceleration values or the maximum rope force can be calculated without any complex simulation.

During this reaction time, the lowering speed is reduced to

$$v_{\text{reak}} = v - a_{\text{reak}} \cdot t_{\text{reak}},$$

$$a_{\text{reak}} = \frac{m_L \cdot g}{m_L + m_{\text{red}}} \quad \text{with } m_{\text{red}} = I_{\text{MBG}} \cdot \left(\frac{\omega_M}{v}\right)^2 + I_T \cdot \left(\frac{\omega_T}{v}\right)^2. \tag{4.76}$$

During the braking process, the braking torque and the load torque, which also brakes, initially act. However, since the rope is very quickly unloaded during the braking process, i. e., the load is "thrown upwards," and thus the load torque can no longer have a braking effect, it is useful in the approximate calculation to set the load torque on the drum to zero in order to determine the braking time and the braking distance – thus, only the braking torque has an approximate effect:

$$a_{brems} = \frac{2 \cdot M_{brems} \cdot \frac{\omega_M}{v}}{m_Q + m_E + m_{red}} s,$$

$$\varepsilon_{brems} = \frac{2 \cdot M_{Brems}}{I_{MBG} + I_{T_{red}} + I_{L_{red}}}, \quad a_{brems} = \frac{v}{\omega_M} \cdot \varepsilon_{brems}.$$

Thus, the motor comes to a standstill after the time t_{brems} and the braking distance s_{brems}:

$$t_{brems} = \frac{v_{reak}}{a_{brems}}, \quad s_{brems} = \frac{v_{reak}^2}{2 \cdot a_{brems}},$$

$$t_{bremsges} = t_{reak} + t_{brems}, \quad s_{bremsges} = s_{reak} + s_{brems}.$$

Then the throw upwards and the subsequent fall of the load into the slack rope is considered. To determine the resulting values for deceleration and rope force, however, the elasticity (spring rate) of the rope must now be taken into account. Therefore, this simplified calculation approach can also be considered relevant for longer rope lengths.

The throwing height with the initial speed v_{reak} until standstill is

$$S_{max} = \frac{v_{reak}^2}{2 \cdot g}.$$

During this time, the rope has moved on by s_{brems}, so that a drop height h remains for the load:

$$h = s_{max} - s_{brems} = \frac{v_{reak}^2}{2} \cdot \left(\frac{1}{g} - \frac{1}{a_{brems}} \right).$$

Taking into account the elasticity of the rope (spring rate c), the rope force and deceleration at the end of the fall process is

$$F_{stör} = m_L \cdot g \left[1 + \sqrt{1 + \frac{2 \cdot h \cdot c}{m_L \cdot g.}} \right], \tag{4.77}$$

$$a_{stör} = \frac{F_{stör}}{m_L} - g. \tag{4.78}$$

The same result can be obtained by applying the formulas for a harmonic oscillation for long rope lengths, similar to the emergency stop in case of lowering. For this purpose, however, the velocity v_0 must first be determined in the zero position on the rope stretched by the load with Δl_0 (see Section 3.10).

With the total strain Δl, the amplitude of the oscillation is calculated to be

$$A = \Delta l - \Delta l_0 = \frac{m_L \cdot g}{c} \left[1 + \sqrt{1 + \frac{2 \cdot c \cdot h}{m_L \cdot g}} \right] - \frac{m_L \cdot g}{c},$$

$$v_0 = A \cdot \omega,$$

$$a_{stör} = a_{max} = v_0 \cdot \omega,$$

$$F_{stör} = m_L \cdot (g + a_{stör}).$$

Sudden braking in the event of a fault during the lowering movement of a small partial load

Figure 4.40 shows that the brakes, which are far oversized for this small partial load, cause very large values of deceleration, whereas the rope forces are naturally very small because of the small load and are not relevant for dimensioning. The fact that the results of the approximate calculation for rigid rope correspond better to the actual values is due to the fact that the smaller the hoisted load, the smaller the influence of the rope's elasticity.

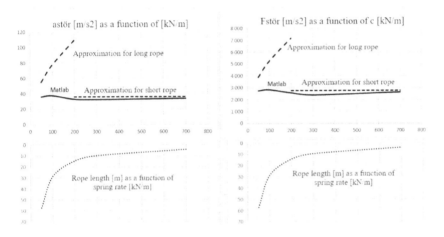

Figure 4.40: Evolution of the deceleration $a_{stör}$ and the rope force $F_{stör}$ in the fault case plotted against the spring rate c when *lowering* a 40 kg *partial load* mass of a 5 kN point hoist.

Sudden braking in the event of a fault during the lifting movement of a small partial load

The results are shown in Figure 4.41 and again exhibit very high values of the delay $a_{stör}$.

4.8.3 Supplementary notes

- In the diagrams shown in Figures 4.38–4.41, a point hoist was investigated. In the case of a load bar hoist, the bar is suspended on z hoisting ropes, usually of slightly different lengths. To simplify matters, in this case an average length of all z ropes can be calculated and the spring rate of the z springs connected in parallel can be expressed as $c = \frac{E_S \cdot A_m}{l} \cdot z$ (see Eqs. (3.102) and (4.7)), i. e., the spring rate tends to somewhat higher values for bar hoists and the system behavior shifts to the range of somewhat shorter rope lengths. However, this simplifying approach assumes that the load mounted on the load bar is actually supported by all ropes. If a point load is hanging under only one of the ropes, for example, this assumption certainly does not apply.

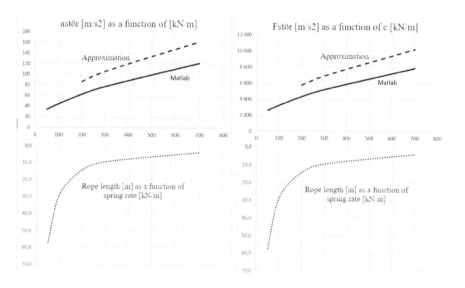

Figure 4.41: Evolution of the deceleration $a_{stör}$ and the rope force $F_{stör}$ in the fault case plotted against the spring rate c when *lifting* a 40 kg *partial load* mass of a 5 kN point hoist.

- In the investigations presented, it was assumed that the two brakes are applied at the same time as the nominal braking torque. In reality, the friction values at the brake pads are subject to a certain bandwidth, and the "nominal brake torque" is to be understood as a minimum value guaranteed under defined operating conditions. The maximum value actually given is, for example, 1.6 times the value specified by manufacturers. As a result, dimensioning must also be carried out assuming these less favorable higher braking torque values, which results in even higher values of deceleration and rope forces (see Section 4.7).
- As noted in Section 4.7, a reduction in the values for deceleration and rope force can be achieved by allowing one of the two brakes to apply with a slight delay. This effect can also be taken into account in the simulation calculation or in the approximate calculation. However, the influence is relatively small, since only a very short period of time is allowed before the second brake is applied. A long delay in the application of the second brake is therefore not permissible because, if the first brake fails, the downward movement is increased too much and stopping the system would result in excessively long braking distances.

4.8.4 Findings from the investigations

From the diagrams shown in Figures 4.37–4.41, but also from investigations on other hoists (bar and point hoists of different load capacities) not presented here, the following general conclusions can be drawn:

- Whereas in the event of a fault during lowering motion the maximum loads occur at medium rope lengths, the maximum values during lifting motion are always given at the shortest possible rope length.
- If we compare the peak values of deceleration and rope force in the event of a fault, we see that the larger critical values during a lifting movement occur with the smallest rope length.
- The peak values of the rope load are, as a rule, almost always greater in the case of an incident than in the case of a double dynamic nominal load. However, it should be noted that the dimensioning from the double nominal load case refers to a fatigue strength verification and that from the load in case of failure to a force fracture/deformation verification.
- In the case of an emergency stop during a lowering movement, the actual loads or Matlab results agree relatively well with the approximate calculation only for very long and for very short rope lengths. The critical maximum values, however, occur at medium rope lengths and are much higher than twice the nominal dynamic load. It is hardly possible to state generally valid maximum values for point and bar hoists, since these depend on a large number of factors. For this reason, the approximate calculations presented, with the appropriate safety factors, can only be used for rough estimates for dimensioning the drives, or calculations can be carried out for the specific case, e. g., with MATLAB-SIMULINK. Of course, measurements with accelerometers can be carried out on the actually realized plant to specify exact values.
- For *partial loads*, the rope forces $F_{stör}$, which are not relevant for dimensioning, are much lower. However, the values for the deceleration $a_{stör}$ are higher than in the nominal load case. This is self-evident, since the brakes must be dimensioned in such a way that they result in acceptable braking distances in the nominal load case (or also at overload test load) and therefore lead to short braking distances with high deceleration values at partial load. This problem must also be taken into account, especially if a hoist is also used as a performer flying. Large deceleration values can cause extremely serious injuries to a person suspended in a flybar system (especially in a horizontal position) (cf. whiplash syndrome in traffic accidents). The smaller the partial load compared to the nominal load, the greater this effect. Therefore, it may make sense to apply an additional load in the case of a very small partial load.
- In the calculations presented for a 5 kN point-hoist winch, it was assumed – as already mentioned – that in the event of a fault both brakes are applied simultaneously with their nominal braking torque guaranteed by the manufacturer, so that twice the braking torque comes into effect. For this reason, reference is made to the explanations in Section 4.7, in which it is explained, among other things, that there are also possibilities of reducing the braking torque acting in the event of a fault. Figure 4.31 shows a brake design in which only one-times the braking torque acts in the event of a fault.

– Behavior at lower speeds: If the speed is reduced on an existing installation where the brakes are, of course, designed for the nominal load and speed, this does not result in any change in the deceleration values if the rotating masses of the winch and the lifting load are rigidly coupled, since there is no change in the magnitude of the effective braking torque and the magnitude of the masses to be braked. In reality, however, the elastic coupling of the masses via the rope can very well cause a certain reduction of the deceleration values. Therefore, it can still make sense to work with lower lifting speeds than the nominal lifting speed for critical lifting movements, e. g., when used as performer fly.

– Since the elasticity of the rope plays a role, the deceleration values can be reduced in the event of a fault by installing spring elements (+ dampers) in the area of the rope guide, which should, however, only take effect if a certain maximum rope force is exceeded. Such an additional spring connected in series to the spring action of the rope results in a reduction of the resulting spring rate or a fictitious increase of the rope length:

$$\frac{1}{c_{res}} = \frac{1}{c_1} + \frac{1}{c_2}, \quad c_{res} = \frac{c_1 \cdot c_2}{c_1 + c_2} \quad \text{(Eq. (3.103))},$$

$$c_1 = c_{Seil} = \frac{A_m \cdot E_S}{l_{Seil}} \quad \text{and} \quad c_2 = c_{Feder} = \frac{A_m \cdot E_S}{l_{spring,fiktiv}}.$$

From this, it is also possible to calculate a fictitious rope length achieved with it:

$$l_{Seil} = \frac{A_m \cdot E_S}{c_{Seil}} \quad \text{and} \quad l_{spring,fiktiv} = \frac{A_m \cdot E_S}{c_{spring}},$$

$$l_{ges,fiktiv} = A_m \cdot E_S \cdot \left(\frac{1}{c_{Seil}} + \frac{1}{c_{spring}} \right).$$

If this is to improve the conditions for dimensioning a hoist, these spring (damper) elements would have to take effect when a certain rope force is exceeded.

Although such a measure would change the conditions for dimensioning the hoist, since the stresses would be reduced in the event of a malfunction, the high values of deceleration would not change in operation with partial loads, because the additional spring system would not take effect at all with partial loads. If, for example, large values of deceleration are to be avoided when the lifting device is used as a performer fly, the effect of the additional spring would have to become effective when a certain deceleration value is exceeded.

– In order to comply with the dimensioning criteria required by EN 17206, the peak values for the load or deceleration occurring in the event of a fault may have to be determined approximately. After completion of the installation, it may be useful to measure the decelerations actually occurring at different loads and rope lengths using accelerometers.

4.8.5 The load limiter from Bosch Rexroth

The task of the *load limiter* developed by Bosch Rexroth is to ensure that, even in the event of a category 0 fault, no loads greater than twice the nominal dynamic load can occur.

The load limiter consists of a short-stroke hydraulic cylinder and a hydraulic accumulator component which, in relation to the geometry of the rod side of the cylinder, is preloaded to a value corresponding to a value between single and double the nominal load. The cylinder is integrated into the rope path via a pulley. The wrap angle influences the pretension equally and must therefore be taken into account.

In normal operation, compression of the prestressing volume is not possible. The cylinder therefore acts as a structurally rigid element within the rope routing and has no function. As soon as the preset pressure in the accumulator component is exceeded in the event of a fault, the cylinder extends, displacing the oil volume of the rod side of the cylinder against the prestressed cushion. The excess energy from the actual hard impact is thus very quickly converted into stored hydraulic energy. As soon as the external force is no longer acting, the stored volume relaxes to the preset value. The cylinder moves to the starting position.

The cylinder stroke is position-monitored to guaranty the safety-relevant functionality. On the rod side, there are two pressure gauges with the help of which the load pressure on the cylinder and thus the suspended load is continuously measured redundantly and the drives are switched off when the set overload is reached (see Figure 4.42).

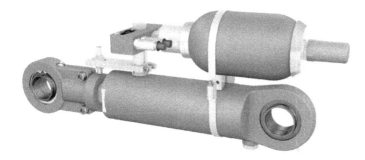

Figure 4.42: Load limiter. Picture credits: Bosch Rexroth.

The classic load limiter is only used for maximum load protection and therefore primarily protects the drive machine of a stage equipment as well as the adjacent supporting structure (e. g., the grid).

As already explained, although no critical force effects occur at the hoist drive with smaller loads, deceleration takes place with high values of deceleration, which represent a very high risk of injury, especially when used as a performer fly. Such deceleration values, some of which are even life-threatening, cannot be prevented with a load limiter of the classic type designed for maximum load.

The variable load limiter was designed for this purpose. Depending on the suspended load, it varies the preload of the storage component in such a way that the damping effect is already effective at low working loads. In addition to maximum load protection, this also ensures protection of the suspended decoration and its connection to the load handling attachment.

When these systems are used as flying structures for artists and performers, the risk of serious injury is dramatically prevented because the maximum deceleration can be limited to 2g in all load cases.

Bosch Rexroth also offers a measuring device called *"Stage Guard."* The basic component is a highly sensitive acceleration sensor from Bosch. For one-dimensional movements, the device offers the possibility of recording the entire course of the acceleration values for the duration of a brake test and converting these into an analog load curve with knowledge of the load. For this purpose, the measuring device is attached to the load handling attachment. The velocity curve can also be read out.

The start of the measurement can be triggered via radio connection by a cell phone, tablet or laptop and then also terminated.

4.9 Trusses and truss systems for rigging purposes

In event technology, "rigging" refers to the assembly and operation of event-specific work equipment for load suspension.

The basic elements are usually trusses available as a modular system, which can be assembled into complex truss systems and used to carry mainly static loads or for purely decorative purposes. They can be hung, placed, fixed or be movable for use.

A *truss* is a multichord lattice girder element made of metallic materials whose different system lengths can be joined together by means of fasteners (screws, bolts) specified by the manufacturer (see Figures 4.43 and 4.44).

A *truss system* (rig) is a complex truss construction created using special elements, such as angle elements (fixed or movable) or arch elements, or from the combination of different system elements and/or from different systems.

Assembly may only be carried out on the basis of assembly instructions and the corresponding user information for the type of truss used and must be carried out in the correct installation position. Particular attention must be paid to the application of forces to the trusses. All live loads should act vertically and be distributed evenly across the main flanges. Additional horizontal loads should be avoided. Load-specific static calculation verifications are required for complex systems.

In order to achieve the lowest possible dead weights, the trusses are designed as plane trusses or as space trusses and are made of aluminum alloys (e. g., AlMgSi1). The standard designation F 28 to F 31 is used to indicate the strength properties.

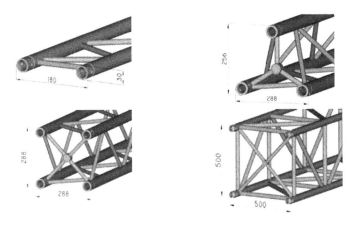

Figure 4.43: Trusses of different design: (top) two-tube truss/double-cone connection system and three-tube truss/double-cone connection system, (bottom) four-tube truss/double-cone connection system and four-tube truss/fork-head connection system. Photo credit: A. T. C. Production & Handels GmbH.

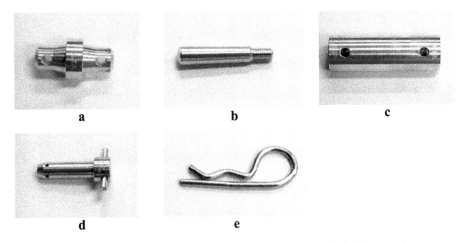

Figure 4.44: Selection of accessories: (a) double cone adapter, (b) stalk pin for double cone adapter, (c) cone sleeve adapter, (d) steel pin for clevis, (e) safety split pin. Image credit: A. T. C. Production & Handles-GmbH.

The truss elements can be connected in various ways to form larger units or truss systems:
- With end plates.
- As tube-in-tube push-in joints – the ends of the beams are open; pipe sockets of smaller diameter are inserted into these open pipes and secured with bolts or screws.
- As double cone connections (the most commonly used connection system) – a tube-element is welded into the ends of the main chords, which is conically shaped on the inside. A double cone is inserted into this conical sleeve as a connecting element.

The double cone is provided with two conical holes into which galvanized pins are usually hammered.
- As clevis connections – turned parts with a long attachment are pushed into the main chords of the truss girder and fastened with dowel pins. The truss girders are provided with "male" turned parts on one side and "female" turned parts on the other.
- As hybrid connection – combination of double cone and clevis system.

Figure 4.45 shows examples of various trusses that can be assembled to form planar or spatial truss systems (constructions with a possible downward and/or upward connection). The picture shows so-called two-, three- and four-point trusses.

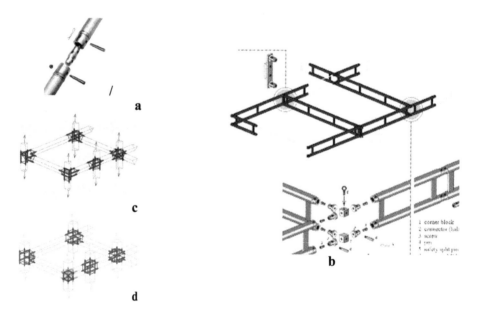

Figure 4.45: Example of assembling trusses into plane and spatial truss systems: (a) installation of a double cone adapter, (b) two-point truss, (c) three-point truss, (d) four-point truss. Photo credit: Eurotruss.

Notes for supplementary literature on Chapters 2, 3 and 4
- Böge, A., Böge, W. (Eds.): *Handbuch Maschinenbau-Grundlagen und Anwendungen der Maschinenbau-Technik*, Springer Fachmedien Wiesbaden, 2021.
- Bosch Rexroth: *Hydraulics Trainer Volumes 1–4*, Bosch Rexroth AG.
- Brosch, P. F.: *Moderne Stromrichterantriebe*, Würzburg: Vogel, 1989 (Kamprath series).
- Bender, B., Göhlich, D. (Eds.): *Dubbel – Taschenbuch für Maschinenbau – Grundlagen und Tabellen – Band 1*, Berlin: Springer, 2020.

- Decker, K.-H.: *Maschinenelemente – Funktion, Gestaltung und Berechnung*, Carl Hanser Verlag, 2018.
- Henn, H.: *Engineering Acoustics. Physical Fundamentals and Application Examples*, Vieweg Teubner, 2008.
- Hoffmann, K., Krenn, E., Stanker, G.: *Fördertechnik*, Bd. 1: *Bauelemente, ihre Konstruktion und Berechnung*, Oldenbourg, 2012.
- Niemann, G., Winter, H.: *Maschinenelemente 1*, 3 Bde., Berlin: Springer, 2019.
- Roloff/Matek (Wittel, H., Spura Chr., Jannasch D.): *Maschinenelemente – Normung, Berechnung, Gestaltung*, Springer Fachmedien Wiesbaden, 2020.

Article in the Bühnentechnische Rundschau:

2013	No. 3	Kirsch, V. – Bosch Rexroth AG: *Test results with consequences*
	No. 5	Lucke, P. – Mayr Antriebstechnik: *Brakes save lives*
2017	No. 1	Lucke, P. – Mayr Antriebstechnik: *New braking system for more safety*
2016	No. SH	Partl. C: *Simulating the emergency stop investigation of loads on winches*
2020	No. 6	Lucke, P. – Mayr Antriebstechnik: *Double safety – single load*

Article in the brochure (ÖTHG)

2014	No. 1	Feurer, F.: *The brake test in stage technology*
2016	No. 2	Partl. C: *Investigation of loads on stage winches during emergency stop*

5 Safety regulations – standards

5.1 Hazards to stage personnel and performers

Efforts to eliminate hazards in stage operations as far as possible have led to the adoption of standards and other regulations, which have to be revised again and again due to advancing technical developments. Regulations and standards are not the same in all countries and are edited by different organizations and authorities depending on the legal structure. In Europe, the new European *EN 17206* is of importance in connection with the content of this book:

EN 17206:
"Entertainment Technology – Lifting and Load-bearing Equipment for Stages and other Production Areas within the Entertainment Industry – Specifications for general requirements (excluding aluminum and steel trusses and towers)"

    EN 17206 was adopted on a Europe-wide basis in 2020 and, as a result, previously existing national standards within the EU will lose their validity.

    This standard replaces:

in Germany:
DIN 56950 "Event technology – Mechanical equipment, safety requirements and testing"

in Austria:
ÖNORM M 9630-1 "Mechanical stage equipment – General information"
ÖNORM M 9630-2 "Mechanical stage equipment – Upper stage machinery"
ÖNORM M 9630-3 "Mechanical stage equipment – Under-stage machinery"
ÖNORM M 9632   "Mechanical stage equipment – Test specifications"

Standard to be mentioned in Germany in this context:
DIN/TS 56951 Event technology – Drive and control for safety-related equipment

The EN 17206 standard defines the detailed safety requirements for machinery used in event technology. It is based on the principle of risk assessment and risk minimization according to EN ISO 12100, EN 62061 and EN ISO 13849.

    For example, according to EN ISO 12100, risk assessment and minimization must be carried out as follows (Steps 1–4 thus concern risk assessment, step 5 risk minimization measures):

1. Determination of the limits of the machine including its intended use.
2. Identification of hazards or hazardous situations.
3. Assessment of the risk for each hazard or hazardous situation.
4. Assessing the risk and making decisions about the need for risk minimization.

https://doi.org/10.1515/9783111366968-005

5. Elimination of the hazard or reduction of the risk associated with the hazard through protective measures.

According to EN 62061, the risk assessment for each hazard should be performed by determining the following risk parameters:
(a) Severity of the damage.
(b) Probability of occurrence of the damage, assessable according to the:
–   Frequency and duration of exposure of persons to the hazard.
–   Probability of the occurrence of a hazard event.
–   Possibilities of avoiding or limiting the damage.

This assessment is also used in particular for the assignment of a required *"Safety integrity level"* – *"SIL"*. It is used to assess electrical/electronic/programmable electronic (E/E/PE) systems with regard to the reliability of safety functions and offers discrete levels from SIL1 to SIL 4 as a classification, whereby SIL 3 is relevant for many applications in stage technology.

The protection goal must be to eliminate or minimize hazards. The basic safety concept in the EN 17206 standard is based on the principle of the
–   *Intrinsic safety*, e. g., by doubling the operating coefficient in calculations (design for double the nominal load) or by redundancy.
–   *Single fault safety*, i. e., simultaneous occurrence of two independent faults is not considered.

In this chapter of the book, only a few basic aspects of safety engineering are presented as examples and typical hazards are listed.

In principle, there are three types of *hazards* for performers, operating personnel and nonoperating personnel who may be in the stage area:
–   Hazards arising from the presence or use of stage equipment (equipment as defined in this book);
–   Hazards due to scenic installations and decorations; and
–   Hazards associated with artistic performances.

Hazards during artistic performances are the responsibility of the artist and are not of interest in this context.

In the context of the subject matter dealt with in this book, the first-mentioned aspect is of particular interest, and the following question arises: What are the resulting special hazardous situations in the stage area?

Stage crew and performers act during set construction and dismantling, during rehearsals, and during a performance:
–   In the travel range of moving equipment; or
–   On moving stage floor elements;

– And stay below a lot of loads suspended on hoists at the grid. In contrast, in normal hoist operations, it is generally forbidden to stay under suspended loads.

The following considerations can be derived from this:

Hazards arise primarily from the fact that there are *moving devices* in the lower and upper stage, such as hoists, lowering devices, stage wagons, turntables, etc., which persons in the immediate vicinity or persons moving with the devices:
– In the execution of intended work movements, as well as
– In the event of unintentional movements due to a malfunction.

Unintentional movements could be caused by malfunctions in open loop control or feedback control systems in hydraulic or electrical or electronic circuits, caused by errors in software or hardware elements, or by mechanical failure of components due to overload, wear, corrosion, etc.

There will also be scenic sequences in which a certain hazardous situation is unavoidable. Then, as far as possible, organizational protective measures must be taken and the scenic procedure must be rehearsed sufficiently often under original conditions (lighting, costume).

Examples of concrete measures to increase safety
In the following, without claiming completeness, a few construction regulations for *stage equipment* are given as examples in order to explain the type of safety thinking involved. However, as already mentioned, it should be borne in mind that the regulations in the individual countries differ in some cases:
– *Higher safety factors for the dimensioning of components especial for lifting equipment* than is usual in general lifting technology:
    – *The wire ropes, chains and steel belts* used as *suspension means* must have a safety factor at nominal dynamic load which is twice as high as specified in Directive 2006/42/EC of the European Parliament and of the Council (dated May 17, 2006 or its new versions and amendments) on machinery and called in short "Machinery Directive." Here, the safety factor is the quotient of the minimum breaking force, related to the prorated rope tensile force at *nominal dynamic load* (see Section 4.1.1 or Eq. (4.4)). Therefore, 10 times the safety factor applies to ropes and steel belts and 8 times the safety factor applies to chains.
    Explanation. The nominal dynamic load consists of the static load components and the additional dynamic loads which relate to operational use, i. e., not to an incident (e. g., power failure). In older standards, sometimes only the static nominal load and the "calculated breaking load" instead of the minimum breaking load was used, but with higher safety factors.
    – When calculating the *structural elements supporting the lifting load in the power flow between the drive motor or brake and the lifting load,* for example, two times the nominal dynamic load must be used for the verification of a determined

finite life fatigue strength. This is to exclude a break in the power flow. The load-bearing elements (e. g., ropes, chains, steel straps) and the load handling attachment (e. g., the load bar) are excluded. For the load in the event of a fault, force fracture or inadmissible deformation must be prevented.

- *Cylinders, pressure piping*, etc., must be sized for twice the maximum working pressure in that area of the equipment where their bursting would cause the load to fall.
- The burst pressure of *hose assemblies* must be at least four times the maximum working pressure.
- *Design requirements and redundancy requirements* to prevent a load from falling or objects from falling:
  - All load-bearing elements of suspension means must be made of noncombustible material. Ropes made of synthetic or natural fibers are only permitted if the risk assessments show that this is safe. As a rule, therefore, only steel wire ropes, steel chains or steel belts can be used as load-bearing elements for lifting equipment.
  - *Driving systems* (power- or hand-operated) must be equipped with at least the following as a safeguard against unintentional movement:
    * *self-locking* from the movement, or
    * *two brakes* acting independently of each other in any operating condition and controlled independently of each other, one of which may be applied with a delay, or
    * in the case of hydraulic cylinder drives, *two independent hydraulically releasable non return poppet valves* (see Section 2.3.1 or Figure 2.17). One poppet valve may be replaced by a *clamping device.*
    While this requirement in the previously valid standards (DIN 56950 and ÖNORM M9630) only applied to lifting devices, this requirement initially applies generally in EN 17206. However, it is supplemented by the following wording: "If a risk assessment has verifiably established that adequate safety can be achieved with only one safety device, one safety device may be sufficient for horizontal movements (e. g., stage wagons, turntables, curtain systems, trolleys) even with only one drive unit."
  - For *sliding spindle drives*, the spindle must have a higher wear resistance than the nut. The support nut must be designed for twice the nominal load. If a safety nut is positioned in addition to the support nut, it is sufficient to design both nuts for one times the nominal load. The wear of the support nut must be monitored by a wear measuring device, e. g., a control nut. This is a non-load-bearing spindle nut that is used to check the wear of the support nut.
- *Falling down objects – avoidance or protection:*
A stage-specific hazard arises from the fact that the stage area is surrounded by working galleries and there is a walk-on grid floor above the stage area. Thus, there is a fundamental risk that objects can fall down, also due to carelessness. Design

measures are intended to reduce the risk. For example, *work galleries* are to be equipped with particularly high baseboards. The counterweights of manual counterweight hoists can also be a particular source of danger. *Counterweights of* manual *counterweight hoists* must be prevented from falling out of the counterweight slide, for example. If there is a traffic route underneath the track of counterweight carriages, as is the case, for example, at the transition from the main to the side stage of doubled counterweight hoists, safety devices must be provided.

– *Determination of the minimum load-bearing capacity of structural elements* to avoid their overloading:

   For example, by the following specifications: In the case of moveable stage platforms, the load capacity per lifting floor level must be at least 500 kg/m^2 when stationary; lower loads are generally assumed when the device is in motion. EN 17206 therefore distinguishes between:

   – ELL (entertainment load limit) maximum load (mass) that an item of lifting equipment is designed to raise, lower or suspend at rest.
   – ELL/R (entertainment load limit at rest) as the maximum load (mass) that an item of lifting equipment is designed to suspend at rest.

      In the ÖNORM valid before the publication of EN 17206, for example, it was stipulated that in the case of lifting platforms a payload of 250 kg/m^2 may be assumed for the driving condition. However, the load capacity may be further reduced if this would result in a load capacity of more than, for example, 5000 kg, but a value of 100 kg/m^2 must not be undercut.

– *Control requirements*:

   Here, too, only a few examples are given in this chapter without claiming completeness. Automatic shutdown occurs, among other things:

   – when passing over safety-defined limits of a travel path,
   – on rope drums to ensure that at least two spare turns remain when the rope is unwound,
   – in the emergence of slack rope,
   – when a specified synchronization tolerance is exceeded.

   If the shutdown takes place in the control circuit, this should be done via a ramp with a defined delay in order to avoid additional hazards like tear-off or overturning of sceneries located on trolleys. If the power supply is interrupted for the purpose of an emergency shutdown or in the event of a power failure, controlled shutdown is generally not possible.

   When automatically shutting down a plant, three types can be distinguished:

   – *Stop category 0.* The machine is brought to an uncontrolled standstill by switching off the power supply to the drive. The stopping distance, the magnitude of the deceleration and the force effects caused are determined by the load holding device used – usually the mechanical brakes. This case therefore occurs in the event of a power failure.

- *Stop category 1.* The machine is brought to a standstill in a controlled manner with a specified braking ramp (i. e., the specified deceleration), whereby stopping distance and deceleration can be specified. After reaching standstill, the power supply is interrupted.
- *Stop category 2.* The system is stopped in the same way as in stop category 1; after the system has come to a standstill, further travel in the dangerous direction is prevented by the safety control system when the energy is switched on, but travel in the safe opposite direction is enabled. This type of stop is required in particular for the safe prevention of consequential damage, e. g., in the event of crushing, slack rope, setting down of inadmissibly lifted loads).
- A special hazard also exists when lifting platforms are moved and performers could, for example, get their foot caught in a shear edge. For this reason, it is generally stipulated to provide a system for automatic shutdown as a *shear edge safety device*.
- *A good overview of* the hazardous area from the control station must be ensured, possibly using auxiliary equipment such as TV cameras or a control panel placed in the viewing area. A mobile or portable control panel can also be used for this purpose.
- *Acoustic signaling devices* for safety curtains in case of emergency shall be provided.
- *Definition of maximum speeds for working movements of movable stage equipment*: In some regulations (EN 17 206 does not specify anything in this respect), maximum speeds were specified for hoisting, travelling and slewing movements. This is intended to prevent tendencies to demand ever higher operating speeds which, for physical reasons, would increase the hazard potential to an unreasonable extent. Up to now, the usual maximum traversing speeds were or are:
  - 1.8 m/s (previously 1.2 m/s) for hoists.
  - 0.7 m/s for lifting platforms.
  - 0.7 m/s for stage wagons.
  - 1 m/s peripheral speed on rotating platforms.

  These values of the maximum speed, which are decisive for the design, may not be utilized operationally in all situations. Therefore, the following speed reductions may be necessary. For example, if there is an access and exit of persons in the case of moving stage wagons or turntables, or a change from one moving part to another, the differential speed at the transition point should not exceed 0.3 m/s.
- *Requirements that the operator should consider*:
  In the event of a malfunction – as already mentioned above – considerably higher decelerations or inertia forces can occur than in normal operation. These extraordinary loads not only stress the mechanical equipment but also the transported loads, i. e., decorative elements or persons suspended in fly device require that the acceleration and deceleration values for the operating and fault cases (in particular, the maximum possible deceleration values for performer fly systems) must be specified in the documentation of the mechanical equipment.

–	*Requirement for special organizational measures*:
	For example, the following organizational measures are required:
	–	Technical and performing personnel must be made aware of hazards within the scope of their activities – in particular, from moving equipment before their first work assignment.
	–	Staying in the movement area of stage equipment and scenery areas during operation and in confusing traffic and transport areas is only permitted if operational or scenic requirements are given.
	–	In the case of hazardous scenic events requiring specific behavior, instruction shall be repeated or adequately rehearsed at appropriate intervals.
	–	Movement processes that can cause hazards may only be carried out if the speed is appropriate to the situation, protective devices are in place to safeguard the hazardous spots or the hazardous points are monitored by the operating personnel and clearly recognizable attention is drawn to the dangerous spots.
	–	At workplaces, scene areas, transportation routes and accesses that are higher than 1 m in relation to adjacent areas, effective devices must be provided to prevent people from falling (handholds, railings).
	–	If, in individual cases, such devices cannot be used for scenic reasons, devices for catching falling persons must be provided instead of them. If it is not possible to use these catching devices on scenic areas, a colored marking must be placed at least 0.5 m in front of the edge of the fall and must be clearly visible under all scenic lighting conditions. Such measures are not required at the stage ramp.
	–	In the case of flies in which persons are suspended, organizational measures must be taken to ensure that they can be lowered in the event of danger.
	–	Operational gaps of more than 20 mm must be covered.
	–	If non-load-bearing surfaces adjoin walkable surfaces, these must be clearly demarcated.
–	Obligatory performance of *acceptance tests and inspections of technical equipment* by experts:
	–	Prior to first use or after substantial changes, the machinery shall be subject to testing in accordance with any specific national safety legislation and meet the standards.
	–	Be reviewed again at regular intervals.
–	*Obligation to evaluate working conditions*:
	Legal regulations on safety at work oblige the operator of a facility, respectively the employer, to carry out an evaluation of the hazards associated with the work for the employees and, if necessary, to provide for measures in the sense of occupational safety. In this sense, stage equipment and its use in specific productions are to be "evaluated" with regard to their hazard potential by means of hazard analyses.

– Consideration must also be given to the competency or suitability of persons on stage.

5.2 Hazards for the spectators

Hazards for spectators exist primarily in two respects. On the one hand, as in all cases where there are large gatherings of people, escape facilities must be provided; emergency exits, emergency lighting, doors that open outward, etc., are required. However, since these regulations do not concern stage technology in the sense of this book, no further statements will be made about them.

On the other hand, hazards result from stage operations. A direct hazard may arise, for example, when spectators are placed on the stage or, in special cases, when aerial movements are performed above spectators.

An indirect hazard for spectators exists in the sense of an increased fire risk due to stage operation. For this reason, a fire protection curtain is mandatory for most of theater and opera houses. Similarly, smoke escape systems and, where appropriate, water spray extinguishing systems are also part of fire protection measures. These facilities were discussed in more detail in Section 1.8.

Index

https://doi.org/10.1515/9783111366968-006

www.ingramcontent.com/pod-product-compliance
Lightning Source LLC
Jackson TN
JSHW051944131224
75386JS00005B/328